BASIC WAVE MECHANICS:
For Coastal and Ocean Engineers

BASIC WAVE MECHANICS:

For Coastal and Ocean Engineers

ROBERT M. SORENSEN
Professor of Civil Engineering
Lehigh University

A Wiley-Interscience Publication

JOHN WILEY & SONS, Inc.

New York • Toronto • Chichester • Brisbane • Singapore

This text is printed on acid-free paper.

Library of Congress Cataloging in Publication Data:

Sorensen, Robert M., 1938–
Basic wave mechanics: for coastal and ocean engineers/Robert M. Sorensen.
p. cm.
Includes bibliographical references and index.
ISBN 0-471-55165-1 (cloth)
1. Ocean waves. 2. Wave mechanics. I. Title.
TC172.S64 1994
627′.042—dc20 93-13309

Printed in the United States of America

10 9 8 7 6 5 4 3 2

Dedicated to
Rita, Jenny, and Jon
With Love

PREFACE

This book is an expansion of my lecture notes on wave mechanics for two courses given to engineers and marine scientists at Lehigh University. It was written because I found no existing texts adequate for the range and level of material I wanted to cover. I had to use selected sections of several texts.

Several texts include a chapter or two on elementary aspects of wave theory and the nature of waves at sea, to support main topics such as coastal engineering or beach processes. There are also a number of quite esoteric texts on wave theory, often with a narrow focus, that are of limited use for the engineer and marine scientist working with ocean waves. This book is written to bridge the gap between these two types of book.

To understand and analyze the behavior and impact of surface water waves requires a comprehension of the processes involved in the generation and transformation of waves as they propagate across the water surface. This includes an understanding of the physical characteristics of the waves and a knowledge of the important theories that define these characteristics and their transformation. A practical understanding of available techniques for wave measurement and prediction, the analysis of wave records, the basis and procedures for the determination of "design" waves for a specific site and type of coastal work, and the procedures for predicting the results of wave interactions with structures and the shore are also needed.

The objectives of this book are to present first a discussion of the physical processes involved in ocean wave mechanics and second the analytical bases of these processes at the level required by the marine engineer and scientist. Emphasis is placed on understanding wave characteristics and the basic techniques for wave analysis, rather than on the derivation and manipulation of equations.

Chapter 1 briefly introduces the subject of sea surface gravity waves and the available literature on waves. Chapter 2 develops the small amplitude or linear wave theory and employs it to describe most of the basic characteristics of surface waves. Chapter 3 then considers the two-dimensional transformations that occur as waves propagate from deep water to the shore. Chapter 4 presents the practically important nonlinear wave theories, recommended conditions for their use, and a discussion of the important wave characteristics best described by these theories. Chapter 5 describes and presents analysis techniques for the three-dimensional transformations that occur as waves propagate toward the shore and past obstructions, as well as related considerations such as ship-generated waves.

Chapters 2–5 only consider monochromatic waves. The characteristics and analysis of the more complex wind-generated waves are presented in Chapter 6. This includes a discussion of wind wave generation and growth; wave record analysis; wave height probability distribution, grouping, and spectral characteristics; and wave prediction. Chapter 7 looks at the determination of design waves, including the wave information required, wave measurement techniques, breaking limits, and return period analysis of wave data. Chapter 8 considers the modification of waves owing to wave-structure interactions. This includes wave reflection, wave runup on slopes, and wave overtopping and transmission over and through structures. In Chapter 9 the long wave equations are developed. They are then applied to typical examples, including the calculation of storm surge, basin oscillations, and Kelvin waves. The final chapter is devoted to a discussion of laboratory wave studies with an emphasis on wave generation techniques.

This book is written to be used as a text for a formal course or for self study as a practical reference for coastal and ocean engineers and marine scientists. The reader only needs a general background in basic fluid mechanics and hydromechanics as well as an understanding of calculus.

I want to acknowledge the outstanding job done by Mrs. Cathy Miller in typing the equations and other related efforts to support the preparation of the original text manuscript. I appreciate the thorough review of the draft manuscript of this text that was provided by Thorndike Saville, Jr. former Technical Director of the U. S. Army Coastal Engineering Research Center. Also, thanks are extended to my graduate students who used the draft manuscript as a text and provided review comments.

R. M. SORENSEN

Lehigh University

CONTENTS

NOTATION

a_c, a_t	Wave crest amplitude; trough amplitude
a_x, a_z	Horizontal and vertical components of acceleration
B	Wave orthogonal spacing; structure crest width
B_0	Wave orthogonal spacing in deep water
C	Wave celerity; Chezy resistance coefficient
C_0	Wave celerity in deep water
C_g	Wave group celerity
C_r	Wave reflection coefficient
C_R	Prototype to model wave celerity ratio
C_s	Wave celerity in still deep water; celerity of the significant wave
C_t	Wave transmission coefficient
C_{t0}	Wave transmission coefficient, over structure
C_{tt}	Wave transmission coefficient, through structure
D	Wave decay distance
D_{50}	Median armor stone diameter
d	Water depth
d^*	Water depth where H_s is maximum
d'	Setup, setdown of the mean water level
d_b	Water depth at wave breaking
d_s	Water depth at structure toe
E, E_k, E_p	Total, kinetic, potential energy per unit crest width
$\overline{E}$	Average wave energy per unit surface area; encounter probability
f	Wave frequency; friction factor; Coriolis parameter
f_p	Peak wave frequency in a spectrum
F	Froude number; wind fetch length; structure freeboard
F^*	Dimensionless structure freeboard
F_{eff}	Effective fetch length
$G(f, \theta)$	Directional spectrum spreading function
g	Acceleration of gravity
H	Wave height
$\overline{H}$	Average wave height
H_b	Wave breaker height
H_D	Significant wave height at end of decay distance
H_F	Significant wave height at end of fetch
H_i, H_r, H_d	Incident, reflected, diffracted wave height
H_i	Individual wave height in a record
H_{max}	Maximum wave height
H_{m0}	Significant wave height based on spectral energy
H_n	Average height of highest n percent of waves
H_0	Wave height in deep water
H_0'	Unrefracted deep water wave height

H_r	Reflected wave height
H_{rms}	Root mean square wave height
H_s	Significant wave height based on individual wave analysis
H_{s0}	Significant wave height in deep water
H_t	Transmitted wave height
I_r	Iribarren number; surf similarity parameter
K	Darcy permeability coefficient
K_b	Bottom stress coefficient
K_d	Diffraction coefficient
K_p	Pressure response function
K_r	Refraction coefficient
K_s	Shoaling coefficient; surface stress coefficient
k	Wave number
L	Wave length; structure lifetime
L_0	Wave length in deep water
L_p	Wave length of spectral peak frequency
L_{p0}	Deep water wave length of spectral peak frequency
L_R	Prototype to model length ratio
l_c	Characteristic linear dimension of a breakwater armor unit
m	Beach slope
m_n	nth moment of wave spectrum
m_0	Zeroth moment of wave spectrum
n	Ratio of wave group to phase celerity; direction along wave crest; Manning's resistance coefficient
M, N	Solitary wave theory coefficients; basin resonance modes
N	Number of waves in a record
P	Wave power per unit crest length; precipitation minus evaporation rate
P_d	Wave power dissipated per unit crest length
p	Pressure; probability of exceedence
p_a	Atmospheric or surface pressure
$p(H)$	Probability of occurrence of H
$p(H, T)$	Probability of occurrence of H, T combination
$p(t)$	Time dependent pressure variation
$P(H)$	Cumulative probability distribution of H
Q	Bernoulli constant; volumetric overtopping rate
Q_a	Average overtopping rate
Q_s	Overtopping rate of significant wave height
Q_p	Spectral peakedness parameter; peak overtopping rate
Q_*	Dimensionless overtopping rate
Q_0^*	Parameter in overtopping equation
q_x, q_y	Volumetric flow rates per unit width
R	Wave runup above the SWL; radius to maximum wind speed in a hurricane; Reynolds number

R_{max}	Maximum runup
R_p	Wave runup associated with a particular probability of exceedence
R_s	Runup of significant wave
R_*	Mean rate of energy dissipation per unit surface area owing to bottom percolation
r	Radial distance; time interval between successive data points for extreme wave analysis; wave runup correction factor
S	Wave generator blade stroke amplitude
S_{ds}	Spectral energy dissipation
S_{in}	Spectral energy input from the wind
S_{nl}	Spectral energy transfer by nonlinear interaction
S_{xx}, S_{xy}, S_{yy}	Radiation stress components
$S(f)$	Frequency spectrum energy density
$S(f, \theta)$	Directional spectrum energy density
$S(T)$	Period spectrum energy density
s	Directional spectrum spreading factor, wave generator blade horizontal position versus time
t	Time
t_d	Wind duration
T	Wave period
$\bar{T}$	Average wave period
T_D	Significant period at end of decay distance
T_F	Significant wave period at end of fetch
T_N, T_{NM}	Resonant periods of basin oscillation
T_p	Peak wave period in a spectrum
T_R	Model to prototype wave period ratio
T_r	Return period or recurrence interval
T_s	Average period of highest one-third of waves
U	Current speed; wind speed; average depth integrated velocity
U_A	Wind stress factor
U_p	Forward speed of a pressure disturbance
U_R	Wind speed at radius to maximum wind in a hurricane
U_{10}	Wind speed at standard 10 m elevation above ground
U_r	Ursell parameter
U_t	Wind speed averaged over a period of time t
U_z	Wind speed measured at elevation z above ground
u	Mass transport velocity
u, v, w	Components of fluid velocity
W	Floating breakwater width in the direction of wave propagation; wind speed; wave generator blade width
V	Volume of mass transport in a solitary wave; vessel speed

V_F	Forward velocity of a hurricane
x, y, z	Coordinate directions
X_p	Breaker travel distance
α	Angle between wave crest and x axis; Phillip's parameter for a wave spectrum; probability distribution shape parameter; arctan of structure slope
β	Orthogonal separation factor; angle to point of interest in diffraction zone; probability distribution scale parameter
Γ	Basin length
γ	Breaker height/water depth at breaking ratio; JONSWAP spectrum parameter; probability distribution location parameter
ΔP	Difference between central and ambient pressures in a hurricane
ϵ	Vertical component of particle displacement; perturbation parameter; spectral width parameter; wave phase lag
ζ	Horizontal component of particle displacement
η	Surface elevation above still water level
θ	Angle between the direction of wave propagation and the x axis; angle between the wind direction and the x axis
ν	Fluid kinematic viscosity; spectral width parameter
ρ	Fluid (water) density
ρ_a	Air density
σ	Wave angular frequency
τ	Viscous shear stress
Φ	TMA spectrum function
ϕ	Velocity potential; latitude
Ψ	Stream function
Ω	Surface tension
ω	Speed of Earth's rotation

1

SEA SURFACE GRAVITY WAVES

For coastal and ocean engineers and for many marine scientists, the single most important phenomenon with which we must contend are the waves on the sea surface. Waves damage shore protection structures and reshape beaches, they sink moored vessels in inadequately protected harbors, they generate flow in to and out of tidal estuaries, and they are a sources of energy through devices for the conversion of wave motion to electrical power. They are relatively simple in their basic form, but complex when we deal with them in the real world.

If the surface of a body of water is disturbed, for example, by the bow of a ship moving through the water, gravitational and surface tension forces act to return the displaced water surface to its equilibrium position. However, the returning surface water now has inertia that causes it to pass the equilibrium position (like a displaced pendulum arm) and establish a surface oscillation. This oscillation, in turn, disturbs the adjacent water surface, causing the surface disturbance to propagate away as a wave.

For the waves of greatest interest to us, surface tension forces normally may be neglected. We commonly refer to these waves as surface gravity waves because gravity is the dominant restoring force. As a wave propagates, the oscillatory water motion in the wave continues because of the interaction of gravity and inertia. Also, since water particles are continuously accelerating and decelerating as the wave propagates, dynamic pressure gradients develop in the water column. These are superimposed upon the normal vertical hydrostatic pressure gradient in the water column.

Waves have potential energy in the form of their surface displacement and kinetic energy in the motion of the water particles. Waves transmit this energy as they propagate. They also transmit a signal in the form of the surface time-history of the waves at a point. And there is a relatively small mass transport in the direction of wave propagation.

Boundary shear stresses at the bottom and at the air–water interface, as well as internal shear stresses, dissipate wave energy. If the bottom is a porous sand so that the wave pressure variations can induce flow in to and out of the bottom, or if the bottom is a soft mud that responds to the wave motion, additional energy dissipation can take place if the water is sufficiently shallow. The cumulative effects of this energy dissipation cause a damping of the waves as they propagate.

If waves encounter a rigid structure or a sloping beach, some of their energy will be reflected and the rest may be dissipated, primarily by wave breaking and subsequent runup on the structure or beach face.

1.1 THE NATURE OF SURFACE WAVES

A casual observer will notice that surface waves have a variety of origins. Most common are the waves generated by the wind. While these waves are still under the action of the wind we refer to them as wind waves or sea. After they travel away from the generating winds they take on a smoother surface form and are called swell. On the quiet waters of a marina ship-generated waves may commonly be observed. And on large bodies of water the tide, a wave generated primarily by gravitational attraction of the sun and moon, will be observed. Less common are the seismically generated surface waves or tsunamis generated by earthquakes that displace the sea bottom in sufficiently shallow water.

Besides differing in their mechanism of generation, these waves can be distinguished by their period. Wind waves and swell have a range of periods from about 1 to 25 sec with dominant ocean storm wave periods being between 5 and 15 sec (e.g., see Thompson, 1980). The periods of ship-generated waves depend on the ship speed and commonly range from 1.5 to 3 sec (Sorensen, 1967). Tsunami waves have periods from 10 min to over an hour and the dominant periods of the tides are around 12 and 24 h.

Ocean wind waves at sea are normally less than 10 ft high and swell tend to be of lesser height. But during major storms, wave heights of greater than 20 ft often occur. Ship wave heights rarely exceed 3 ft and are normally less than 2 ft. In the deep ocean, tsunami wave heights are believed to be about 2 ft or less, but as the waves approach shore they can increase in height to in excess of 10 ft, depending on the nature of the nearshore hydrography (Wiegel, 1964). Likewise, midocean tide wave heights (tide ranges) are relatively low, but coastal tide ranges in excess of 20 ft occur at several locations.

1.2 WAVE THEORIES AND WAVE CHARACTERISTICS

A starting point and building block for dealing with complex sea wave conditions are the theories that describe the important characteristics of a single wave. Some of the characteristics of interest are demonstrated in Figure 1.1,

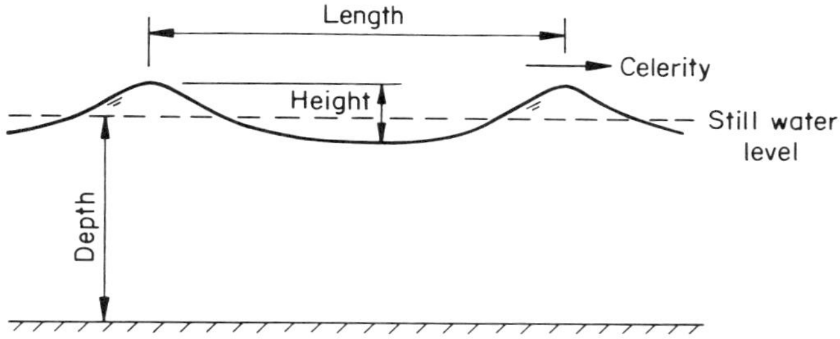

Figure 1.1 Two-dimensional wave profile.

which shows a two-dimensional profile view of a wave that is traveling from left to right. A wave is completely specified by its height, the water depth, and the wave period or wave length. Other wave characteristics of interest can be theoretically derived from these basic quantities. These include (1) the water surface profile; (2) the forward speed or celerity of the wave form; (3) wave water particle velocities, accelerations, and motion paths; (4) the dynamic pressure field in the wave; (5) the wave kinetic and potential energy; and (6) the wave power and momentum flux.

Water is a viscous fluid, but fortunately over most of the wave form gravity, pressure and inertial forces dominate. Only in the thin boundary layers at the bottom and surface do viscous forces become significant. Consequently, wave theories can be developed using the inviscid potential or irrotational flow hydrodynamic theory. An outline of the irrotational flow requirements for the development of wave theories that are presented in subsequent chapters will be given. For more background on irrotational flow the reader should see such basic references as Rouse (1961), Streeter (1948), and Vallentine (1967).

Besides defining the important characteristics of a wave at a specific water depth, we must be able to determine the changes that occur as the water depth changes owing to wave propagation toward the shore. And we must be able to predict the effects of three-dimensional phenomena such as refraction and diffraction.

1.3 WIND-GENERATED WAVES

The wave theories that we consider in Chapters 2 and 4 define the characteristics of a monochromatic wave, that is, a wave train that has the same height and length from wave to wave. A record of a wind-generated wave surface profile is much more irregular and complex than a monochromatic wave profile. So are the surface profiles of ship-generated waves, tsunamis, and the tide (but to a lesser extent).

The primary concern of this book is wind-generated waves. To deal with wind waves effectively, we will have to consider both the statistical distribution of wave heights and periods and the wave spectrum. Basically, a wave spectrum is a plot of the distribution of wave energy versus wave period or its inverse, wave frequency. The wave profile shown in Figure 1.1 is not symmetrical around the still water line. But this profile can be described, through Fourier analysis, by multiple sine waves having the same phase. Likewise, the more complex wind wave profile can be described by multiple sine waves of different amplitude, period, and phase. The cumulative energies of these sine components then define the wave energy spectrum.

1.4 DESIGN WAVES

A major objective of engineers and marine scientists is to be able to determine the appropriate design wave conditions for a specific location or to forecast near-term wave conditions for marine operations. As a background to developing these capabilities, effective wind wave measurement, analysis, and prediction techniques are needed. Also site-specific wave data are needed for a sufficient duration to make predictions of long return period wave conditions. These site-specific data are often determined by a combination of direct measurements and hindcasts made from longer term meteorological data.

Depending on the design requirement, different types of design wave information are needed. A single wave might seriously damage a rigid structure, whereas the stability of structures like stone mound breakwaters is more dependent on the level of repeated attack by many high waves. The performance of floating breakwaters and other floating structures is very dependent on the period of the higher incident waves. And beach processes depend on the annual wave climate variation, including the directions of the incoming waves.

And, given a design wave height and period or a design wave spectrum at an offshore point for a specific coastal site, analyses must be performed to quantify changes that occur as these waves propagate to the shore. This analysis might include the need to quantify the effects of wave breaking and wave transformations across the surf zone.

1.5 SURFACE WAVE LITERATURE

Several books dealing with surface waves have been written during the past three decades; most can be grouped in two broad categories. Those in one group contain one or a few chapters on wave mechanics as a prelude to the main topic of coastal and oceanographical engineering, beach processes, port engineering, or offshore structure design. Included in this group are Horikawa (1978), Horikawa (1988), Ippen (1966), Sarpkaya and Isaacson (1981), Sorensen (1978), U.S. Army Coastal Engineering Research Center (1984), and Wiegel (1964). The other group is only concerned with the mechanics of water waves. Books in this group generally have a very analytical orientation and most are restricted in their coverage. This group includes Dean and Dalrymple

(1984), Kinsman (1965), LeBlond and Mysak (1978), LeMehaute (1976), Mei (1983), Tucker (1991) and Whitham (1974).

Another important source of information, especially on the more applied aspects of surface waves such as field measurement programs, wave statistics, and the development of design criteria, are the proceedings of various conferences. Of particular value are those of the biannual International Conferences on Coastal Engineering and speciality conferences such as Ocean Waves Measurement and Analysis (Waves 74) and Directional Wave Spectral Applications, all published by the American Society of Civil Engineers.

Several journals are of interest. These include *Applied Ocean Research, Coastal Engineering*, the *Journal of Fluid Mechanics*, the *Journal of Geophysical Research*, the *Journal of Physical Oceanography*, the *Journal of the Waterway, Ports, Coastal and Ocean Engineering Division of the American Society of Civil Engineers* and *Ocean Engineering.*

Finally, research and project reports from major research laboratories such as the Danish Hydraulic Institute, Delft Hydraulics, the Canadian National Research Council Hydraulics Laboratory, Hydraulics Research Limited in England, Japan's Port and Harbor Research Institute and the U.S. Army Coastal Engineering Research Center are a worthwhile source of information on both fundamental and applied wave research.

REFERENCES

Dean, R. G. and Dalrymple, R. A. (1984), *Water Wave Mechanics for Engineers and Scientists*, Prentice-Hall, Englewood Cliffs, NJ.

Horikawa, K. (1978), *Coastal Engineering: An Introduction to Ocean Engineering*, University of Tokyo Press.

Horikawa, K. (1988), *Nearshore Dynamics and Coastal Processes: Theory, Measurement, and Predictive Models*, University of Tokyo Press.

Ippen, A. T. (1966), *Estuary and Coastline Hydrodynamics*, McGraw-Hill, New York.

Kinsman, B. (1965), *Wind Waves*, Prentice-Hall, Englewood Cliffs, NJ.

LeBlond, P. H. and Mysak, L. A. (1978), *Waves in the Ocean*, Elsevier Scientific, New York.

LeMehaute, B. (1976), *An Introduction to Hydrodynamics and Water Waves*, Springer-Verlag, Dusseldorf.

Mei, C. C. (1983), *The Applied Dynamics of Ocean Surface Waves*, Wiley, New York.

Rouse, H. (1961), *Fluid Mechanics for Hydraulic Engineers*, Dover, New York.

Sarpkaya, T. and Isaacson, M. (1981), *Mechanics of Wave Forces on Offshore Structures*, Van Nostrand Reinhold, New York.

Sorensen, R. M. (1967), ''Investigation of Ship-Generated Waves,'' *J. Waterw. Harbors Div. Am. Soc. Civ. Eng.*, February, 85–99.

Sorensen, R. M. (1978), *Basic Coastal Engineering*, Wiley, New York.

Streeter, V. L. (1948), *Fluid Dynamics*, McGraw-Hill, New York.

Thompson, E. F. (1980), "Energy Spectra in Shallow U. S. Waters," Technical Paper 80-2, U. S. Army Coastal Engineering Research Center, Ft. Belvoir. VA.

Tucker, M. J. (1991),. *Waves in Ocean Engineering—Measurement, Analysis, Interpretation*, Ellis Horwood, New York.

U. S. Army Coastal Engineering Research Center (1984), *Shore Protection Manual*, U. S. Government Printing Office, Washington, DC.

Vallentine, H. R. (1967), *Applied Hydrodynamics*, Butterworths, London.

Whitham, G. B. (1974), *Linear and Nonlinear Waves*, Wiley, New York.

Wiegel, R. L. (1964), *Oceanographical Engineering*, Prentice-Hall, Englewood Cliffs, NJ.

2

SMALL AMPLITUDE WAVE THEORY AND CHARACTERISTICS

In this chapter the simplest theory for two-dimensional progressive surface gravity waves, known as the small amplitude or linear theory, is given. This presentation includes a statement of the fundamental flow equations and boundary conditions for gravity wave theories, the simplifying assumptions made to derive the small amplitude theory, and an outline of the derivation of this theory. Also, the important kinematic and dynamic characteristics of two-dimensional surface waves is developed using the small amplitude theory. An important objective is to provide a strong physical understanding of the characteristics and behavior of an individual wave.

Small amplitude theory is extensively used in practice for a variety of reasons. It is much easier to apply than the more complex finite amplitude theories that are available, while still yielding useful results for many applications. Some of the characteristics of surface waves are easier to comprehend when described by the small amplitude theory. And this theory is most useful as a building block for dealing with more complex phenomena such as reflecting waves, wave refraction and diffraction, and wave spectra.

It is important, however, to have a clear picture of the limitations of this theory. The derivation assumes that the wave amplitude is small relative to the wave length and the water depth. Since water particle velocities are related to the wave amplitude and the wave celerity depends on the water depth and the wave length, this implies that particle velocities must be small compared to the wave celerity. Consequently, for high waves at sea or for waves propagating in shallow nearshore areas where these assumptions do not strictly hold, the small amplitude wave theory is of more limited accuracy. However, experiments have shown that the small amplitude theory yields surprisingly realistic results for conditions that markedly deviate from the small amplitude assumption.

The small amplitude wave theory was first developed by Airy (1845). Useful presentations of this theory can also be found in Dean and Dalrymple (1984), Ippen (1966), Lamb (1945), Sorensen (1978), U.S. Army Coastal Engineering Research Center (1984), and Wiegel (1964).

2.1 TWO-DIMENSIONAL SURFACE WAVES

Figure 2.1 is a definition sketch for a wave propagating in water of constant depth d in an x, z coordinate system. The x axis is the still water position and the wave is shown with its crest at the origin of the x axis. The bottom is at $z = -d$ and the wave surface profile is $z = \eta$, where η is a function of x and time t. The surface profile propagates forward at speed or celerity C and is defined by the wave length L and height H. Since the wave travels a distance of one wave length in a time equal to the wave period T, we have $C = L/T$.

Arrows indicate the direction of water particle motion at key points along the surface of the wave. Note that as the wave propagates one wave length in the x direction, a surface water particle would move forward, down, backward, and then up to complete a clockwise orbit. Similar but smaller orbital paths would be followed by particles down in the water column. An orbit for one of these particles is shown on the figure. The position of the particle at any instant during its orbit is given by its horizontal and vertical coordinates ζ and ϵ, respectively, referenced to the center of the orbit. The horizontal and vertical components of the water particle velocity are u and w respectively. At any instant, the water particle is located at a distance $d - (-z) = d + z$ from the bottom.

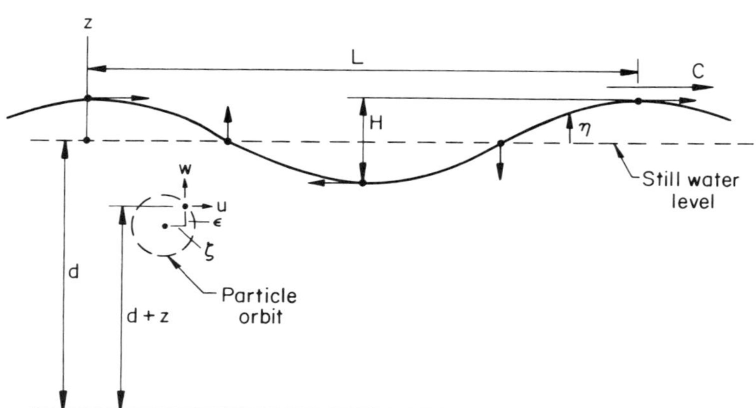

Figure 2.1 Definition of wave parameters.

In the material that follows, two useful dimensionless parameters are used:

$$k = 2\pi/L \text{ (wave number)}$$

$$\sigma = 2\pi/T \text{ (wave angular frequency)}$$

From this we see that $C = L/T = \sigma/k$. Also, we employ the phrase "wave steepness," defined as the wave height divided by the wave length (i.e., H/L), to describe a wave. And we classify wave characteristics according to the "relative depth," defined as the water depth divided by the wave length (i.e., d/L).

2.2 GOVERNING EQUATIONS, ASSUMPTIONS, AND BOUNDARY CONDITIONS

As discussed in Chapter 1 gravity, pressure, and inertia dominate wave motion outside of the thin boundary layers at the air–water interface and at the bottom. So the small amplitude and other wave theories we consider are based on the assumption of irrotational flow. Assuming flow is irrotational throughout the wave implies that there is no shear stress at the air–water interface or at the bottom. The wave theories will thus be valid throughout the wave except at the thin surface and bottom boundary layers.

Basically, we want to obtain the periodic velocity potential that satisfies the Laplace equation (which is an expression of continuity for irrotational flow) and the appropriate bottom and surface boundary equations. The Laplace equation for two-dimensional flow is

$$\frac{\partial^2 \phi}{\partial x^2} + \frac{\partial^2 \phi}{\partial z^2} = 0 \tag{2.1}$$

where ϕ is the velocity potential. The velocity potential we seek should be a function of the basic wave-defining parameters H, d, T, or L, the spatial position x and z, and time t.

Other governing assumptions include:

1. The water is homogeneous and incompressible and surface tension forces are negligible. This precludes internal gravity waves or pressure waves affecting the flow. And wave lengths must be longer than about 3 cm so that surface tension effects are unimportant. (In Section 2.5.5 we will briefly consider the effect of surface tension on gravity wave motion.)

2. The bottom is horizontal, impermeable, and stationary. Thus there is no vertical component of flow at the bottom and the bottom is not adding or removing energy from the wave or causing wave reflection. If the bottom slope is sufficiently small, as for example when waves propagate toward shore in the

nearshore region, the wave theories developed herein are adequate for most purposes.

3. The pressure along the air–water interface is constant. Thus no wind pressure would exist and the vertical aerostatic pressure difference between the wave crest and trough is negligible.

Three boundary conditions will apply, one at the bottom and two at the free surface. Since there is no flow normal to the bottom, the bottom boundary condition (BBC) is

$$w = \frac{\partial \phi}{\partial z} = 0 \quad \text{at } z = -d \tag{2.2}$$

At the surface there is a kinematic boundary condition (KSBC) that relates the vertical component of water particle velocity at the surface to the surface position:

$$w = \frac{\partial \eta}{\partial t} + u \frac{\partial \eta}{\partial x} \quad \text{at } z = \eta \tag{2.3}$$

And, there is a dynamic surface boundary condition (DSBC) based on the Bernoulli equation for unsteady flow which is

$$\frac{1}{2}(u^2 + w^2) + gz + \frac{p}{\rho} + \frac{\partial \phi}{\partial t} = 0 \tag{2.4}$$

where p is the pressure and ρ is the fluid density. At the surface, where the pressure is zero, the DSBC becomes

$$\frac{1}{2}(u^2 + w^2) + g\eta + \frac{\partial \phi}{\partial t} = 0 \quad \text{at } z = \eta \tag{2.5}$$

2.3 SMALL AMPLITUDE THEORY DERIVATION

Thus, to develop a theory for surface waves, we need to find a solution to the Laplace equation that satisfies the BBC, KSBC, and DSBC. Some difficulties immediately appear. The BBC is linear and it applies at a well-defined location—the bottom. However, the KSBC and the DSBC are nonlinear and they are defined at the unknown free surface. If we assume that the wave height is small relative to the wave length and the water depth, we can linearize these two boundary conditions and apply them at the still water level rather than at the water surface. Thus the linearized KSBC becomes

$$w = \frac{\partial \eta}{\partial t} \quad \text{at } z = 0 \tag{2.6}$$

and the linearized DSBC becomes

$$g\eta + \frac{\partial \phi}{\partial t} = 0 \quad \text{at } z = 0 \tag{2.7}$$

Employing the Laplace equation, the BBC, and the linearized DSBC, the desired velocity potential for small amplitude linear waves can be derived (see Ippen, 1966 or Sorensen, 1978). A useful form of this velocity potential is

$$\phi = \frac{gH}{2\sigma} \frac{\cosh k(d + z)}{\cosh kd} \sin (kx - \sigma t) \tag{2.8}$$

2.4 SMALL AMPLITUDE WAVE CHARACTERISTICS

With the velocity potential, we have a complete definition of the flow field for a small amplitude progressive monochromatic wave from which a description of most of the important wave characteristics can be developed. Also, the Laplace equation is linear so the sum or difference of any two solutions is also a solution. This allows us, for example, to add the solutions for two waves propagating in opposite directions to develop the solution for a standing wave or to add the solutions for a wave and a current to obtain a solution for wave–current interaction.

2.4.1 Surface Profile

Inserting the velocity potential into the linearized DSBC and letting $z = 0$ yields the equation for the wave surface profile:

$$\eta = \frac{H}{2} \cos (kx - \sigma t)$$

or, in expanded form,

$$\eta = \frac{H}{2} \cos 2\pi \left(\frac{x}{L} - \frac{t}{T} \right) \tag{2.9}$$

Thus the small amplitude theory defines the surface profile as a cosine form, which is reasonable for low-amplitude waves. But as the wave amplitude increases, wave forms become vertically asymmetric, as shown in Figure 1.1, where the wave crest has become sharper and the trough flatter.

2.4.2 Wave Celerity

Combining the DSBC and the KSBC yields

$$\frac{\partial^2 \phi}{\partial t^2} + g\frac{\partial \phi}{\partial z} = 0 \quad \text{at } z = 0$$

and, inserting the velocity potential, differentiating and rearranging yields

$$\sigma^2 = gk \tanh kd$$

or

$$C = \frac{\sigma}{k} = \sqrt{\frac{g}{k} \tanh kd} \tag{2.10}$$

Equation 2.10 is the classic dispersion equation that relates the wave celerity, the wave length, and the water depth. According to the small amplitude wave theory, the wave celerity is independent of the wave height. Finite amplitude wave theories, which are discussed in Chapter 4, show that there is a small but noticeable dependence of the wave celerity on the wave height, especially for steeper waves.

Using the relationship $C = L/T$, Eq. 2.10 can be converted to the alternate useful forms

$$C = \frac{gT}{2\pi} \tanh \frac{2\pi d}{L} \tag{2.11}$$

and

$$L = \frac{gT^2}{2\pi} \tanh \frac{2\pi d}{L} \tag{2.12}$$

Note that if the wave period and the water depth are known, the wave length and the wave celerity can be determined by trial and error from Eqs. 2.12 and 2.11 or 2.10 respectively. Also, tables are available (see U.S. Army Coastal Engineering Research Center, 1984) to directly determine the wave length from the wave period and depth. It can easily be shown (Ippen, 1966) that as a wave propagates across water of changing depth (e.g., from deep water toward the shore) the wave period will remain constant since the number of waves per unit time passing sequential points must be equal. Other characteristics such as the height, length, celerity, surface profile, internal pressure field, and particle kinematics change as a wave propagates across water of changing depth.

2.4.3 Wave Classification by Relative Depth

As a wave propagates from deep water toward the shore the wave length decreases (Eq. 2.12), but at a slower rate than the depth decreases. So the relative depth (d/L) decreases as the depth decreases. The relative depth is a valuable parameter for classifying waves because for certain ranges of relative depth the equations that describe wave characteristics are significantly simplified. And these ranges define the limits for some unique wave behavior patterns. We can define three depth ranges—deep, intermediate or transitional, and shallow water.

When the relative depth for a wave is greater than approximately 0.5 we have deep water. The tanh term in Eqs. 2.10–2.12 is approximately equal to unity and these equations respectively become

$$C_0 = \sqrt{\frac{gL_0}{2\pi}} \tag{2.13}$$

$$C_0 = \frac{gT}{2\pi} \tag{2.14}$$

and

$$L_0 = \frac{gT^2}{2\pi} \tag{2.15}$$

Note that the deep water wave length and celerity are denoted by the subscript zero, but no such designation is needed for the wave period because the period does not change as a wave propagates from deep to shallow water.

In deep water, the wave length and celerity are independent of the water depth; they depend only on the wave period. This is because the water particle orbit diameters decrease with increasing depth to near zero at $-z/L = 0.5$, so the bottom has no effect on the wave. The waves do not "feel" the bottom.

When the relative depth becomes less than 0.5 the waves enter the intermediate range. The water particle motion and other wave characteristics now depend on both the wave period and the water depth (Eqs. 2.10–2.12 must be used). Dividing Eq. 2.14 by Eq. 2.11 and Eq. 2.15 by Eq. 2.16 yields

$$\frac{C}{C_0} = \frac{L}{L_0} = \tanh\frac{2\pi d}{L} \tag{2.16}$$

a useful relationship that will be employed in a later chapter. The intermediate or transitional depth range continues until the relative depth decreases to less than approximately 0.05, where the shallow water range begins. In shallow water, tanh kd is approximately equal to kd and Eq. 2.10 becomes

$$C = \sqrt{gd} \tag{2.17}$$

Now, the wave celerity depends only on the water depth, and the wave period and length are related by $L = CT$ or

$$L = \sqrt{gd}\, T \tag{2.18}$$

It is important to emphasize that this classification is based on the local relative depth d/L. A tide wave having a period of around 12 h is so long that it is a shallow water wave in the deepest part of the ocean. The relative depth limits of 0.5 and 0.05 for deep and shallow water are somewhat arbitrarily chosen. At this deep water limit, using Eq. 2.13 rather than Eq. 2.10 to calculate wave celerity results in an error of only 0.4%. At the shallow water limit, use of Eq. 2.17 rather than Eq. 2.10 to calculate wave celerity results in an error of 1.6%.

Since the wave period is constant as a wave propagates toward the shore and is easier to measure than the wave length, it is sometimes more convenient to express the relative depth in the terms of the depth and wave period. From the dispersion equation, the limits for deep and shallow water become $d/gT^2 > 0.08$ and $d/gT^2 < 0.0025$ respectively. Thus, for example, for an 8-sec wave, depths greater than 50.2 m are deep water and depths less than 1.5 m are shallow water.

Note that the celerity of deep water waves is period dependent, whereas the celerity of shallow water waves is dependent on the water depth but not the wave period. Thus, in deep water a spectrum of waves is dispersive because the longer period (length) waves travel faster than, and move out ahead of, the shorter period waves. But small amplitude waves in shallow water are not dispersive, they all have the same celerity, which only depends on the water depth.

2.4.4 Particle Velocity, Acceleration, and Orbit Geometry

The horizontal and vertical components of the water particle velocity in a wave can be obtained from the velocity potential where $u = \partial\phi/\partial x$ and $w = \partial\phi/\partial z$. After inserting the dispersion relationship and some algebraic manipulation, we have

$$u = \frac{\pi H}{T}\left[\frac{\cosh k(d + z)}{\sinh kd}\right]\cos(kx - \sigma t) \tag{2.19}$$

$$w = \frac{\pi H}{T}\left[\frac{\sinh k(d + z)}{\sinh kd}\right] \sin (kx - \sigma t) \tag{2.20}$$

These equations give the particle velocity components at a point $(x, -z)$ as time passes and different fluid particles pass through that point.

Note that each velocity component consists of three parts: (1) the surface deep water particle speed $\pi H/T$ (which is easily seen as the particle circular orbit circumference πH divided by the time required to complete an orbit, T); (2) a hyperbolic function of z which causes an exponential decrease in particle velocity with increasing distance below the free surface, and which accounts for modifications caused by the wave moving into transitional and shallow water; and (3) a phasing term that defines the cyclic velocity variation through a wave phase (as expected, the components are 90° out of phase with each other). Remember that the linearized surface boundary conditions were applied at $z = 0$, so the velocity component equations are only valid between the still water level and the bottom.

Calculation of the force exerted by a wave on a vertical cylindrical pile, for example, requires a calculation of both the water particle velocity and acceleration for the given wave height and period and the water depth. The total acceleration has both convective and local components. The horizontal component of acceleration is given by

$$a_x = u\frac{\partial u}{\partial x} + w\frac{\partial u}{\partial z} + \frac{\partial u}{\partial t}$$

where the first two terms on the right side are the convective acceleration and the third term is the local acceleration. The magnitude of the local term is of the order of the wave steepness (H/L) and the magnitude of the convective term is of the order of the wave steepness squared. Since the wave steepness for most conditions is small, it is common to neglect the convective term when determining the wave acceleration. Consequently, the horizontal and vertical components of particle acceleration are closely approximated by

$$a_x = \frac{2\pi^2 H}{T^2} \frac{\cosh k(d + z)}{\sinh kd} \sin (kx - \sigma t) \tag{2.21}$$

$$a_z = \frac{2\pi^2 H}{T^2} \frac{\sinh k(d + z)}{\sinh kd} \cos (kx - \sigma t) \tag{2.22}$$

Note that the hyperbolic decay/shoaling term is the same as for the respective particle velocity equations, but that there is a 90° phase difference between the respective velocity and acceleration terms. This phase difference is easily seen by considering a water particle as it transits around its orbit. The particle velocity is tangent to the orbit and the particle acceleration is at right angles to the velocity, being directed toward the center of curvature of the orbit.

As a water particle traverses its orbit, the particle horizontal and vertical displacement components (see Fig. 2.1) can be determined by integrating the velocity components (with time). This yields

$$\zeta = -\frac{H}{2}\frac{\cosh k(d + z)}{\sinh kd} \sin (kx - \sigma t) \tag{2.23}$$

$$\epsilon = \frac{H}{2}\frac{\sinh k(d + z)}{\sinh kd} \cos (kx - \sigma t) \tag{2.24}$$

for the horizontal and vertical displacements from the mean position. The displacements are evaluated for the particle orbit relating to the instantaneous particle position, but the small amplitude assumptions allow us to assume that the coordinates x, z in Eqs. 2.23 and 2.24 apply to the orbit center position.

A plot of typical orbit geometries for deep and shallow water is shown in Figure 2.2. In deep water the orbits are circular with a diameter at the surface equal to the wave height, and decreasing exponentially to a diameter of approximately 4% of the wave height at a depth equal to half of the wave length. In shallow water the particle orbits are elliptical. The ellipses become flatter with distance down in the water column in both transitional and shallow water owing to wave contact with the bottom. The horizontal particle displacements increase (at any particular $-z$) as a wave propagates from deep to shallow water. At the bottom in transitional and shallow water, the particle motion is strictly horizontal as required by the BBC. Since the decay/shoaling term is the same for the respective velocity, acceleration, and displacement equations, the particle velocity and acceleration components have the same spatial relative behavior as do the particle displacements.

The small amplitude wave theory yields a sinusoidal surface profile which, as previously stated, is satisfactory for waves of relatively low height in deep water. However, as a wave propagates into transitional and shallow water, the

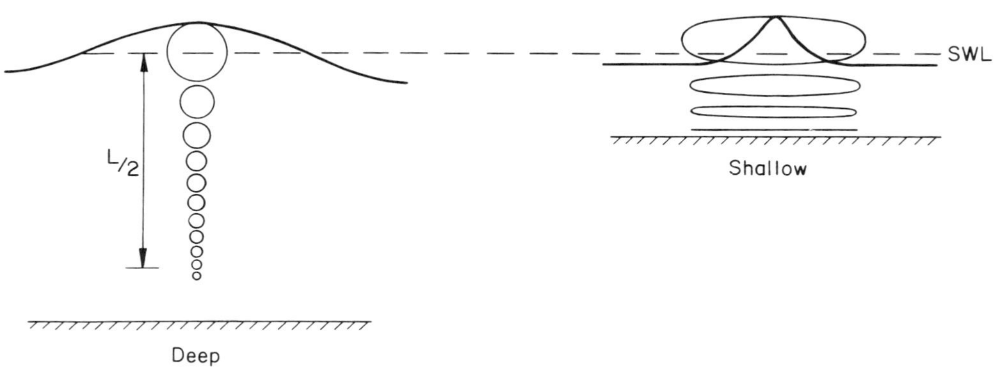

Figure 2.2 Transformation of particle orbits and surface profile, deep to shallow water.

surface profile becomes trochoidal with long flat troughs and shorter peaked crests, as shown in Figure 2.2. Consistent with this profile change, the amplitude of the wave crest increases while the amplitude of the trough decreases. In transitional and shallow water, particles still move in essentially closed orbits. Because they must therefore travel the same distance forward under the crest as they do back under the trough, but in less time, owing to the trochoidal profile, peak velocities under the crest will significantly exceed those under the trough in shallower water. These surface profile and particle velocity asymmetries are not accounted for by the small amplitude theory.

As was done for the dispersion equation, it is useful to look at the limits reached by the particle velocity, acceleration, and displacement equations in deep and shallow water. At these limits, it can be shown that for

$$\text{Deep water:} \quad \frac{\cosh k(d+z)}{\sinh kd} \approx \frac{\sinh k(d+z)}{\sinh kd} \approx e^{kz} \tag{2.25}$$

$$\text{Shallow water:} \quad \frac{\cosh k(d+z)}{\sinh kd} \approx \frac{1}{kd} \tag{2.26}$$

$$\frac{\sinh k(d+z)}{\sinh kd} \approx 1 + \frac{z}{d} \tag{2.27}$$

Inserting the results of Eq. 2.25 into Eqs. 2.19–2.24 provides equations that define the exponential decay in the particle velocity, acceleration, and displacement for deep water. Substitution of Eqs. 2.26 and 2.27 into Eqs. 2.19 and 2.20 yields the following equations for the particle velocity in shallow water:

$$u = \frac{\pi H}{T}\left(\frac{1}{kd}\right)\cos(kx - \sigma t) \tag{2.28}$$

$$w = \frac{\pi H}{T}\left(1 + \frac{z}{d}\right)\sin(kx - \sigma t) \tag{2.29}$$

Note that in shallow water the horizontal component of particle velocity is independent of the distance below the still water line ($-z$), but the vertical component decreases from a maximum at the still water level to zero at the bottom. Similar equations and related statements can be developed for the particle accelerations and displacements in shallow water. As a wave just enters shallow water $d/L = 1/20$, so $1/kd = 10/\pi = 3.18$. And at the still water line $(1 + z/d) = 1$, so the horizontal to vertical ratio of the particle velocity, acceleration, and displacement components is 3.18 : 1 at the entrance to shallow water.

2.4.5 Pressure Field

Substitution of the velocity potential into the linearized form of the unsteady Bernoulli equation (Eq. 2.4) yields an equation that defines the pressure field in a wave:

$$p = -\rho g z + \frac{\rho g H}{2} \frac{\cosh k(d + z)}{\cosh kd} \cos(kx - \sigma t) \tag{2.30}$$

The first term on the right is the normal hydrostatic pressure increase with distance below the still water level that occurs in a static fluid. The second term on the right is the dynamic pressure owing to fluid particle acceleration in the wave motion. These static and dynamic pressure components are plotted in Figure 2.3 for vertical sections under the wave crest and trough of a deep water wave. Under the wave crest, water particles are accelerating downward so a downward pressure gradient is required. This adds to the static pressure to yield the total pressure under the wave crest. Under the wave trough water particles are accelerating upward, resulting in an upward dynamic pressure gradient that subtracts from the static pressure. Halfway between the crest and trough the water is accelerating horizontally so no vertical dynamic gradient develops and the vertical pressure distribution is hydrostatic. Again, owing to the application of the surface boundary conditions at $z = 0$, Eq. 2.30 is only valid below the still water level. Above this level under the wave crest, the pressure must regularly decrease to zero at the water surface.

A pressure sensor placed in the water at a distance less than about $L/2$ below the still water line will detect a constant static pressure plus a fluctuating dynamic pressure. This pressure sensor can be used as a wave gauge. The period of the pressure fluctuation is the wave period, from which the value of kd may be calculated using the dispersion equation. The dynamic pressure component of Eq. 2.30 can then be used to determine the wave height.

2.4.6 Wave Energy

The energy in a wave is important to the understanding of several phenomena including the generation of waves by wind, the changes that occur as a wave

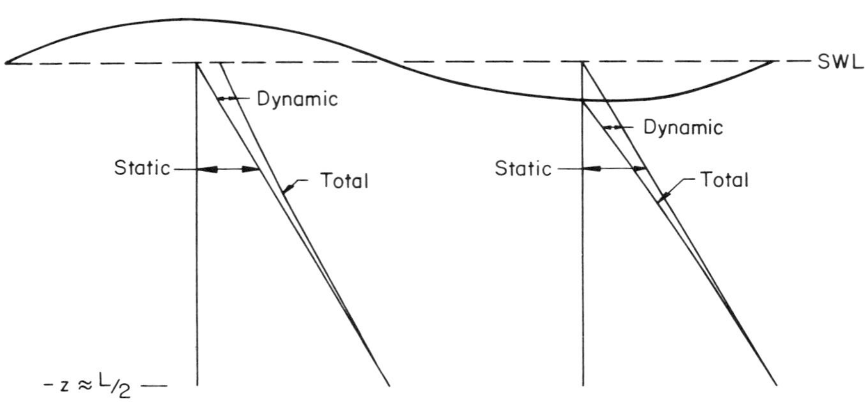

Figure 2.3 Pressure field, deep water wave.

propagates from deep to shallow water, and the spectral characteristics of waves. The total energy in a wave is the sum of its kinetic energy manifest by the water particle motion and its potential energy owing to the water surface displacement from the still condition. Equations for each may be derived by considering Figure 2.4.

The kinetic energy per unit width of wave crest and for one wave length (E_k) is equal to the integral over one wave length and the water depth of the kinetic energy of the small element *dx dz*, or

$$E_k = \int_0^L \int_{-d}^{0} \tfrac{1}{2} \rho dx\, dz\, (u^2 + w^2)$$

Inserting the velocity components (Eqs. 2.19 and 2.20), integrating, and performing the required algebraic manipulations yields

$$E_k = \frac{\rho g H^2 L}{16}$$

The potential energy per unit crest width and for one wavelength (E_p) can be found by subtracting the potential energy of the still water from the potential energy of the wave surface profile or

$$E_p = \int_0^L \rho g (d + \eta) \left(\frac{d + \eta}{2}\right) - \rho g d L \left(\frac{d}{2}\right)$$

Inserting the surface elevation as a function of x (Eq. 2.9, let $t = 0$) and integrating yields

$$E_p = \frac{\rho g H^2 L}{16}$$

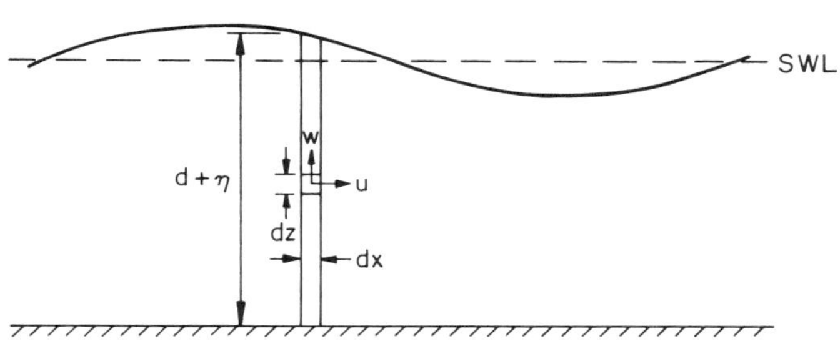

Figure 2.4 Wave energy parameters definition.

Thus the kinetic and potential energies are equal and the total energy E is the sum of the two or

$$E = \frac{\rho g H^2 L}{8} \tag{2.31}$$

Note that the energy is a function of the wave height squared. For example, a floating breakwater that dissipates and reflects a total of 50% of the incident wave energy (thus allowing 50% energy transmission), would only reduce the wave height in the lee of the breakwater by 29%.

The energy is variable from point to point along a wave length. (The kinetic energy is essentially uniform along a horizontal line, but the potential energy varies from a maximum at the wave crest to a minimum at the wave trough). A useful concept is the average energy per unit surface area:

$$\overline{E} = \frac{E}{(1)L} = \frac{\rho g H^2}{8} \tag{2.32}$$

This is usually known as the specific energy or energy density of a wave. Equations 2.31 and 2.32 apply for deep to shallow water within the limits of the applicability of small amplitude wave theory.

2.4.7 Energy Flux and Group Celerity

An important property of waves is that they transmit energy. The energy flux or power P of a wave is the average energy per unit time and per crest width transmitted in the direction of wave propagation. The power can be written as the product of the force acting on a vertical plane situated normal to the direction of wave propagation times the particle flow velocity across this vertical plane. This force is the product of the dynamic pressure and the area of the vertical plane. Thus

$$P = \frac{1}{T} \int_0^T \int_{-d}^0 (p + \rho g z) u \, dz \, dt$$

where the term in parentheses is the dynamic pressure. Inserting the dynamic pressure from Eq. 2.30 and the horizontal particle velocity from Eq. 2.19 and integrating yields

$$P = \frac{\rho g H^2 L}{16T} \left(1 + \frac{2kd}{\sinh 2kd}\right)$$

or

$$P = \frac{E}{T}\left[\frac{1}{2}\left(1 + \frac{2kd}{\sinh 2kd}\right)\right] \tag{2.33}$$

Letting

$$n = \frac{1}{2}\left(1 + \frac{2kd}{\sinh 2kd}\right) \tag{2.34}$$

Equation 2.33 becomes

$$P = \frac{nE}{T} = n\bar{E}C \tag{2.35}$$

The term n is a function of kd or the relative depth d/L. Its value varies from 0.5 in deep water to 1.0 in shallow water. Equation 2.35 indicates that n can be interpreted as the fraction of the energy in a wave that is transmitted forward each wave period.

An important phenomenon related to wave energy flux is the celerity of a group of waves. Consider a long wave tank in which a small group of a few deep water waves is generated and then the wave generator is stopped. As the waves travel along the tank, waves in the front of the group will gradually decrease in height and, if the tank is long enough, will disappear in sequence starting with the first wave. As the waves in the front of the group are decreasing in height, new waves will appear at the back of the group and commence to grow. One new wave will appear each wave period so the total number of waves in the group will continually increase. But since no energy is being added to the group, the average height of the waves in the group will continuously decrease. Also, the wave group celerity is less than the celerity of the individual waves in the group.

An explanation for this phenomenon can be found by considering Eq. 2.35, which shows that only a fraction n of the wave energy is transmitted forward each period. Thus, for example, as the first wave of a group of waves traveling in deep water advances one wave length into still water, only half of its energy is transmitted forward. Consequently this front wave decreases in energy by a half each successive wave period. Its height would decrease by 29% of the height it had one wave length back during each period. The last wave in the group leaves half of its energy behind during each succeeding period to form a new wave at the end of the group.

A practical consequence of the group celerity being less than the celerity of individual waves in the group is that when waves are generated by a storm, prediction of their arrival time at a given location must be based on the group celerity.

The top portion of Figure 2.5 shows two wave trains, one having a slightly larger celerity and length than the other, traveling in the same direction. The resulting surface configuration, shown in the bottom portion of the figure, is the sum of the individual surface elevations at each point along the still water line. The result is a beating effect in which the two component waves are alternately in and out of phase. This produces the highest wave when the two components are in phase, reducing to zero wave height where the component

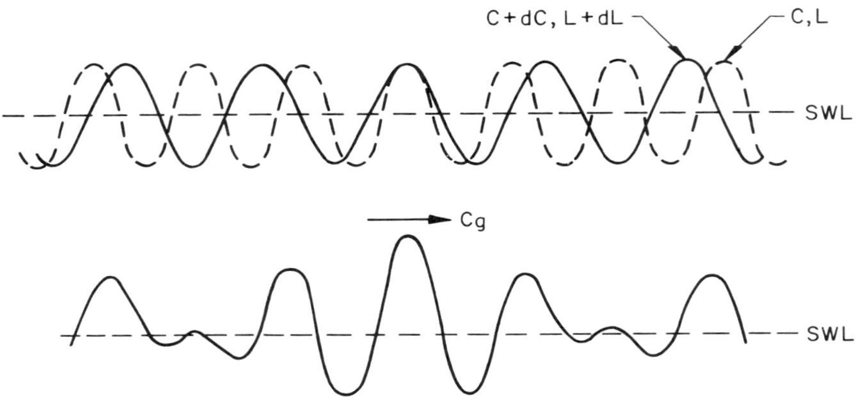

Figure 2.5 Wave group development.

waves are exactly out of phase. The result is a wave group propagating forward at a celerity C_g. As you follow an individual wave phase in the group its amplitude increases to a peak, then decreases as it moves through the group and disappears at the front. This is the group phenomena discussed above.

It can be shown (see Ippen, 1966 and Sorensen, 1978) that the group celerity is related to the length and celerity of the individual waves in the group by

$$C_g = C - L\frac{dC}{dL} \tag{2.36}$$

In shallow water, small amplitude waves are nondispersive ($dC/dL = 0$) and the group celerity equals the phase celerity. In deep water $dC/dL = C/2L$ (from Eq. 2.13) so the group celerity is half the phase celerity.

A general relationship can be obtained by inserting Eq. 2.10 into Eq. 2.36 yielding

$$C_g = \frac{C}{2}\left(1 + \frac{2kd}{\sinh 2kd}\right) \tag{2.37}$$

or

$$C_g = nC \tag{2.38}$$

Equation 2.38 shows that n is the ratio of the group celerity to the phase celerity as well as the fraction of the energy in a wave being transmitted forward. Another way to look at this is that the wave energy is propagated forward at the group celerity.

2.4.8 Momentum Flux, Radiation Stress

For basic fluid flow, some problems are best solved using the energy equation and others (such as a hydraulic jump where there is a strong concentrated dissipation of energy) the impulse–momentum principle. Similarly, for waves, some problems are best addressed considering the flux of momentum. This approach was first applied to waves by Longuett–Higgins and Stewart (1960, 1964) who introduced the term "radiation stress," which they defined as "the excess flow of momentum due to the presence of waves."

The instantaneous horizontal flux of momentum at a given location consists of the pressure force on a vertical plane plus the transfer of momentum through that vertical plane. The latter is the product of the momentum in the flow and the flow rate across the plane. Dividing by the area of the vertical plane yields the momentum flux for the x direction which is

$$p + \rho u^2$$

The resulting radiation stress S_{xx} for a wave propagating in the x direction is

$$S_{xx} = \overline{\int_{-d}^{\eta} (p + \rho u^2)\, dz} - \int_{-d}^{0} \rho g z dz$$

where the subscript xx denotes the x-directed momentum flux across a plane defined by x = constant. In the equation shown above p is the total static plus dynamic pressure so that the static pressure must be subtracted to obtain the radiation stress for the wave. The overbar denotes that the term is averaged over the wave period. Inserting the pressure and particle velocity terms from the small amplitude theory yields

$$S_{xx} = \frac{\rho g H^2}{8}\left(\frac{1}{2} + \frac{2kd}{\sinh 2kd}\right) = \overline{E}\left(2n - \frac{1}{2}\right) \tag{2.39}$$

Likewise S_{yy}, the y-directed momentum flux across a plane defined by y = constant (for a wave traveling in the x direction) is

$$S_{yy} = \frac{\rho g H^2}{8}\left(\frac{kd}{\sinh kd}\right) = \overline{E}\left(n - \frac{1}{2}\right) \tag{2.40}$$

The radiation stress components S_{xy} and S_{yx} are both zero. Note, in deep water

$$S_{xx} = \frac{\overline{E}}{2} \qquad S_{yy} = 0$$

and in shallow water

$$S_{xx} = \frac{3\overline{E}}{2} \qquad S_{yy} = \frac{\overline{E}}{2}$$

If a wave is propagating in a direction that is at an angle with the x direction, the radiation stress components become

$$\begin{aligned} S_{xx} &= \overline{E}\left[n(\cos^2\theta + 1) - \frac{1}{2}\right] \\ S_{yy} &= \overline{E}\left[n(\sin^2\theta + 1) - \frac{1}{2}\right] \\ S_{xy} &= \frac{\overline{E}}{2} n \sin^2\theta = \overline{E}n \sin\theta \cos\theta \end{aligned} \tag{2.41}$$

The radiation stress components presented above are useful for analyzing a number of wave phenomena, including mean water level set up in the surf zone, wave–current interaction, and the alongshore currents generated in the surf zone by waves that obliquely approach the shore.

2.5 OTHER WAVE CHARACTERISTICS

There are other wave characteristics that are important for developing a basic understanding of small amplitude progressive waves, but that are not directly derived from the linear wave velocity potential and do not involve more complex phenomena than progressive waves in still water. In this section we briefly consider the mass transport that accompanies wave propagation, the breaking limits on deep and shallow water waves, standing waves and wave reflection, wave–current interaction, and the effects of surface tension on gravity waves.

2.5.1 Mass Transport

When dye is injected below the water surface in a wave tank and waves are generated, an observer will see the dye trace out the expected particle orbits. It will also be observed that the dye slowly drifts in the direction of wave propagation. The drift speed of the dye will be maximum near the surface and decrease with increasing distance below the surface.

This mass transport is predicted by finite amplitude wave theories, but not by the small amplitude theory. If the small amplitude theory horizontal component of particle velocity at a particular distance below the water surface is averaged over a wave period, the resulting average velocity is zero. This is true for points below the wave trough or, from the small amplitude assumptions, below the water surface since the vertical distance between the crest and trough is assumed to be very small.

The small amplitude theory can be used to gain some insight into the nature of mass transport by employing the Taylor theorem to approximate the particle velocity at the water surface for a wave of finite amplitude (see Dean and Dalrymple, 1984). The resulting surface velocity is greater at the crest than at the trough. If this surface particle velocity for a deep water wave is then averaged over a wave period, the result is

$$\overline{u_{z=\eta}} = \frac{\pi^2}{2}\left(\frac{H}{L}\right)^2 C \tag{2.42}$$

Equation 2.42 indicates that the surface drift speed is a function of the wave steepness (H/L) squared. For a relatively steep wave, say $H/L = 0.05$, the surface drift speed is 1.2% of the wave celerity. If this wave had a period of 7 sec, its celerity (Eq. 2.14) would be 10.9 m/s and the surface drift speed would be 0.13 m/s.

2.5.2 Wave Breaking Limits

When a wave reaches a certain height it will become unstable and break. In deep water breaking is primarily dependent on the wave steepness. As a rule of thumb, the maximum deep water wave steepness H_0/L_0 is about 1/7 or 0.14. In shallow water the maximum wave steepness is less and depends on the relative depth d/L and the beach slope. The ratio of the wave height to water depth at breaking for common beach slopes and wave periods is between 0.8 and 1.2.

Wave breaking is a complex phenomena. A common simplified approach to determining wave breaking limits is to assume that the wave crest particle velocity equals the wave celerity at the breaking point. Of course, applying the small amplitude theory to an analysis of this type gives unrealistic results owing to the significant inaccuracy of the small amplitude theory for steep waves.

2.5.3 Standing Waves, Wave Reflection

Consider two waves of identical height and period that are propagating in opposite directions along the x axis. The resulting motion from the superposition of these two waves would be a standing wave, illustrated in Figure 2.6*a*. The water surface oscillates between the two surface envelope positions shown during each time interval $T/2$. Water particles oscillate in a horizontal plane beneath the node and in a vertical plane beneath the antinode. When the water surface is at one of the envelope positions, water particles instantaneously come to rest and all of the wave energy is potential; halfway between the envelope positions the water surface is level and all energy is kinetic. The total energy

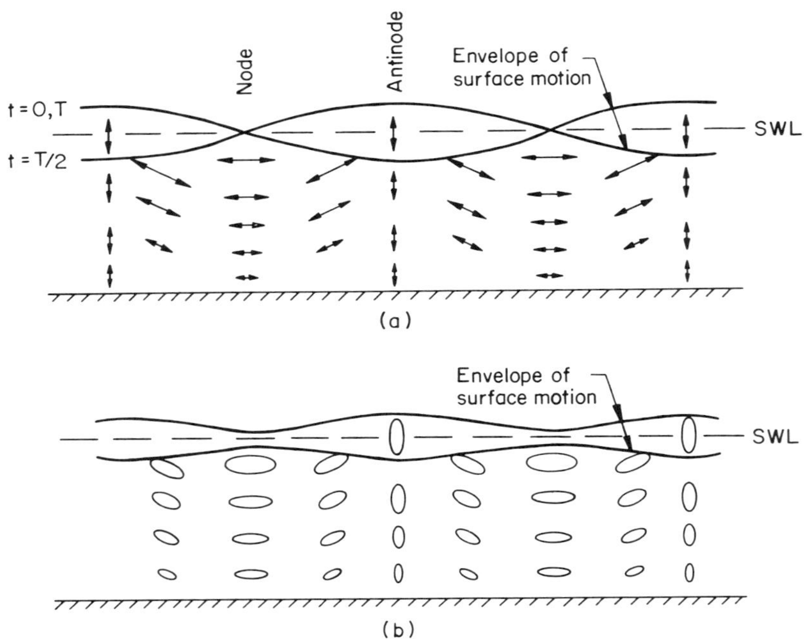

Figure 2.6 Standing, reflected waves. (a) Standing wave resulting from superposition of two waves moving in opposite directions. (b) Reflectioned wave height is less than incident wave height.

in a standing wave is the sum of the two component wave energies and the net energy flux is zero.

The velocity potential for a standing wave can be obtained by adding the velocity potentials for the two component waves moving in opposite directions. If each of the component waves has a height H, the standing wave height is $2H$. The result is

$$\phi = \frac{gH}{\sigma} \frac{\cosh k(d + z)}{\cosh kd} \cos kx \sin \sigma t$$

With this, the various characteristics can be determined as was done for progressive waves. The water surface profile then becomes

$$\eta = H \cos kx \cos \sigma t \tag{2.43}$$

and the particle velocities are

$$u = \frac{\pi H}{T} \frac{\cosh k(d + z)}{\sinh kh} \sin kx \sin \sigma t \tag{2.44}$$

and

$$w = \frac{\pi H}{T} \frac{\sinh k(d + z)}{\sinh kd} \cos kx \sin \sigma t \tag{2.45}$$

The pressure becomes

$$p = -\rho g z + \rho g H \frac{\cosh k(d + z)}{\cosh kd} \cos kx \cos \sigma t \tag{2.46}$$

and the horizontal and vertical particle displacements are

$$\zeta = -H \frac{\cosh k(d + z)}{\sinh kd} \sin kx \cos \sigma t \tag{2.47}$$

$$\epsilon = H \frac{\sinh k(d + z)}{\sinh kd} \cos kx \cos \sigma t \tag{2.48}$$

It is interesting to compare Eqs. 2.44 and 2.48 with their progressive wave counterparts. The hyperbolic decay/shoaling terms are the same for progressive and standing waves. But, at a given point x, $-z$, the horizontal and vertical velocity and displacement components are in phase, rather than having a 90° phase lag as is the case for progressive waves. The total pressure is hydrostatic under a node and the dynamic pressure fluctuates from positive to negative under an antinode.

The total standing wave energy per unit crest width (for one wave length) is

$$E = \frac{\rho g H^2 L}{4} \tag{2.49}$$

which consists of kinetic and potential energy components given by

$$E_k = \frac{\rho g H^2 L}{4} \sin^2 \sigma t \tag{2.50}$$

and

$$E_p = \frac{\rho g H^2 L}{4} \cos^2 \sigma t \tag{2.51}$$

so, as mentioned above, at $t = 0, T/2, T, \ldots E = E_p$ and at $t = T/4, 3T/4, \ldots E = E_k$.

A standing wave can be caused by a progressive wave reflecting from a perfectly reflecting (i.e., frictionless, impervious, inelastic) vertical wall located at an antinode. The water surface position on the wall, the particle velocity along the wall, and the pressure distribution on the wall would be given by Eqs. 2.43–2.46 with $\cos kx = 1$.

As the wall slope decreases or the wall roughness or permeability increase, the reflected wave height will decrease. Also, for a given obstacle, wave reflection decreases with an increase in incident wave steepness, particularly when wave breaking occurs. The water surface envelopes and the water particle motions for the case where the reflected wave height H_r is less than the incident wave height H_i (i.e., the reflection coefficient $C_r = H_r/H_i$ is less than 1) are shown in Figure 2.6*b*. Water particles move in flat elliptical paths with an orientation somewhat similar to a pure standing wave. As the reflection coefficient decreases from 1 to 0, the particle trajectories change from the pure standing wave pattern to the orbital pattern of a progressive wave. It can be shown (see Ippen, 1966) that the envelope height at the antinode equals $H_i + H_r$ and the envelope height at the node equals $H_i - H_r$. The reflection coefficient equals the difference between the two envelope heights divided by the sum of the two envelope heights. When tests are being run with monochromatic waves in a wave tank, the wet mark on the side of the tank displays the upper envelope shown in Figure 2.6*b* and is a good indicator of the amount of wave reflection. A wave gauge mounted on a movable carriage and moved along the tank will allow a more precise evaluation of the wave reflection by measuring the node and antinode envelope heights.

2.5.4 Wave–Current Interaction

To this point we have only considered waves propagating over still water. When such a wave encounters flowing water, for example, when waves from the relatively still ocean enter a restricted tidal channel having a high flood or ebb velocity, the wave characteristics are significantly changed. Since the number of waves per unit time passing a point in the ocean must equal the number of waves per unit time passing a fixed point in the channel, the wave period remains the same.

If the waves encounter an ebb current, the wave celerity relative to the land must decrease so the wave length will also decrease ($C = L/T$). A reduced wave length causes an increased wave height and a more significant increase in wave steepness. Wave breaking may result. If the wave encounters a flooding current, the wave length will increase and the height and steepness will decrease.

By a simple analysis using small amplitude wave theory Unna (1942) developed the following relationship between the wave celerity in deep still water, C_s, and the celerity and length at a point where there is a uniform current speed U at a depth d.

$$C = \frac{C_s}{2} \tanh \frac{2\pi d}{L} \left[1 \pm \sqrt{1 + \frac{4U}{C_s} \coth \frac{2\pi d}{L}} \right] \tag{2.52}$$

This equation and $C = L/T$ allows one to calculate the resulting change in wave celerity and length owing to an opposing $(-U)$ or following $(+U)$ current.

If the channel current is also in deep water, Eq. 2.47 reaches a limit at $U = -C_s/4$. Waves cannot penetrate a current having a higher opposing speed than this. At this condition $C = C_s/2 = -2U$; that is, the wave celerity in the ebb current is twice the current speed so the wave group celerity would be equal to the current speed. Wave energy, which travels at the group celerity, would "pile up," causing the waves to break. This breaking would occur no matter what amplitude the incident waves have. For finite amplitude waves Perigrine (1976) found that the "stopping current" velocity was slightly higher $(U = -0.3C_s)$.

Perigrine (1976) also evaluated the effect of a current flowing in the direction of wave propagation on the calculated wave height from bottom pressure measurements (see Section 2.5). For common current speeds and wave periods this effect can be significant.

2.5.5 Capillary Waves

One assumption made when we derived the small amplitude theory was that surface tension is negligible. This assumption is reasonable if the wave length exceeds about 3 cm—a requirement that has little practical impact on most everyday uses of the small amplitude theory. But it is of academic interest to look briefly at the affect of surface tension on small amplitude waves.

Including surface tension Ω, the dynamic surface boundary condition (see Eq. 2.7) becomes

$$g\eta + \frac{\partial\phi}{\partial t} + \frac{\Omega\partial^2\eta}{\rho\partial x^2} = 0 \quad \text{at } z = 0$$

The bottom and kinematic surface boundary conditions do not change.

Solving the boundary value problem yields

$$\phi = \frac{H\left(g - \dfrac{k\Omega^2}{\rho}\right)}{2\sigma} \frac{\cosh k(d + z)}{\cosh kd} \sin (kx - \sigma t)$$

Which is similar to Eq. 2.8 except for the surface tension term. The dispersion equation becomes

$$C = \sqrt{\left(\frac{g}{k} + \frac{\Omega k}{\rho}\right) \tanh kd} \tag{2.53}$$

Waves that have significant surface tension effects will be short, so the deep water form of Eq. 2.53 will represent most capillary waves. This becomes

$$C = \sqrt{\frac{g}{k} + \frac{\Omega k}{\rho}} \tag{2.54}$$

Note that the gravity component is proportional to the wave length, but the surface tension component is inversely proportional.

We can employ Eq. 2.54 to plot the two components of the wave celerity as a function of wave length. This is Figure 2.7, where the dashed lines represent the gravity (1) and surface tension (2) components and the solid line is the combined effect.

There is a minimum celerity $C_{\min}$ at a wave length $L_{\min}$. The celerity increases for decreasing wave length below $L_{\min}$. Taking the partial derivative of C with L (Eq. 2.54) and setting this equal to zero yields

$$L_{\min} = 2\pi\sqrt{\frac{\Omega}{\rho g}} \tag{2.54}$$

$$C_{\min} = \sqrt[4]{\frac{4g\Omega}{\rho}} \tag{2.56}$$

For water at 20°C, $\Omega = 0.073$ N/m, which yields $L_{\min} = 1.7$ cm, $C_{\min} = 23$ cm/sec and $T_{\min} = L_{\min}/C_{\min} = 0.074$ sec.

Another interesting characteristic of capillary waves is their group celerity.

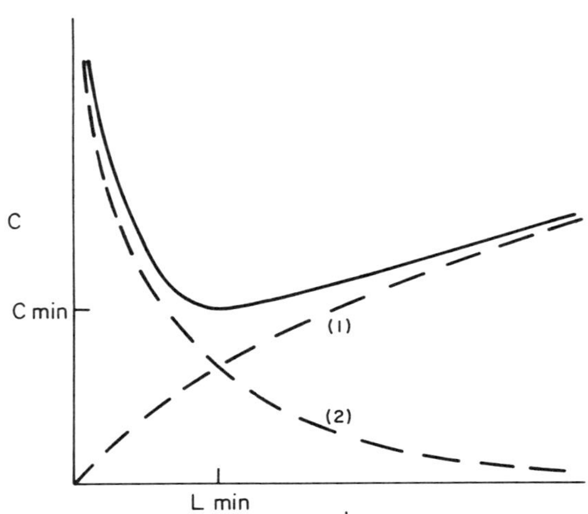

Figure 2.7 Dispersion relationship for capillary-gravity waves.

Neglecting the gravity component of Eq. 2.54 and using this to solve Eq. 2.36, we find that for deep water capillary waves $C_g = 3C/2$ rather than $C/2$ for pure gravity waves.

REFERENCES

Airy, G. B. (1845), "On Tides and Waves," *Encyclopedia Metropolitan*, London, pp. 241–396.

Dean, R. G. and Dalrymple, R. A. (1984), *Water Wave Mechanics for Engineers and Scientists*, Prentice-Hall, Englewood Cliffs, NJ.

Ippen, A. T. (1966), *Estuary and Coastline Hydrodynamics*, McGraw-Hill, New York.

Lamb, H. (1945), *Hydrodynamics*, Dover, New York.

Longuet-Higgins, M. S. and Stewart, R. W. (1960), "Changes in the Form of Short Gravity Waves on Long Waves and Tidal Currents," *J. Fluid Mech.*, **8,** 565–583.

Longuet-Higgins, M. S. and Stewart, R. W. (1964), "Radiation Stress in Water Waves: A Physical Discussion, with Applications," *Deep Sea Res.*, **11,** 529–549.

Peregrine, D. H. (1976), "Interaction of Water Waves and Currents," *Advances in Applied Mechanics*, Vol. 16, Academic, New York, pp. 9–117.

Sorensen, R. M. (1978), *Basic Coastal Engineering*, Wiley, New York.

Unna, P. J. H. (1942), "Waves and Tidal Streams," *Nature*, 219–220.

U. S. Army Coastal Engineering Research Center (1984), *Shore Protection Manual*, U. S. Government Printing Office, Washington, DC.

Wiegel, R. L. (1964), *Oceanographical Engineering*, Prentice-Hall, Englewood Cliffs, NJ.

3

TWO-DIMENSIONAL WAVE TRANSFORMATION

In Chapter 2 the basic characteristics of two-dimensional surface waves were presented using the small amplitude wave theory. In this chapter we expand on the description of wave characteristics by considering in more detail the transformations that occur as a train of waves propagate from deep to shallow water, break, and run up the beach face.

Typically, wave conditions are forecast for deep water, or wave conditions are known from offshore measurement programs. Results of these forecasts or measurement programs are usually then summarized as representative wave heights, periods, and directions for important design storm conditions or for daily or more frequent time intervals throughout the year. Given these deep water conditions, it is important to be able to determine subsequently the changes that occur as the waves propagate toward the shore, break, and run up on the beach face. Of related interest are the changes in mean water level near and in the surf zone owing to the incident waves and wave reflection from shore structures or a beach.

Waves typically approach the shore from an oblique direction and nearshore bottom conditions are usually irregular, so a complete analysis of wave transformation requires a three-dimensional approach. However, it is convenient to consider first two-dimensional (x, z) wave transformation and to add the three-dimensional effects at a later time (see Chapter 5). First, a brief physical discussion of the changes that occur as a wave train propagates toward shore is presented. Then, some important features of this transformation, not covered in the presentation of the small amplitude theory, are given.

3.1 PHYSICAL DESCRIPTION OF WAVE TRANSFORMATION

In Chapter 2 the limits for the deep, intermediate, and shallow water regions were defined in terms of the local relative depth, and some of the basic wave characteristics for these regions were described employing the small amplitude theory. It is appropriate at this point to present a physical description of all the important transformations that occur as a train of regular waves moves toward the shore from deep water. Key features of these transformations are given in more detail, using the small amplitude wave theory where appropriate, in the sections that follow.

The wave period is constant from deep water in to the breaking point. As the water depth decreases toward the shore, both the wave phase celerity and the wave length decrease at the same rate (i.e., $L/C = T$), as defined by the dispersion equation. The group celerity increases as a fraction of the local phase celerity from a half in deep water to unity in shallow water. But since the phase celerity decreases with decreasing water depth, the absolute group celerity at first increases and then decreases. This variation in the wave group celerity can be worked out from Eqs. 2.11, 2.12, and 2.37.

If the energy in a wave were constant as a wave train propagates from deep to shallow water, the continuous decrease in wave length should cause the wave height to continuously increase as the depth decreases. But, owing to the behavior of the group celerity discussed above, the wave height decreases slightly through much of the intermediate depth range and then starts to increase rapidly as the wave approaches and passes through the shallow water range. Of course, any energy input from the wind or any dissipation or reflection of wave energy as the wave train propagates toward the shore can alter this behavior accordingly. Also, for very steep waves, the wave steepness may reach its limit and wave breaking may reduce the wave energy and height.

Very low steepness waves in deep water have a surface profile that is approximately sinusoidal. But for steeper waves in deep water, or as the wave steepness increases with decreasing water depth toward the shore, the surface profile develops increasingly sharper crests and flatter troughs. The crest elevation becomes increasingly more than half of the wave height. And the duration of the wave crest is increasingly less than half of the wave period. Water particle orbits similarly become asymmetric about their horizontal axis, being flatter below this axis and rounder above.

The small amplitude theory indicates that the horizontal component of the water particle velocity, at any particular point below the surface, increases significantly as the depth decreases, particularly in shallow water (see Eq. 2.28 where kd is decreasing and H is increasing). The vertical component of the water particle velocity at the surface decreases slightly then increases, as does the wave height as a wave propagates toward the shore. In deep water both the horizontal and vertical components decay exponentially with increasing distance below the surface. In shallow water, the horizontal velocity component is constant from the surface to bottom, while the vertical velocity component decreases linearly to zero at the bottom (see Eqs. 2.28 and 2.29).

The discussion of particle velocities is based on the small amplitude theory. Surface profile asymmetry causes a significant difference between the particle velocities under the wave crest and trough. Shorter crest durations mean that a water particle has significantly less time for the forward portion of its orbit than for the return portion under the trough. And the larger crest amplitude increases the travel distance owing to the vertical asymmetry of the particle orbit. Thus for steep waves, particularly in shallow water, the particle velocity under the wave crest will be significantly greater than the particle velocity under the trough.

As a wave train approaches shallow water before breaking, consideration of wave momentum flux over a sloping bottom shows that the mean water surface elevation is not constant but decreases slightly in the shoreward direction. Landward of the breaking point, in the surf zone where energy is being dissipated, momentum flux considerations show that the mean water surface elevation increases in the shoreward direction.

Also, in the surf zone oscillatory wave motion is transformed to translatory motion of the water particles causing a runup of water on the beach face. The relative elevation of this runup depends primarily on the beach slope and surface condition, and the steepness of the incident waves.

3.2 WAVE HEIGHT CHANGE

Figure 3.1 is a schematic representation of a section profile oriented normal to the shore, with a wave train propagating in from deep water. As a wave travels from point 1 to point 2, the ratio of the group to phase celerity and the fraction of energy in the wave being transmitted forward increases owing to the decrease in water depth. Also, the energy content of the waves may be increased or decreased by the wind blowing over the water surface. And energy may be dissipated at the bottom by bottom friction, by percolation of water into and out of the porous bottom, and by movement of the bottom if it is, for example,

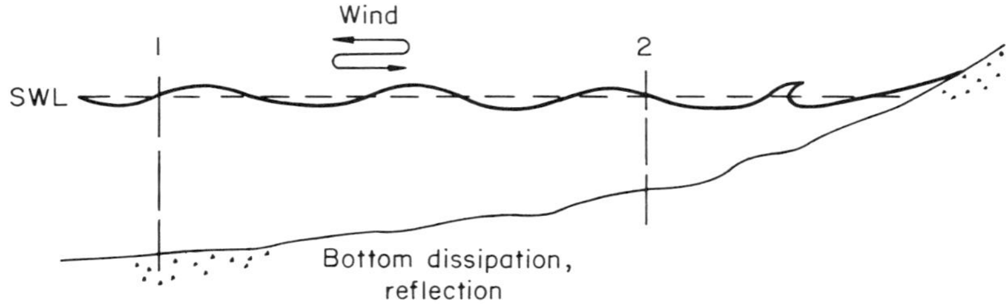

Figure 3.1 Shore-normal section profile.

a mud slurry. The sloping bottom or larger bottom irregularities may reflect wave energy seaward.

Often, considering the accuracy to which the incident wave height and period are known and the accuracy of the small amplitude theory, it is appropriate to calculate wave height changes as a wave train propagates toward the shore by neglecting the energy transfer to and from the waves owing to surface and bottom effects and the losses owing to wave reflection. This is particularly so when the distance over which the waves propagate is relatively short. Doing so, the wave energy per unit time passing 1 is equal to the wave energy per unit time passing 2, or $P_1 = P_2$.

If we construct lines normal or orthogonal to the wave crests with a spacing along the crest equal to B, then the total power in the wave between these orthogonal lines is BP. And if, as the wave train advances, no energy flows along the crest (i.e., across the orthogonal lines),

$$BP = \left(\frac{BnE}{T}\right)_1 = \left(\frac{BnE}{T}\right)_2$$

Inserting the wave energy from Eq. 2.31 and rearranging yields

$$\frac{H_1}{H_2} = \sqrt{\frac{n_2 L_2}{n_1 L_1}}\sqrt{\frac{B_2}{B_1}} \tag{3.1}$$

The first term on the right in Eq. 3.1 is a function of the relative depth (d/L) and represents the effect of depth change from 1 to 2 on the change in wave height. It is normally called the shoaling coefficient, K_s. The second term on the right would equal unity for our two-dimensional case as $B_1 = B_2$. But when three-dimensional waves approach the shore and refraction occurs (Chapter 5) orthogonal lines can converge or diverge, the value of this term will be greater or less than unity, respectively. This term is often called the refraction coefficient K_r. Thus $H_1/H_2 = K_s K_r$.

For two-dimensional wave propagation from deep water to some transitional or shallow water depth, Eq. 3.1 becomes

$$\frac{H}{H_0'} = \sqrt{\frac{L_0}{2nL}} = \sqrt{\frac{C_0}{2C_g}} \tag{3.2}$$

where the prime denotes wave shoaling without refraction. Table 3.1 is a tabulation of H/H_0' versus d/L and d/L_0. For reference, $C/C_0 = L/L_0$, C_g/C_0, n, and $(H/L)/(H_0'/L_0)$ are included in the table. Note that H/H_0' decreases to a minimum of 0.9129 at $d/L = 0.1891$ $(d/L_0 = 0.1570)$ and then continuously increases (until breaking) for decreasing relative depths. Since the wave length decreases as the depth decreases, the wave steepness (H/L)

TABLE 3.1 Tabulated Wave Parameters

d/L	d/L_0	H/H_0'	C/C_0, L/L_0	C_g/C_0	n	$(H/L)/(H_0'/L_0)$
0.02	0.0025	2.005	0.1250	0.1244	0.9947	16.04
0.03	0.00559	1.648	0.1863	0.1841	0.9884	8.85
0.04	0.00985	1.440	0.2462	0.2411	0.9795	5.85
0.05	0.01521	1.303	0.3042	0.2947	0.9685	4.28
0.07	0.02895	1.134	0.4136	0.3891	0.9409	2.74
0.09	0.04608	1.037	0.5120	0.4646	0.9074	2.03
0.10	0.05569	1.005	0.5569	0.4952	0.8892	1.80
0.15	0.1105	0.9254	0.7367	0.5839	0.7930	1.26
0.20	0.1700	0.9134	0.8500	0.5993	0.7049	1.075
0.30	0.2865	0.9445	0.9550	0.5605	0.5870	0.989
0.40	0.3948	0.9749	0.9870	0.5260	0.5330	0.988
0.50	0.4981	0.9903	0.9962	0.5098	0.5117	0.994
1.00	1.0000	1.0000	1.0000	0.5000	0.5000	1.000

first decreases slightly and then increases rapidly as a wave moves toward the shore. For a more extensive tabulation of the values given in Table 3.1 refer to the Shore Protection Manual (U.S. Army Coastal Engineering Research Center, 1984).

The term H_0' is commonly known as the equivalent deep water wave height. It is the deep water height an intermediate or shallow water wave would have if it had propagated in from deep water without refraction or energy losses. Hence H_0' is useful, for example, in wave tank research with monochromatic waves where the height measurement is made at some intermediate or shallow water depth in the tank. Rather than denote the depth at which each height measurement was made (since height changes with depth), it is better to calculate (and use in data plots) the equivalent deep water height for each measured height.

The small amplitude theory is most valid for waves having a small height relative to their length and the water depth. This leads to two questions concerning the accuracy of a wave theory in predicting wave characteristics: (1) How accurate is the theory in calculating the change in a specific wave characteristic (e.g., wave height) as the wave propagates from deep to intermediate or shallow water? (2) How accurate is a given theory for predicting the particle velocity or surface profile for a wave of given height and period at a particular water depth. The small amplitude theory may be less successful than the appropriate finite amplitude theory for the latter if H/L and H/d are relatively large, but may be relatively useful for the former. As we shall see in Chapter 4, the various finite amplitude theories are most appropriately applied in specific relative depth ranges, but they can be less useful for calculating changes that occur as a wave travels from deep to shallow water.

Ippen (1966) summarizes data from Wiegel (1950) and Eagleson (1956) on wave tank experiments to evaluate Eq. 3.1. The wave tank bottoms were smooth, but not frictionless, and the experimental slopes were between 1 : 10.8 and 1 : 20 but, the amount of wave reflection from these slopes was not evaluated. The small amplitude theory consistently underpredicted (given H_0 and T) the wave height as the wave train propagated through the intermediate depths to breaking. For the steeper bottom slopes and larger deep water incident wave steepnesses (H_0/L_0), the underprediction was greater. On the steepest slope, the wave height was underpredicted by about 20% when the wave reached a d/L of 0.07. However, the small amplitude theory was better at predicting the wave steepness at small values of d/L because it underpredicted both the wave height and length. For d/L approaching 0.1, the wave steepness was only underpredicted by 3%.

3.3 WAVE ATTENUATION WHILE SHOALING

Equation 3.2 neglects energy dissipation in predicting the change in wave height that occurs as a wave shoals. It has been mentioned that energy transfer to and from the waves owing to surface and bottom effects can affect the consequent change in height as a wave travels in transitional and shallow water. These surface and bottom effects are briefly discussed in this section.

3.3.1 Wind Effects

In most situations the component of the wind velocity in the direction of wave propagation will cause the waves to grow rather than decay. This phenomenon is discussed in Chapter 6. Energy transfer from the wind to a wave train depends, in basic terms, on the wind speed relative to the wave phase speed and the water surface speed (surface particle velocity), the duration of the wind action, the wind boundary layer profile, and the wave surface profile. Wind speed greater than the phase speed transfers energy to a wave by form drag, and wind speed greater than the surface water particle velocity transfers energy to the wave by surface shear. An opposing wind would have a greater attenuating impact than a following wind of the same velocity and duration of action would have in causing wave growth. Because the occurrence of strong opposing winds is not a matter of frequent concern, little research is available on this subject. An opposing wind will increase the attenuation of swell propagating over a long distance. Note that even for swell propagating through a windless atmosphere, the waves must do a small amount of work on the air and thus lose some energy.

3.3.2 Bottom Friction

In intermediate and shallow water, wave interaction with the bottom causes an unsteady oscillatory boundary layer to develop. In a laboratory wave tank this boundary layer may be laminar, but at sea for typical bottom conditions the boundary layer is practically always rough turbulent (Jonsson, 1966). Wave energy dissipation will occur owing to the turbulent motion and resulting bottom shear stress in this boundary layer as well as by viscous dissipation outside the boundary layer. The former may be significant but the latter is usually always negligible.

The typical approach used to describe the bottom shear stress and, from this, to determine the rate of energy dissipation in the wave is to employ a bottom stress definition that is analogous to the classical pipe flow definition of boundary shear stress. The shear stress is written as a function of a friction factor and the square of the water particle velocity just outside the boundary layer (calculated by the small amplitude wave theory). Wave and oscillatory flow tunnel experiments are then employed to relate this friction to a wave Reynolds number and bottom relative roughness (only the latter for rough turbulent flow). The relative roughness is defined as the horizontal amplitude of water particle motion at the edge of the boundary layer divided by the bed roughness dimension (à la Nikuradse). For results of these studies see Jonsson (1966), Kamphius (1975), and Jonsson and Carlson (1976).

Grosskopf (1980) gives the results of calculations of bottom friction attenuation using small amplitude wave theory and a friction factor diaphragm presented by Kamphius (1975). The results only deviated on average by 6% (under) from measurements made at two offshore wave gauges spaced 2200 m apart. For example, calculations showed that a 10-sec period, 2-m high wave traveling 2200 m from a depth of 18 m to a depth of 10 m was reduced to a height of 1.91 m with the effects of shoaling included. Shoaling alone (Eq. 3.1) would have reduced the wave height to 1.97 m; the difference was attributed primarily to bottom friction.

Svendsen and Jonsson (1976) present results of calculations using the work of Jonsson that further indicate the significance of bottom friction effects. For an 8-sec period, 2-m high wave in water 7 m deep and a bottom roughness height of 5 cm, the wave would lose 1.7% of its energy by bottom friction and only 0.00004% of its energy by internal viscous dissipation while traveling one wave length. For the same wave height and bottom roughness but a depth of 10 m, the boundary layer thickness was calculated as a function of wave period. The results are shown in Table 3.2. This demonstrates that for normal wind waves the bottom boundary layer is very small compared to the water depth. But as the wave period increases, the boundary layer size increases significantly. For tide waves in 10 m water depth, the boundary layer would envelope the entire water depth.

TABLE 3.2 Computed Boundary Layer Thicknesses[a]

Wave period	10 sec	1 min	6 min	30 min
Boundary layer thickness (m)	0.04	0.11	0.46	1.86

[a] Data from Svendsen and Jonsson (1976).

3.3.3 Bottom Percolation

The bottom boundary condition used to develop the small amplitude wave theory required that there be no flow normal to the bottom. However, if the bottom is porous, the horizontal pressure variation along the bottom (which is above hydrostatic under the wave crest and below hydrostatic under the wave trough) will generate an unsteady flow into and out of the bottom. Normally this flow is insufficient to modify the wave length and celerity. But, if the bottom is sufficiently permeable to a sufficient depth and over a sufficient lateral distance, the flow in and out of the bottom will dissipate a noticeable amount of the wave energy, resulting in an attenuation of the wave height as the wave propagates forward.

The solution to the problem of a small amplitude wave propagating over a horizontal porous bed is given by Reid and Kajiura (1957). They matched the potential flow solution for the wave motion and the Darcy flow solution for the porous bed, which was assumed to be homogeneous and of infinite vertical extent. The pressure and the vertical component of flow were assumed to be continuous at the water–soil interface. The resulting mean rate of energy dissipation per unit surface area R_* is given by

$$R_* = \frac{\pi \rho g^2 K H^2}{4 L \nu \cosh^2 kd} \tag{3.3}$$

where K is the Darcy permeability coefficient and ν is the water kinematic viscosity.

An interesting feature of this solution is that the maximum rate of decay for a given permeability and water depth occurs when $d/L_0 = 0.13$. Longer and shorter waves have a smaller rate of decay because for a given water depth and wave height the wave period defined by this relationship has the maximum horizontal pressure gradient at the bottom. Longer waves have a larger horizontal crest to trough pressure difference, but this difference is spaced over a longer distance and vice versa. Thus, for a spectrum of waves in water 5 m deep and propagating over a porous bed, there would be selective attenuation with the greatest effect being on those waves having periods around 5 sec.

For a coarse sand, say $K = 0.001$ m/s, and $d = 7$ m, $T = 8$ sec, and $H = 2$ m (as in the example in Section 3.3.2), the energy loss per wave length would be 0.06% versus the 1.7% loss from bottom friction. However, if the bottom consisted of gravel, say $K = 0.01$ m/s, the loss would be comparable to the loss from bottom friction.

The analysis by Reid and Kajiura (1957) assumes a porous medium of infinite vertical extent. However, in practice Eq. 3.3 can be used if the vertical extent of the porous medium is greater than about 30% of the wave length. Laboratory measurements of wave energy loss owing to bottom percolation collected by Savage (1953) were in good general agreement with Eq. 3.3.

3.3.4 Bottom Movement

If a wave train propagates over a bottom consisting of a soft viscous material of sufficient thickness so that the wave pressure field can set the soft bottom in motion, wave energy is transferred to the bottom where it is dissipated. Tubman and Suhayda (1976) reported wave measurements from two stations on the Mississippi River Delta that showed a wave height attenuation more than an order of magnitude greater than might be accounted for by bottom friction or percolation. They measured wave surface profiles and internal pressure fluctuations as well as the wave-induced movement of the mud bottom. Calculations of the energy transfer to the bottom account for the wave attenuation observed. They suggest that this process is important wherever shallow coastal areas with extensive fine-grained bottom sediments exist, such as the Guianas, the north coast of China, and southwest India.

Tubman and Suhayda (1976) found that the bottom-induced motions dissipate relatively greater amounts of wave energy in deeper intermediate water depths than bottom friction would dissipate. Consequently, such coasts are quite effective in protecting their shoreline by dissipating incident wave energy.

Assuming that the bottom is a viscoeleastic material, Hsiao and Shemdin (1980) derived a complex equation (which had to be solved numerically) for the bottom-motion-induced wave attenuation. Using estimates for the mud viscosity in the area where wave measurements were made by Tubman and Suhayda (1976) they were able to calculate wave height attenuations that were consistent with measured attenuations. For a given water depth and wave period, wave attenuation is found to increase with increased mud viscosity to a certain point and then to decrease with further increases in viscosity. This behavior occurs because dissipation increases as the mud viscosity increases, but dissipation also requires the mud to move in response to the wave motion and this mud motion decreases with increased viscosity.

3.4 WAVE PROFILE ASYMMETRY

As the deep water wave steepness (H_0/L_0) increases or as a wave propagates into shallow water, wave surface profile asymmetries develop and grow. Initially a vertical asymmetry around the still water line appears, but soon thereafter horizontal asymmetries around a vertical line through the wave crest also appear (Adeyemo, 1968; Ramberg and Griffin, 1987). Figure 3.2 is a typical surface profile showing these asymmetries.

The wave crest amplitude a_c exceeds half of the wave height. The front face slope of the wave is steeper than the rear face slope. And the forward horizontal distance from the wave crest to the still water line and the preceding trough are less than the rearward horizontal distances to the still water line and the following trough. This asymmetry builds to the point at which the wave becomes unstable and breaks. It is also consistent with the development of relatively high crest particle velocities leading to wave breaking.

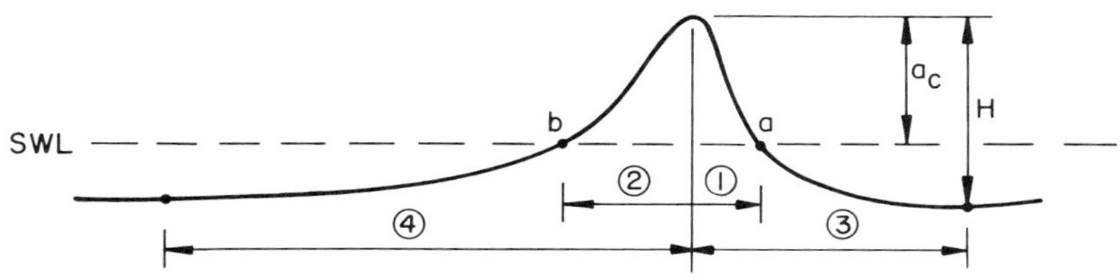

Figure 3.2 Wave surface profile asymmetry.

The surface profile vertical asymmetry will be satisfactorily defined by the commonly used finite amplitude wave theories in their appropriate ranges of application (see Chapter 4), but these theories do not show the horizontal asymmetries. The horizontal asymmetry has been shown by numerical simulations of wave motion prior to breaking (Longuet-Higgins and Cokelet, 1976; New et al., 1985).

Adeyemo (1968) conducted extensive wave tank experiments on surface profile asymmetry for waves propagating in intermediate water depths on slopes from 1 : 18 to 1 : 4. He considered four asymmetries, defined as follows (see Fig. 3.2):

Vertical asymmetry $= a_c/H$
Slope asymmetry $= 0.5$ (slope a + slope b)
Horizontal asymmetry (1) = distance 1/distance 2
Horizontal asymmetry (2) = distance 3/distance 4

The slopes were measured in radians with slope b positive and slope a negative.

For all of the slopes investigated, the vertical asymmetry continuously increased to a maximum at breaking. At breaking, vertical asymmetries varied from 0.62 to 0.74 for the slopes investigated. In the shallower water depths ($d/L < 0.10$) wave vertical asymmetry was greater for the flatter slopes. The slope and horizontal asymmetries also continuously increased as the depth decreased and (as opposed to the vertical asymmetry) steeper slopes caused greater slope and horizontal asymmetries. All of the asymmetries increased most rapidly for $d/L < 0.15$.

3.5 WAVE BREAKING

In simplest terms, for a given water depth and wave period a wave will break when the wave height grows to reach a certain limiting height. As the wave height and horizontal asymmetry increase so does the crest particle velocity, which approaches and becomes equal to the wave phase speed at breaking. This was demonstrated by Iverson (1952), who made particle velocity measurements in breaking waves by taking motion pictures of neutrally buoyant colored particles in the waves. The fine details of the breaking mechanism are complex (see Cokelet, 1977 and Melville, 1982), involving such matters as the interaction with subharmonic instability waves at the vicinity of the wave crest and the rate and nature of the surface profile asymmetry growth.

Commonly, breaking waves have been classified into four different types based on the physical changes of the surface profile during the breaking process. They have been given the names spilling, plunging, collapsing, and surging (Patrick and Wiegel, 1955 and Galvin, 1968) as they progress from one form to the next. These four types of breakers are depicted in Figure 3.3 and described as follows:

Spilling: Turbulence and foam first appear at the wave crest and spread down the front face of the wave as the wave propagates forward. It appears as if the wave is "plowing" the foam as it moves forward. The turbulence is uniformly dissipating wave energy, resulting in a continual decrease in the wave height as the wave propagates forward.

Plunging: The wave crest sharpens and then curls forward over the front face to plunge at the base of the front face of the wave. The breaking process

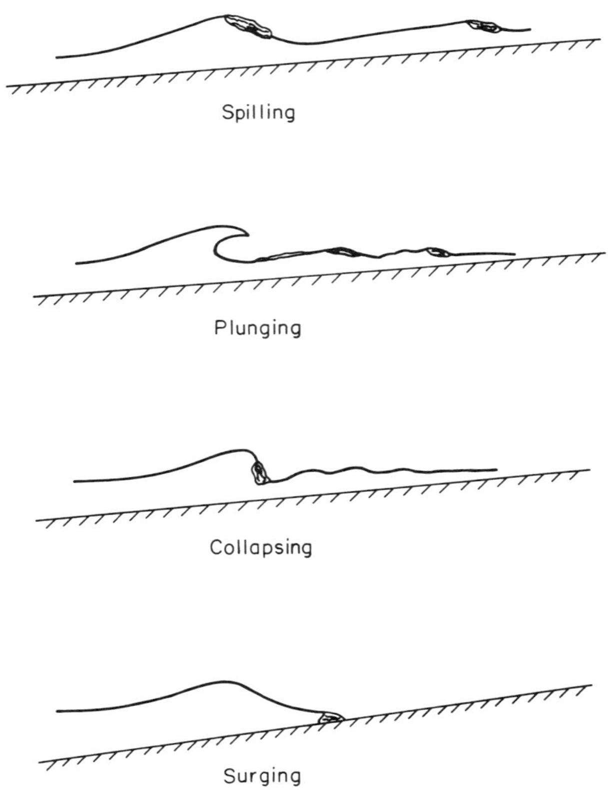

Figure 3.3 Wave breaker types.

and energy dissipation is more confined than for a spilling breaker. The plunging jet may regenerate smaller more irregular waves that propagate forward.

Collapsing: As the front face of the wave steepens at incipient breaking, the lower portion of the face plunges forward and the wave collapses. The collapsing breaker is an intermediate form between the plunging and surging form and is not as clearly defined as the others. (Some authors exclude this form from their classification.)

Surging: The crest and front face of the wave retain a fairly stable shape as they "surge" up the beach slope and return. This is a progression toward a standing or reflecting wave.

All four types of breakers can occur in shallow water, but only spilling and plunging breakers occur in deep water and they are most common in shallow water. Spilling breakers are most common in deep water and, if accompanied by a strong wind, "whitecapping" occurs.

The type of breaker that occurs can have several important consequences. For example, the stability of rubble mound structures is very dependent on the type of breaker to which the structure is exposed. And the rate of energy dissipation across the surf zone and the resulting water motion and wave runup on a beach face depend on the breaker type.

The type of breaker that occurs in shallow water depends on the wave steepness and the beach slope. At large wave steepnesses and flat bottom slopes, spilling breakers occur. There is a progression through the plunging and collapsing forms to surging breakers as the wave steepness decreases and the bottom slope increases. A number of authors have proposed a parameter consisting of the wave steepness divided by the beach slope squared to classify breaker types. Both the wave deep water steepness and the steepness at breaking have been used. From Galvin (1972) we have the following transition points:

TABLE 3.3 Wave Breaker Type Transition Points (Galvin, 1972)

Parameter	Spilling/Plunging	Plunging/Collapsing-Surging
$H_0/(L_0 m^2)$	4.8	0.09
$H_b/(gmT^2)$	0.068	0.003

In Table 3.3 m is the beach slope and H_b is the wave height at breaking. It should be emphasized that these transitions are gradual and the stated transition points are strongly dependent on the judgment of the observer. Also, the presence of an offshore bar that triggers wave breaking will lower the demarcation point between the spilling and plunging types, and an onshore or offshore wind will affect both transition points.

In addition to the breaker type, it is often important to know the height a wave will reach at breaking, and in shallow water, the depth at which breaking will occur. The former depends on the wave period and the latter on the period and beach slope. Simple approximate rules of thumb for the maximum deep water wave steepness and maximum wave height to water depth ratio in shallow water were given in Section 2.5.2.

Horikawa (1988), from a review of several sources, indicates that the ratio of wave height to water depth for nearshore breaking varies from 0.8 to 1.0 for spilling breakers and from 1.0 to 1.2 for plunging breakers, is just above 1.2 for collapsing breakers, and is difficult to define for surging breakers since the break point is not well defined. Considering the depth at which nearshore waves break and the slopes that relate to the different breaker types, spilling breakers tend to create the widest surf zone with a continual decrease in surf zone width with the transition of breaker type to surging breaker.

Several authors have presented empirical relations, based on wave tank studies of deep and shallow water breaking, that define the breaking height and (for shallow water) depth. Ramberg and Griffin (1987), employing their results and data from three other sources, found that the deep water breaking height is best represented by

$$H_b = 0.021gT^2 \tag{3.4}$$

which yields only a slightly lower limiting height than that given by the rule of thumb.

Empirical relations for shallow water breaking conditions have been given by LeMehaute and Koh (1967), Weggel (1972), and Singamsetti and Wind (1980). They follow the general forms H_b/H_0, $H_b/d_b = fct(m, H_0/L_0)$ where d_b is the water depth below the still water line at the wave crest at incipient breaking. From these relationships, given the wave period and deep water height and the beach slope, one can determine the wave height and water depth at breaking. These relationships assume two-dimensional wave transformation toward the shore. Details of these relationships and their application in a design context are given in Chapter 7.

None of the abovementioned laboratory experiments considered the effect of wind on the breaker type, height, and water depth. Douglass (1990) conducted shallow water wave breaking experiments for the conditions of two different onshore and offshore wind speeds ranging from about 1.5 to 3.5 times the wave celerity at breaking. The wind and wave propagation directions were in line. He compared the results to those for similar waves without any wind.

Although his data set was small, some of the qualitative conclusions are noteworthy. Offshore wind retarded the increase in wave height toward the shore and consequently caused the waves to break in shallower water than for the no-wind conditions. Onshore wind had the opposite result, but to a less dramatic extent. The resulting breaker heights were less for the offshore wind than for the onshore wind. But H_b/d_b was greater for offshore winds than for onshore winds. For the same waves, onshore winds caused spilling breakers, whereas offshore winds caused the breakers to plunge. The author concludes that wind has a significant impact on surf zone geometry and wave breaking characteristics.

During storm wave attack, most beach profiles will develop one or more submerged offshore bars. These bars trigger wave breaking and will thus impact on the shallow water breaking height and depth relationships discussed above. Also, there is often a trough in the lee of the bar crest where breaking waves may reform and propagate toward shore to break again. Wave tank tests by McNair and Sorensen (1970), employing a simulated offshore bar that looked like the top portion of an airfoil, showed that for a range of depths over the bar and a range of incident wave steepnesses, the reformed waves had from about 10 to 70% of the incident wave energy. The percentage of energy dissipation correlated best with and increased with a decrease in the ratio of the depth over the bar divided by the incident wave height. Most of the reformed wave energy had the same period as the incident wave, but a small portion of the reformed energy shifted to shorter harmonic periods.

Using the same experimental setup, Chandler and Sorensen (1972) studied the transformation of nonbreaking waves over the simulated bar. Although the waves did not break, instability caused a transfer of some of the wave energy to shorter free harmonic periods. Again, with a decrease in the ratio of depth over the bar to incident wave height there was an increase in the transfer of incident wave energy to the shorter harmonic periods. In some of the tests, the harmonic periods had as much as 60% of the energy as the transmitted incident wave period.

Smith and Kraus (1991) also conducted wave tank tests for waves breaking over a beach profile having a bar section. Bar profiles simulated measured profiles from movable bed wave tank tests and from field experiments. All breakers were spilling, plunging, or collapsing. The breaker-type transition points (see Table 3.3) were shifted so that the range for spilling breakers was very slightly smaller and the range for plunging breakers was much smaller, that is, collapsing breakers were more common. An empirical relationship for $H_b/d_b = fct(m, H_0/L_0)$ was given where m was now the front face slope of the bar; H_b/d_b were in the general range of 0.7–1.2.

Additional discussions of wave breaking are to be found in Chapter 6, which deals with irregular waves, and Chapter 7, which considers breaking limits on design wave heights.

3.6 NEARSHORE SETDOWN AND SETUP

Consider a wave train that propagates toward the shore, breaks in shallow water, and dissipates its energy in breaking and moving across the surf zone. Seaward of the breaker line the mean water level will be depressed or set down below the still water level. This setdown is due to an increase in the radiation stress with decreasing water depth as the waves propagate shoreward. Any wave energy dissipation or reflection between deep water and the breaker line would diminish the increase in radiation stress and consequently reduce the resulting setdown. The setdown is maximum at a point just seaward of the breaker line, but still relatively small in magnitude compared to the wave height.

Shoreward of the breaker line, in the surf zone, the decrease in radiation stress owing to wave energy dissipation is much stronger than the increase owing to decreasing water depth, so a setup of the mean water level occurs. The setup continues to increase toward the shore and is significantly larger than the setdown outside the surf zone.

Longuet-Higgins and Stewart (1964) derived equations for wave-induced setdown and setup by considering a horizontal momentum balance for a wave train propagating normal to the shoreline. Net bottom shear stresses are neglected. Consider Figure 3.4, which shows a segment of the nearshore of length dx situated normal to the shore line. The setup or setdown of the mean water level is d' and the forces and change in the radiation stress at the boundaries are shown. Writing the momentum balance for a segment of unit width along the shore,

$$\frac{\rho g}{2}(d + d')^2 - \frac{\rho g}{2}\left(d + d' + \frac{\partial d'}{\partial x}\,dx\right)^2 = \frac{\partial S_{xx}}{\partial x}\,dx$$

where the second term on the left includes the hydrostatic force along the landward vertical boundary and the bottom. Assuming $d \gg d'$, neglecting higher-order terms and rearranging yields

$$\frac{dS_{xx}}{dx} + \rho g d\,\frac{dd'}{dx} = 0 \qquad (3.5)$$

which relates the change in radiation stress to the mean water level slope. This will be evaluated for the regions seaward and shoreward of the breaker line.

Seaward of the breaker line, assuming the wave energy flux or power is constant, we can employ Eqs. 2.39 and 2.35 to integrate Eq. 3.5 to yield the setdown of the mean water level, giving

$$d' = -\frac{1}{8}\,\frac{H^2 k}{\sinh 2kd} \qquad (3.6)$$

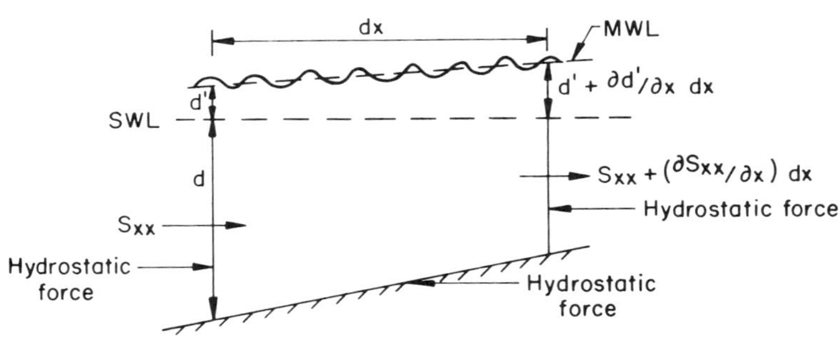

Figure 3.4 Shore-normal coastal segment.

Inspection of Eq. 3.6 indicates that the setdown is zero in deep water irrespective of the wave height. In shallow water, the setdown becomes $d' = -H^2/16d$. Thus, for a 2-m high incipient breaker in 2.5 m water depth the mean water level setdown is 0.1 m.

Shoreward of the breaker line, the mean water level setup depends on the rate of energy dissipation and conversion of wave energy to translatory motion across the surf zone. It is unlikely that this rate is uniform; but this is a complex and only marginally understood phenomenon (see Battjes, 1988 and Svendsen et al. 1978).

An analysis can be performed by making two simplifying assumptions. The first is that shallow water conditions prevail so $S_{xx} = 3E/2$, and the second is that the wave height is proportional to the water depth below the mean water level or $H = \gamma(d + d')$. The latter basically assumes that any reformed waves in the surf zone cannot exceed a simple breaker criteria as given in Section 2.5.2. It also implicitly assumes that the water depth does not increase at any point toward the shore. With these assumptions, the solution of Eq. 3.5 becomes

$$\frac{dd'}{dx} = \left(1 + \frac{8}{3\gamma^2}\right)^{-1} \frac{dd}{dx} \tag{3.7}$$

which gives the local slope of the mean water level as a function of the bottom slope in the surf zone. Continuing our example, assume that $\gamma = 0.8$ and that the bottom slope in the surf zone is 0.02, then from Eq. 3.7 the mean water level slope is 0.004. At the still water line contour on the beach (125 m shoreward from the breaker line) the mean water level would be $-0.1 + 125 (0.004) = +0.4$ m above the still water level.

Note that Eqs. 3.6 and 3.7 indicate that the setdown is a function of the incident wave height but that the slope of the setup is not. However, for higher incident waves the surf zone will extend further seaward so the actual setup at a point in the surf zone will be higher.

Two other factors should be kept in mind. First, the setup/setdown equations were based on the small amplitude theory, which is of limited accuracy in very shallow water. However, the calculations given above should still give a good estimate of expected conditions. Comparison of calculated setup and setdown with measurements made in a large wave tank by Saville (1961) yielded good results. Second, the equations apply to waves approaching normal to the shore. If the waves have a significant along-shore component, only the shore normal component of momentum will generate setup and setdown.

When the train of waves approaching the shore consists of groups of higher and lower waves, the surf zone setup fluctuates with the period of the groups. This is one of the causes of "surf beat" and the generation of longer period free waves.

3.7 WAVE REFLECTION

When a wave hits a vertical rigid impermeable wall it completely reflects from the wall. But waves that approach a flat slope will break. Is there a particular slope where the change from reflection to breaking occurs? Actually, whether a wave reflects or breaks depends on both the wave length and the slope. A useful parameter for considering whether breaking occurs is mL/d. The numerator mL is the change in depth that occurs in one wave length. If this is large compared to the depth (i.e., mL/d large) reflection will occur and vice versa.

The above parameter is implicit in the discussion of wave breaking types. For a given wave steepness (or wave length), as the bottom slope increases, the breaker type experiences a transition from spilling to surging. For a spilling breaker there would be negligible reflection from the beach slope, whereas for a surging breaker some reflection would occur and the motion in a surging breaker is trending toward the motion that occurs when a wave reflects from a sloping wall.

Conversely, for a given beach slope a relatively short wave will break while a long wave will reflect, and the longer the wave the greater the reflection. The flat continental slope and beach will cause wind waves to break, but the tide and other long waves will completely reflect.

Seelig (1983) presents the results of wave tank experiments of wave reflection from smooth nonporous plane slopes. The reflection coefficient C_r is given by

$$C_r = \frac{I_r^2}{5.5 + I_r^2} \tag{3.8}$$

where I_r is known as the Iribarren number or the surf similarity parameter and is defined as

$$I_r = \frac{m}{\sqrt{H/L_0}} \tag{3.9}$$

In Eq. 3.9 H is the incident wave height at the toe of the slope. Note the similarity between I_r and the parameter used to define breaker type. Seelig (1983) and Allsop and Hettiarachchi (1988) present an extensive collection of wave reflection information for various slope and structure shapes all in the general form of C_r as a function of I_r.

When a wave reflects from a slope, the water level at the slope is raised slightly to provide the pressure force to reverse the wave momentum flux. From radiation stress considerations Longuet-Higgins and Stewart (1964) showed that the rise in the mean water level at a vertical wall that has a reflection coefficient of unity (pure standing wave) is given by

$$d' = \frac{\pi H^2}{4L} \coth 2kd \tag{3.10}$$

For example, for a 1-m high, 5-sec period wave reflecting from a vertical wall in water 5 m deep, the rise in mean water level at the wall would be 0.03 m. The rise given by Eq. 3.10 can be thought of as an upper limit value because slopes having lower reflection coefficients would have lower increases in mean water level, for given incident wave characteristics.

More detailed information on wave reflection is given in Chapter 8.

3.8 WAVE RUNUP

In the surf zone a portion of the incident wave oscillatory motion will be converted by the wave breaking process to forward translation of the water mass. This results in the formation of a bore that ''runs up'' the face of a beach or shore structure. We define the runup R as the maximum vertical elevation above the still water level to which the water rises on the beach or structure face (Fig. 3.5). A knowledge of the expected wave runup is, of course, important for a variety of concerns, including the determination of the optimum crest elevation for a structure or the location of a beach setback line for limiting construction.

The relative runup R/H (where H is the deep water height or some other incident height) depends primarily on the incident wave steepness H/L_0 and the beach or structure slope, as well as on the slope characteristics, including surface roughness and porosity, and slope geometry if other than planar. The envelope curves for a classic plot of runup data for monochromatic waves breaking on a smooth plane slope and based on laboratory tests by Saville (U.S. Army Coastal Engineering Research Center, 1984) are shown in Figure 3.6. Also denoted on the plot is the region where nonbreaking reflecting waves occur. Note that many beaches have face slopes that are flatter than 1:30, the right-hand limit of the figure.

Figure 3.6 demonstrates that for breaking waves on a given beach slope the

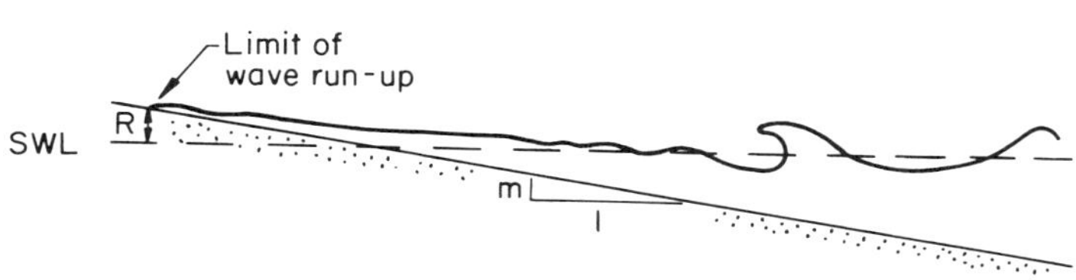

Figure 3.5 Wave runup.

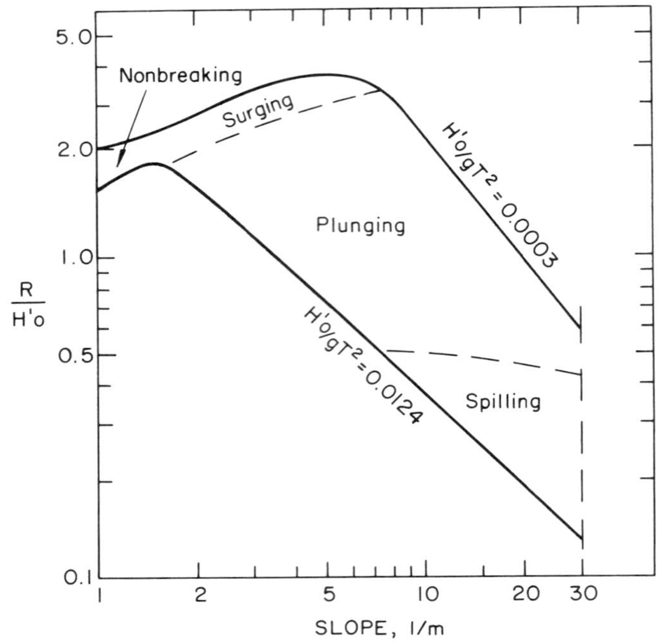

Figure 3.6 Wave runup, smooth plane slopes.

relative runup increases with decreasing wave steepness. That is, longer period waves will run up much higher than shorter period waves having the same height. Conversely, for a given incident breaking wave height and period, the steeper the slope the higher the wave runup. Naturally, runup will be lower on beach or structure faces with greater roughness or permeability.

Wave runup is related to the type of breaker that occurs, which in turn is related to the beach slope and incident wave steepness as discussed in Section 3.6. The ranges for breaker types (taken from Table 3.3) are also denoted on Figure 3.6. Spilling breakers generate the lowest relative runup, producing runup elevations of typically less than half of the incident wave height (i.e., less than the incident wave crest amplitude). Spilling breakers dissipate their energy over a wider surf zone and transfer only a small portion of their energy to forward motion of the water. As the breaker type transforms to plunging, and collapsing/surging the relative runup increases, typically to values in excess of 2.

More information on wave runup for specific conditions is given in Chapter 8.

REFERENCES

Adeyemo, M. D. (1968), "Effect of Beach Slope and Shoaling on Wave Asymmetry," *Proceedings, 11th Conference on Coastal Engineering*, American Society of Civil Engineers, London, pp. 145–172.

Allsop, N. W. H. and Hettiarachchi, S. S. L. (1988), "Reflections from Coastal Structures," *Proceedings, 21st International Conference on Coastal Engineering*, American Society of Civil Engineers, Malaga, Spain, pp. 782–794.

Battjes, J. A. (1988), "Surf Zone Dynamics," *Annual Review of Fluid Mechanics*, Vol. 20, Annual Reviews, Palo Alto, pp. 257–293.

Chandler, P. L. and Sorensen, R. M. (1972), "Transformation of Waves Passing a Submerged Bar," *Proceedings, 13th Conference on Coastal Engineering*, American Society of Civil Engineers, Vancouver, pp. 385–404.

Cokelet, E. D. (1977), "Breaking Waves," *Nature*, pp. 769–774.

Douglass, S. L. (1990), "Influence of Wind on Breaking Waves," *J. Waterw. Port Coastal Ocean Eng. Div., Am. Soc. Civ. Eng.*, November, 651–663.

Eagleson, P. S. (1956), "Properties of Shoaling Waves by Theory and Experiment," *Trans., Am. Geophys. Union*, **37,** 565–572.

Galvin, C. J. (1968), "Breaker Type Classification on Three Laboratory Beaches," *J. Geophys. Res.*, 3651–3659.

Galvin, C. J. (1972), "Wave Breaking in Shallow Water," *Waves on Beaches and Resulting Sediment Transport*, R. E. Myers, Ed., Academic, New York, pp. 413–451.

Grosskopf, W. G. (1980), "Calculation of Wave Attenuation Due to Friction and Shoaling: An Evaluation," Technical Paper 80-8, U. S. Army Coastal Engineering Research Center, Ft. Belvoir, VA.

Horikawa, K. (1988), *Nearshore Dynamics and Coastal Processes: Theory, Measurement and Predictive Models*, University of Tokyo Press, Tokyo.

Hsiao, S. V. and Shemdin, O. H. (1980), "Interaction of Ocean Waves with a Soft Bottom," *J. Phys. Oceanography*, **10,** 605–610.

Ippen, A. T. (1966), *Estuary and Coastline Hydrodynamics*, McGraw-Hill, New York.

Iversen, H. W. (1952), "Waves and Breakers in Shoaling Water," *Gravity Waves*, Circular 521, National Bureau of Standards, Washington, DC, pp. 1–12.

Jonsson, I. G. (1966), "Wave Boundary Layers and Friction Factors," *Proceedings, 10th Conference on Coastal Engineering*, American Society of Civil Engineers, Tokyo, pp. 127–148.

Jonsson, I. G. and Carlsen, N. A. (1976), "Experimental and Theoretical Investigations in an Oscillatory Rough Turbulent Boundary Layer," *J. Hydraulic Res.*, **14,** 45–60.

Kamphius, W. J. (1975), "Friction Factor Under Oscillatory Waves," *J. Waterw. Harbors Coastal Eng. Div. Am. Soc. Civ. Eng.*, May, 135–144.

LeMehaute, B. and Koh, R. (1967), "On the Breaking of Waves Arriving at an Angle to the Shore," *J. Hydraulic Res.*, **5,** 67–88.

Longuet-Higgins, M. S. and Cokelet, E. D. (1976), "The Deformation of Steep Surface Waves on Water. I. A Numerical Method of Computation," *Proc. R. Soc. London, Series A*, 1–26.

Longuet-Higgins, M. S. and Stewart, R. W. (1964), "Radiation Stress in Water Waves: A Physical Discussion, with Applications," *Deep Sea Res.*, **11,** 529–549.

McNair, E. C. and Sorensen, R. M. (1970), "Characteristics of Waves Broken By a Longshore Bar," *Proceedings, 12th Conference on Coastal Engineering*, American Society of Civil Engineers, Washington, DC, pp. 415–434.

Melville, W. K. (1982), "The Instability and Breaking of Deep Water Waves," *J. Fluid Mech.*, **115,** 165–185.

New, A. L., McIver, P., and Peregrine, D. H. (1985), "Computation of Overturning Waves," *J. Fluid Mech.*, **150,** 233–251.

Patrick, D. A. and Wiegel, R. L. (1955), "Amphibian Tractors in the Surf," *Proceedings, 1st Conference on Ships and Waves*, Council on Wave Research and American Society of Naval Architects and Marine Engineers, pp. 397–422.

Ramberg, S. E. and Griffin, O. M. (1987), "Laboratory Study of Steep and Breaking Deep Water Waves," *J. Waterw. Port Coastal Ocean Eng. Div., Am. Soc. Civ. Eng.*, September, 493–506.

Reid, R. O. and Kajiura, K. (1957), "On the Damping of Gravity Waves Over a Permeable Sea Bed," *Trans. Am. Geophys. Union*, **38,** 662–666.

Savage, R. P. (1953), "Laboratory Study of Wave Energy Losses By Bottom Friction and Percolation," Technical Memorandum 31, U. S. Army Beach Erosion Board.

Saville, T. (1961), "Experimental Determination of Wave Set-up," Proceedings, 2nd Conference on Hurricanes, U. S. Department of Commerce National Hurricane Research Project, Report 50, pp. 242–252.

Seelig, W. N. (1983), "Wave Reflection from Coastal Structures," *Proceedings, Coastal Structures '83*, American Society of Civil Engineers, pp. 961–973.

Singamsetti, S. R. and Wind H. G. (1980), "Characteristics of Shoaling and Breaking Periodic Waves Normally Incident to Plane Beaches of Constant Slope," Report M1371, Delft Hydraulics Laboratory.

Smith E. R. and Kraus, N. C. (1991), "Laboratory Study of Wave Breaking over Bars and Artificial Reefs," *J. Waterw. Port Coastal Ocean Eng. Div. Am. Soc. Civ. Eng.*, July/August, pp. 307–325.

Svendsen, I. A. and Jonssen, I. G. (1976), *Hydrodynamics of Coastal Regions*, Technical University of Denmark, Lyngby, Denmark.

Svendsen, I. A., Madsen, P. A., and Buhr Hansen, J. (1978), "Wave Characteristics in the Surf Zone," *Proceedings, 16th Conference on Coastal Engineering*, American Society of Civil Engineers, Hanover, pp. 520–539.

Tubman, M. W. and Suhayda, J. N. (1976), "Wave Action and Bottom Movements in Fine Sediments," *Proceedings, 15th Conference on Coastal Engineering*, Honolulu, pp. 1168–1183.

U. S. Army Coastal Engineering Research Center (1984), *Shore Protection Manual*, U. S. Government Printing Office, Washington, DC.

Weggel, J. R. (1972), "Maximum Breaker Height," *J. Waterw. Harbors Coastal Eng. Div., Am. Soc. Civ. Eng.*, November, 529–548.

Wiegel, R. L. (1950), "Experimental Study of Surface Waves in Shoaling Water," *Trans. Am. Geophys. Union*, **31,** 377–385.

4

FINITE AMPLITUDE WAVE THEORY

The linear or small amplitude wave theory was summarized in Chapter 2; in this chapter we present the most useful and most used two-dimensional non-linear or finite amplitude wave theories. Mathematically, there is no general solution to the basic conservation of mass and momentum equations for gravity waves. All wave theories require one form of approximation or another. In the small amplitude theory we linearized the free surface boundary conditions and applied them at the still water level rather than at the water surface. This required that H/d and H/L be small compared to unity. Consequently, the small amplitude theory could be applied over the complete range of relative water depths (d/L) provided the wave height was sufficiently low.

The ideal wave theory would be accurate in all water depths and for all wave heights and periods (and it would be easy to apply). Such a wave theory does not exist. But we can relax the requirement that one of the parameters H/d or H/L be small and thus develop a finite amplitude theory useful for waves of large amplitude and a specific range of relative water depths. Allowing H/L to be finite results in a wave theory for steep deep water waves and allowing H/d to be finite results in a theory useful for shallow water.

There are two general types of finite amplitude wave theories in use: analytical theories in which a power series is used and numerical theories. In the former the velocity potential (and other parameters such as the surface amplitude and the wave celerity) is defined by a power series, with successively smaller terms defined by a small perturbation parameter raised to a higher power in each succeeding term. Truncation of the power series at, for example, the third term yields the third-order solution and so forth. The complexity of the solutions increases dramatically as the order of the solution is increased. The perturbation parameter is developed in terms of H/L for deep water and H/d for shallow water.

In the numerical theories, a power series solution to the Laplace equation is defined and the free surface boundary conditions are used to iteratively optimize coefficients in this power series by using numerical techniques. Solutions are obtained by computer and tabulated. Again, solutions may be truncated at a certain order.

Specifically, we will consider the Stokes theory for deep water waves, the cnoidal and solitary waves theories for shallow water, and the Dean stream function numerical wave theory, which is applicable throughout the entire range of water depths. The development of these theories is extremely complex as are the results. Only a brief overview of the development of these theories is presented. Results are given in as much detail as is reasonably possible. Extensive references to each theory are given. Good general discussions of finite amplitude wave theory are given in Ippen (1966) and Sarpkaya and Isaacson (1981).

It is important to known which theory to employ for a particular combination of water depth and wave height and period. The choice may be made on theoretical grounds such as which theory best satisfies the free surface boundary conditions for the H, T and d combination of interest. Or the choice may be based on a comparison of theoretical predictions of physical characteristics, such as the surface profile or water particle velocities, with experimentally measured values of these characteristics. The difficulty of choice is compounded by the fact that one characteristic may be best predicted by one theory and another by a different theory. A further complication is the increased difficulty in employing certain theories that may yield better results but may not justify the increased effort because the input conditions are not precisely known or only a preliminary analysis is required. Both theoretical and analytical approaches to defining ranges of application for the various theories are discussed.

Because some theories are best applied for specific ranges of relative depth, a question arises when calculations of wave transformation over a wide range of water depths must be made. (For example, wave hindcasts may yield deep water design wave heights and periods. Refraction and shoaling calculations must then be made to define the wave height at a structure located in shallow water.) Some of the approaches that have been employed for finite amplitude wave shoaling calculations are presented.

4.1 GENERAL FORMULATION OF ANALYTICAL FINITE AMPLITUDE THEORIES

The basic two-dimensional wave boundary value problem is stated in Section 2.2. The Laplace equation (Eq. 2.1) is to be solved given the bottom, kinematic surface, and dynamic surface boundary conditions (Eqs. 2.2, 2.3, and 2.5). For the finite amplitude wave theories these equations must be solved in their given nonlinear form and with the surface boundary conditions applied at the water surface rather than the still water line.

To simplify the solution, the coordinate system is usually modified by moving it in the direction of the wave at a speed equal to the wave celerity. This yields a steady flow situation and removes the unsteady flow terms from the equations. The Laplace equation and the bottom condition are not changed. The KSBC becomes

$$w = (u - C)\frac{\partial \eta}{\partial x} \quad \text{at } z = \eta \tag{4.1}$$

since the water surface elevation no longer changes with time and the horizontal particle velocity is now $u - C$. The DSBC becomes

$$\tfrac{1}{2}[(u - C)^2 + w^2] + g\eta = Q \quad \text{at } z = \eta \tag{4.2}$$

The unsteady velocity potential term disappears but the Bernoulli constant Q, which is normally incorporated in the velocity potential, must be retained. Here Q is the total energy with reference to the free surface elevation.

For the Stokes and cnoidal wave theories, the perturbation method is used to solve Eqs. 2.1, 2.2, 4.1, and 4.2. The dependent variables (e.g., velocity potential, surface amplitude, wave celerity) are written as power series. For example,

$$\phi = \epsilon\phi_1 + \epsilon^2\phi_2 + \cdots$$

where ϵ is a perturbation parameter. For the Stokes theory ϵ is proportional to the wave steepness H/L and for the cnoidal theory ϵ is proportional to H/d. The power series terms are then substituted into the basic boundary value equations and terms of equal order of ϵ are gathered to solve for ϵ. This is carried out to the order desired to yield the wave theory order desired. In the power series, the last term of the solution for each successive order basically adds a correction to the previous order solution. Since the perturbation parameter is small and raised to successively higher powers, the correction applied by successively higher terms should be increasingly smaller.

The solution is straightforward conceptually, but very complex algebraic manipulations are required and they result in very complex sets of relationships representing the final solutions. Different approaches in the details lead to somewhat different final solutions from different authors.

4.2 STOKES FINITE AMPLITUDE THEORY

Stokes (1847) developed a second-order wave theory for finite amplitude waves using a power series based on H/L. The results require that H/d not be large and thus are applicable for deep water and much of the intermediate depth range. (Truncation of the second-order theory at the first order yields the small amplitude wave theory.) Since it is only a second-order wave theory, results diminish in accuracy as the wave steepness increases toward breaking. For large wave steepnesses, higher-order wave theories are more appropriate. Skjelbreia (1959) presented a third-order Stokes theory and Skjelbreia and Henderson (1961) presented a fifth-order Stokes theory with tabulated formulas for the calculation of basic wave characteristics. A fifth-order Stokes theory is also given by Fenton (1985), which corrects some errors in Skjelbreia and Henderson (1961) and is easier to apply. Fenton (1985) also tabulates the necessary formulas for wave calculation. For most applications, the second-order, or if necessary, the fifth-order theory is used. The second-order theory is presented herein.

The velocity potential to the second order is

$$\phi = \frac{gH}{2\sigma}\frac{\cosh k(d+z)}{\cosh kd}\sin(kx-\sigma t) + \frac{3\pi CH}{16}\left(\frac{H}{L}\right)\frac{\cosh 2k(d+z)}{\sinh^4(kd)}\sin 2(kx-\sigma t) \tag{4.3}$$

Note that the first-order term is identical to the small amplitude theory velocity potential (see Eq. 2.8). The magnitude of the second-order term is dependent on the wave steepness and its frequency is twice that of the first-order term.

Wave celerity, as given by the dispersion equation, is the same for the second-order theory as for the first-order or small amplitude theory (Eq. 2.10). Thus, to second order the wave celerity is unaffected by wave steepness. From the third-order theory,

$$C^2 = \frac{g}{k}\tanh kd\left[1 + \left(\frac{\pi H}{L}\right)^2\left(\frac{9 + 8\cosh^4(kd) - 8\cosh^2(kd)}{8\sinh^4(kd)}\right)\right] \tag{4.4}$$

This indicates that wave celerity increases with wave steepness. For deep water and the limiting steepness ($H_0/L_0 = 1/7$), the wave celerity is approximately 20% higher than that given by the small amplitude dispersion equation. Consequently, for a given wave period, the third-order theory wave length would also be larger.

The second-order surface profile is given by

$$\eta = \frac{H}{2}\cos(kx-\sigma t) + \frac{\pi H}{8}\left(\frac{H}{L}\right)\frac{\cosh kd(2+\cosh 2kd)}{\sinh^3 kd}\cos 2(kx-\sigma t) \tag{4.5}$$

The second-order term, having twice the frequency as the first-order term, increases the surface amplitude at the crest (i.e., components are in phase) and decreases it at the trough (i.e., components are out of phase). And, the vertical profile asymmetry increases as the wave steepness increases. In deep water Eq. 4.5 reduces to

$$\eta = \frac{H}{2}\cos(kx - \sigma t) + \frac{\pi H}{4}\left(\frac{H}{L}\right)\cos 2(kx - \sigma t) \tag{4.6}$$

The first- and second-order components and the resulting wave profile for a relatively steep 6-m high, 7-sec deep water wave are plotted (with an exaggerated vertical scale) in Figure 4.1 to demonstrate the effect of the second-order term.

From Eq. 4.6 for deep water the profile asymmetry can be written as

$$\frac{a_c}{H}, \frac{a_t}{H} = \frac{1}{2} \pm \left(\frac{\pi}{4}\right)\frac{H}{L} \tag{4.7}$$

where a_c and a_t are the wave crest and trough amplitudes. For the limiting deep water wave steepness this yields $a_c = 0.611H$ and $a_t = 0.389H$ (or $a_c/a_t = 1.57$). Higher-order Stokes theories would yield a somewhat greater deep water wave asymmetry at the limiting steepness. Also, from Eq. 4.5 the asymmetry would increase as the wave propagates into intermediate depths.

As the wave steepness increases for a given relative depth, the second-order term in Eq. 4.5 increases in size relative to the first-order term. As a consequence, the trough becomes flatter and flatter until a point is reached where

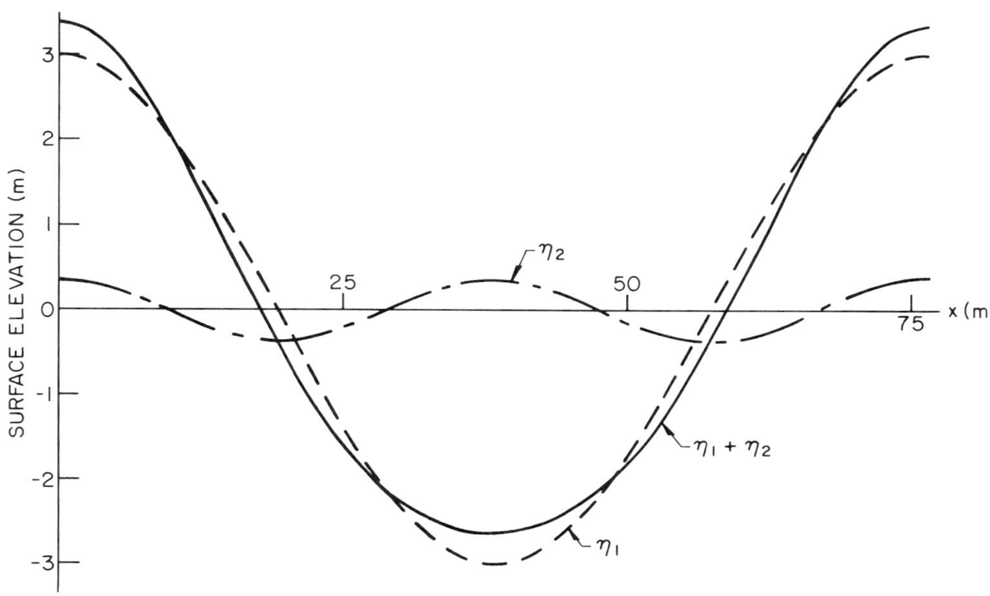

Figure 4.1 Wave surface profile, Stokes second order.

the water surface at the trough becomes horizontal. For further increases in the wave steepness a "hump" starts to appear at the trough. (For very steep waves in shallow water the hump can grow to be larger than the main or first-order wave amplitude.) However, the second-order term should be small compared to the first-order term so this appearance of a hump (which is not a real wave phenomena) is an indication that the theory is being applied beyond its appropriate range of applicability. Setting the limit for the range of applicability at the point where the surface just becomes horizontal at the trough yields the following limit for the wave steepness:

$$\frac{H}{L} = \frac{\sinh^3 kd}{\pi \cosh kd(2 + \cosh 2kd)} \tag{4.8}$$

For deep water Eq. 4.8 yields a limiting steepness greater that $1/7$, so no application limit is placed on the second-order theory in deep water. For an intermediate water relative depth of say $d/L = 0.1$, however, the limit set by Eq. 4.8 on the wave steepness is 0.021. This is a significant restriction.

For reference, Stokes second-order equations for the other important wave characteristics are listed below. (Note that the following shorthand is used: $\omega = (kx - \sigma t)$ and $(\ )_*$ is used to denote the first-order or small amplitude term for the respective characteristic as given in Chapter 2.)

Particle velocity

$$u = (u)_* + \frac{3(\pi H)^2}{4TL} \frac{\cosh 2k(d + z)}{\sinh^4(kd)} \cos 2\omega \tag{4.9}$$

$$w = (w)_* + \frac{3(\pi H)^2}{4TL} \frac{\sinh 2k(d + z)}{\sinh^4(kd)} \sin 2\omega \tag{4.10}$$

Particle acceleration

$$a_x = (a_x)_* + \frac{3\pi^3 H^2}{T^2 L} \frac{\cosh 2k(d + z)}{\sinh^4(kd)} \sin 2\omega \tag{4.11}$$

$$a_z = (a_z)_* - \frac{3\pi^3 H^2}{T^2 L} \frac{\sinh 2k(d + z)}{\sinh^4(kd)} \cos 2\omega \tag{4.12}$$

Particle displacement

$$\zeta = (\zeta)_* + \frac{\pi H^2}{8L \sinh^2(kd)} \left[1 - \frac{3 \cosh 2k(d + z)}{2 \sinh^2(kd)} \right] \sin 2\omega + \frac{\pi H^2}{4L} \frac{\cosh 2k(d + z)}{\sinh^2(kd)} (\sigma t) \tag{4.13}$$

$$\epsilon = (\epsilon)_* + \frac{3\pi H^2}{16L} \frac{\sinh 2k(d + z)}{\sinh^4(kd)} \cos 2\omega \tag{4.14}$$

Pressure

$$p = (p)_* + \frac{3\pi\rho g H^2}{4L \sinh 2kd} \left[\frac{\cosh 2k(d + z)}{\sinh^2(kd)} - \frac{1}{3} \right] \cos 2\omega - \frac{\pi\rho g H^2}{4L \sinh 2kd} (\cos 2k(d + z) - 1) \tag{4.15}$$

The Stokes second-order specific energy and energy flux are the same as for the small amplitude wave theory (Eqs. 2.32 and 2.33).

The second-order terms in the particle velocity and acceleration (Eqs. 4.9–4.12) likewise lead to particle kinematic asymmetries. The particle velocity and acceleration are greater under the wave crest than under the trough and this asymmetry increases with an increase in the wave steepness. As a result of the horizontal velocity asymmetry the particle orbits are not closed. This leads to a small particle drift or mass transport in the direction of wave motion which is evident in the horizontal particle position given by Eq. 4.13. Note that the last term in Eq. 4.13 is not periodic but increases with time. Dividing this term by time yields the mass transport velocity,

$$\bar{u} = \frac{\pi^2 H^2}{2TL} \frac{\cosh 2k(d + z)}{\sinh^2(kd)} \tag{4.16}$$

This is similar to the surface mass transport velocity from Section 2.5.1.

Figure 4.2 is a plot of the mass transport velocity versus depth for the 6-m high, 7-sec deep water wave considered above (Fig. 4.1). For comparison, this wave has a celerity of 10.91 m/sec. The mass transport is maximum at the water surface and then decreases sharply with distance below the surface.

The shoreward mass transport will cause a continuing accumulation of water at the shore, unless this accumulation is relieved by a return flow seaward and/or a flow of water along the shore. In a confined two-dimensional wave flume, continuity of mass would require a return flow equal to the forward wave mass transport and this return flow pattern would be superimposed on the forward mass transport velocity profile. Longuet-Higgins (1953) has developed equations for the mass transport velocity profile for the two-dimensional enclosed situation and these equations have been experimentally verified by Russell and Osorio (1958).

The vertical kinematic asymmetries also produce another interesting feature demonstrated by the second-order theory equation for dynamic pressure (Eq. 4.15). Note that the last term in this equation is not cyclic with time, but varies from zero at the bottom to increasing positive numbers as one moves up the water column toward the surface. At the bottom the boundary requires that

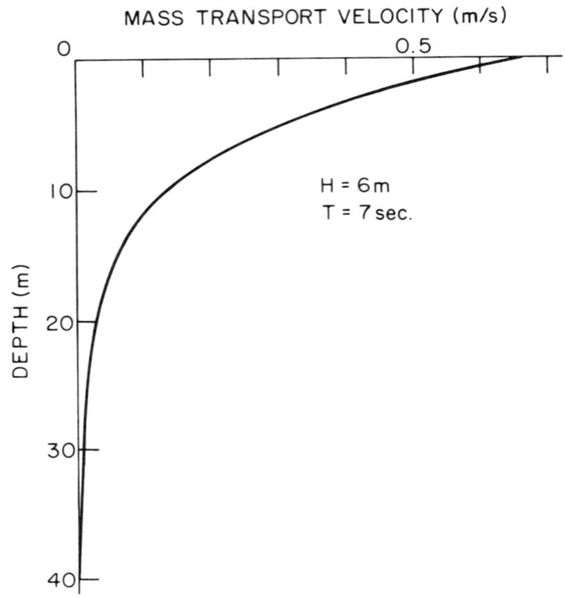

Figure 4.2 Mass transport velocity.

flow and acceleration be horizontal, so there is no vertical momentum flux and the time average pressure must balance the time average weight of water above. Away from the bottom, however, there is a time average vertical momentum flux so the time average dynamic pressure is not zero (as it is for the small amplitude theory).

4.3 CNOIDAL WAVE THEORY

Stokes theory is valid for deep water, but becomes less and less applicable in intermediate water depths as the relative depth decreases owing to the increase in the relative magnitude of the higher-order terms. A commonly quoted (Keulegan, 1950) limiting relative depth for Stokes waves is $d/L = 0.1$, but the more realistic cutoff point depends on the wave steepness, as discussed in the previous section. For finite amplitude waves in shallower water, a theory based on an expansion of the relative depth is needed.

These shallow water theories are generally based on work done by Kortweg and de Vries (1895), commonly known as cnoidal wave theory. Stokes theory results in a series of trigonometric functions, but cnoidal theory involves Jacobian elliptical functions, designated cn, thus the name cnoidal is applied to this wave theory. The most commonly used cnoidal wave theory is to the first order of approximation, but it is capable of describing waves of finite height in shallow water. An interesting feature of this first-order cnoidal theory is that its deep water limit is the small amplitude wave theory and its shallow water limit is the solitary wave theory (see Section 3.4).

Cnoidal wave theories have been presented by several people. See Keulegan and Patterson (1940), Keller (1948), and Laitone (1960) for a first-order approximation, and Laitone (1960) and Chappelear (1962) for higher-order approximations. In all cases the results are extremely difficult to apply—to the extent that some authors recommend extending the range of use of the small amplitude, Stokes, and solitary theories into the relative depths where cnoidal theory is most applicable.

When cnoidal wave theory is used, the results presented by Wiegel (1960, 1964) are most commonly employed. He synthesized the results of Kortweg and de Vries (1895), Keulegan and Patterson (1940), and Keller (1948) and presented the results in as practical a form as possible, employing both equations and graphs. From this it is possible, given the wave height and period and the water depth, to determine the wave length, celerity, surface profile, particle velocity, and particle acceleration. Laitone (1960) showed that, to the first approximation (but not to the second), the pressure variation is hydrostatic, using the distance below the water surface to calculate the pressure. A portion of Wiegel's results will be given herein to demonstrate some of the cnoidal wave characteristics; the reader should consult Wiegel (1960, 1964) to make more thorough cnoidal wave calculations. A slight modification to Wiegel's results presented in U. S. Army Coastal Engineering Research Center (1984) is used to further simplify the presentation.

Cnoidal theory is presented in terms of two parameters—k^2 (or m for some authors) and $U_r = L^2 H/d^3$. The first is a parameter related to the wave length and height and the water depth, that is, one of the independent variables in the elliptical function. It varies from 0 for the small amplitude limit wave to 1.0 for the solitary wave limit. The term U_r, commonly known as the Ursell parameter (Ursell, 1953), is a dimensionless parameter containing the three dependent variables [or the wave steepness and relative depth, $(H/L)/(d/L)^3 = L^2 H/d^3$] that define the range of application of the various wave theories. Generally (Hardy and Kraus, 1987) cnoidal theory is applicable for $U_r > 25$ and Stokes theory is applicable for $U_r < 10$. Both theories are equally valid for the range $U_r = 10 - 25$.

Figures 4.3 and 4.4. (originally from Wiegel, 1960 with modifications) define cnoidal wave characteristics. Given the wave period and height and the water depth, the dimensionless wave period and H/d can be calculated to determine k^2 from Figure 4.3. And the value of k^2 yields U_r (using the dashed line). From U_r the wave length can be calculated and the wave celerity follows from $C = L/T$ since cnoidal waves are periodic and of permanent form. If the wave length is known instead of the wave period, Figure 4.3 can be used in reverse by calculating U_r, determining k^2, and then with H/d determining the wave period.

Given the value of k^2, the surface profile can be determined from Figure 4.4. This is a plot of the water surface amplitude η above the trough elevation

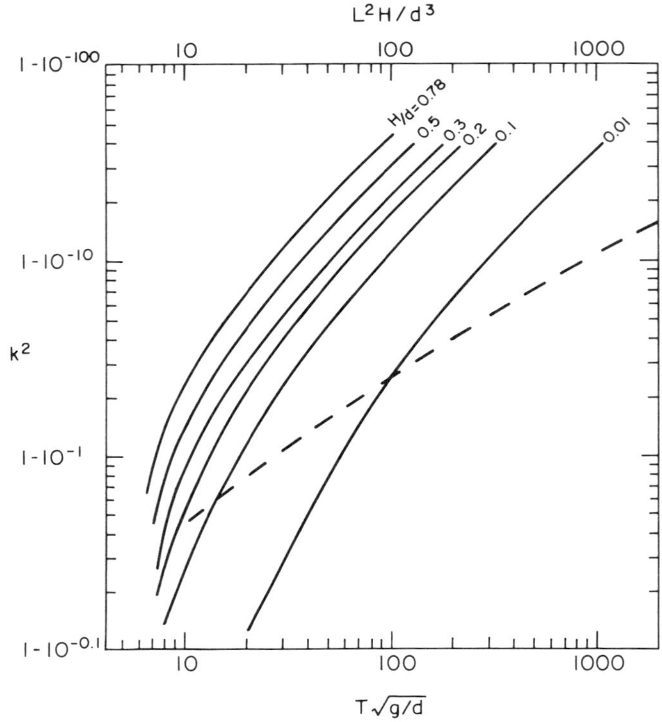

Figure 4.3 Cnoidal wave parameters.

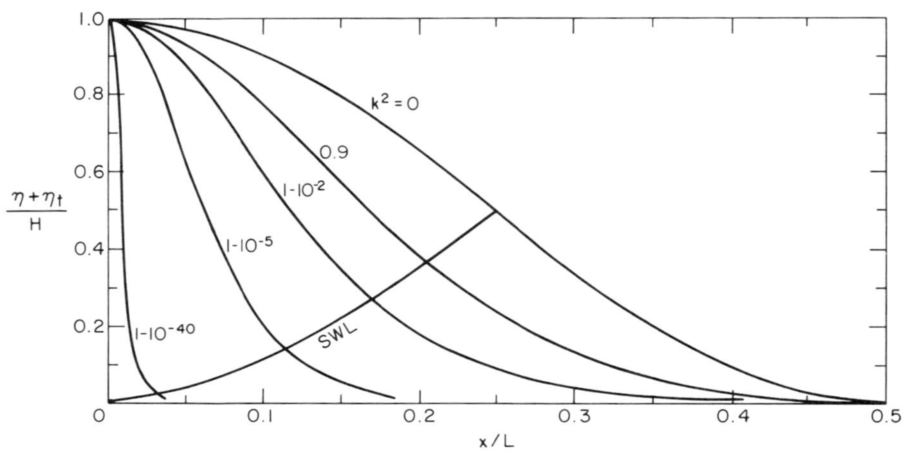

Figure 4.4 Cnoidal wave surface profiles.

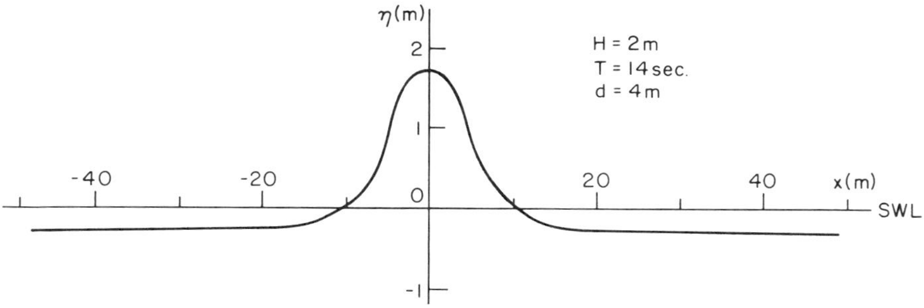

Figure 4.5 Cnoidal wave surface profile.

$-\eta_t$ (i.e., $\eta - (-\eta_t) = \eta + \eta_t$) versus horizontal distance from the wave crest x in dimensionless form. The still water level (SWL) is also given. Note that the surface profile for $k^2 = 0$ is a cosine curve, the surface profile given by the small amplitude wave theory. As k^2 increases, the surface profile becomes increasingly asymmetric with a sharper crest and flatter trough and a larger crest amplitude and smaller trough amplitude. These figures thus allow one to determine the cnoidal theory wave length, celerity, and surface profile. As mentioned earlier, one should see Wiegel (1960, 1964) for procedures to calculate other wave properties.

As an example, consider a 14-sec period, 2-m high wave in water 4 m deep. Thus, $H/d = 0.5$ and the dimensionless period is 21.9. From Figure 4.3, $k^2 = 1 - 10^{-5.3}$ and $U_r = 300$. This yields a wave length of 98.0 m and a celerity of 7 m/sec. The Ursell number of 300 indicates that the cnoidal theory is applicable. The small amplitude theory yields a wave length of 86.5 m and a celerity of 6.2 m/sec (a difference of 11.7%).

With the value of k^2, the surface profile can be determined from Figure 4.4. Figure 4.5 is a plot of this profile (with a 10:1 vertical scale exaggeration), which has a significant vertical asymmetry ($a_c/H = 0.86$).

Finally, although the small amplitude theory showed waves to be period dispersive in deep and intermediate water depths, Stokes third- and higher-order theories and the cnoidal theory show that waves are both period and amplitude dispersive throughout the entire range from deep to shallow water.

4.4 SOLITARY WAVE THEORY

A solitary wave is a wave having a surface displacement that is completely above the still water line; that is, it has a crest but no trough (see Fig. 4.6). The waves considered to this point were oscillatory as the water particles move in orbits. A solitary wave is translatory because the water particles move only in the direction of wave advance as the wave passes by; there is no return flow. The wave period and length are infinite.

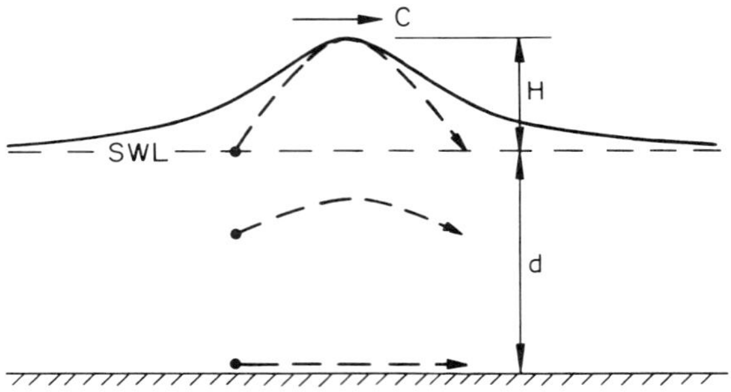

Figure 4.6 Solitary wave surface profile and particle paths.

If a vertical faced paddle in a wave tank is moved forward and then stopped, a solitary wave will be generated. It is difficult to generate a pure solitary wave, particularly for larger wave heights. Instability causes a group of successively smaller waves [Madsen and Mei (1969), Zabusky and Galvin (1971)] to trail behind the main solitary wave. These waves are amplitude dispersive, and accordingly separate as they move along the wave flume. (This is discussed further in Chapter 10.)

The solitary wave is the limit of the cnoidal wave as the relative depth decreases. For an infinite wave length and period, the Ursell number becomes infinite and k^2 becomes unity (see Fig. 4.3). And Figure 4.4 shows that as k^2 approaches unity, the wave trough approaches the still water line, resulting in the wave form depicted in Figure 4.6.

A long period wave in shallow water will approach the solitary form, but breaking will likely occur before it closely approximates this form. Tsunami waves (which have periods of several to tens of minutes) may be approximated by the solitary wave form. In any case, the cnoidal theory is still most applicable to these shallow water waves. However, the difficulty of employing the cnoidal theory equations has lead some investigators to use solitary wave theory for waves in very shallow water.

Good summaries of the solitary theory equations are given by Munk (1949) and Wiegel (1964). These mainly devolve from the cnoidal first approximation theory. A second-order solitary theory is given by Laitone (1960).

When $k^2 = 1$, the cnoidal surface profile reduces to

$$\eta = H \operatorname{sech}^2\left[\sqrt{\frac{3H}{4d^3}}\,(x - Ct)\right] \tag{4.17}$$

where the celerity is

$$C = \sqrt{gd}\left(1 + \frac{H}{2d}\right) \tag{4.18}$$

At breaking (say $H/d = 1.0$) the solitary wave celerity would thus be 50% larger than the celerity given by the small amplitude theory for shallow water. Other equations for wave celerity that are slightly different from Eq. 4.18 and based on experiments or the second-order theory have been given (Wiegel, 1964).

Typical paths of fluid particles as a solitary wave passes are shown in Fig. 4.6. Experiments show (Dailey and Stephan, 1953) that the best equations for determining the particle velocity components are those given by McCowan (1891). These are

$$u = NC \frac{1 + \cos\left(M \dfrac{z + d}{d}\right) \cosh\left(\dfrac{Mx}{d}\right)}{\left[\cos\left(M \dfrac{z + d}{d}\right) + \cosh\left(\dfrac{Mx}{d}\right)\right]^2} \tag{4.19}$$

$$w = NC \frac{\sin\left(M \dfrac{z + d}{d}\right) \sinh\left(\dfrac{Mx}{d}\right)}{\left[\cos\left(M \dfrac{z + d}{d}\right) + \cosh\left(\dfrac{Mx}{d}\right)\right]^2} \tag{4.20}$$

where N and M are determined from

$$\frac{H}{d} = \frac{N}{M} \tan \frac{1}{2}\left[M\left(1 + \frac{H}{d}\right)\right]$$

$$N = \frac{2}{3} \sin^2\left[M\left(1 + \frac{2}{3}\frac{H}{d}\right)\right]$$

The water particle paths cause a forward mass transport which is most easily determined by integrating the water surface elevation above the still water level (Eq. 4.17) from minus to plus infinity. For a unit crest width this yields

$$V = \left(\tfrac{16}{3}\, d^3 H\right)^{1/2} \tag{4.21}$$

This volume of water would move forward at the wave celerity. Bruun (1963), for example, used the solitary wave mass transport at incipient breaking to develop an early method for predicting longshore current velocity in the surf zone.

The energy in a solitary wave is approximately half potential and half kinetic.

For a unit crest width this is

$$E = \frac{8}{3\sqrt{3}} \rho g (Hd)^{3/2} \tag{4.22}$$

The wave power would be the product of the wave energy and wave celerity.

Several authors (see Galvin, 1972) have used solitary wave theory to determine the breaking wave height for a given water depth by equating the wave celerity and crest particle velocity. Typically, the limiting values for H/d range from 0.73 to 0.83 with a value of 0.78 commonly used.

4.5 NUMERICAL WAVE THEORIES

The Stokes and cnoidal-solitary wave theories (even when carried to very high orders) are somewhat deficient in accurately describing wave characteristics for large wave steepnesses near breaking. Use of higher orders does not result in solutions that converge when extreme wave steepnesses are considered (Schwartz, 1974). And extreme difficulties are encountered in attempting to develop these theories to higher than a few orders. Consequently, using numerical techniques and a computer, several successful efforts have been made in which the theory has been carried to extremely high orders.

Although these efforts generally provide the most accurate description of wave characteristics for steep and near breaking conditions, they produce results that are not easily applicable to widespread engineering use. The numerical theory developed by Dean (1965) has the most available tabulated solutions and has been evaluated versus the other theories in a variety of ways. Consequently, it is probably most used in engineering practice. These numerical approaches are discussed in general and a summary of Dean's stream function wave theory is presented.

Numerical wave theories fall into two broad categories. Schwartz (1974) employed complex variables and conformal transformation to transform the x–z wave plane to an annular ring in which the solution was achieved. It was achieved with a different perturbation parameter than Stokes used, and with computer-aided algebra solutions carried to the 112th order for deep water and the 48th order for finite depths. Longuet-Higgins and Fenton (1974), Longuet-Higgins (1975), and Cokelet (1977) used modifications of this approach and different perturbation parameters to develop other solutions for steep waves in all water depths.

None of these approaches have been presented in a form for easy engineering application. However, since they yield the most precise definition of wave characteristics for very steep waves, they have been used to evaluate the results of more commonly used wave theories. And they give greater insight to the behavior of very steep waves. An interesting feature resulting from these solutions involves the characteristics of waves as the wave steepness approaches breaking. With an increase in steepness to the limit of stability, the wave celerity and other integral properties such as energy and momentum flux increase to a peak value and then decrease before the stability limit is reached (i.e., the highest wave is not the fastest).

Recently Williams (1985) presented extensive tables, based on a conformal transformation–numerical perturbation approach, that allow us to calculate finite amplitude wave characteristics for deep to shallow water. The wave field is transformed into a complex potential field in terms of the stream function and velocity potential, the transformation being written in the form of a Fourier series. Solutions are carried to a sufficient order so that an additional term will not change the solution when carried to four decimal places. The tables allow fairly direct hand calculation of the surface profile, particle velocities and accelerations, and pressure as a function of position within the wave, as well as the wave celerity, energy, power, and radiation stress. Results in the tables are presented for 224 selected cases of relative depth d/L and wave height divided by the breaking height H/H_b. A set of coefficients for a total of 2224 cases of relative depth and dimensionless wave height is also presented. These coefficients are used with a computer program to calculate wave properties. So one may use hand calculations for the 224 cases or a more involved computer solution using one of the 2224 cases. To the author's knowledge, the results presented in these tables have not been evaluated in detail versus other theories or experimental data.

A different numerical approach has been presented by Chappelear (1961) and Dean (1965). Chappelear wrote the two velocity components and the surface profile in Fourier series form employing terms similar to those in the Stokes equations (see Eqs. 4.5, 4.9, and 4.10) with Fourier coefficients that are functions of d, H, and T. He then numerically evaluated the Fourier coefficients by an iterative least-squares procedure employing the Bernoulli equation and the KSBC which contain the velocity components at the surface and the surface amplitude. The velocity components and surface profile were calculated for five d, H, T conditions (steep waves, intermediate depth) carrying the numerical technique to the fifth order. Results were compared to similar calculations using the Stokes fifth-order theory. Both approaches gave very similar results.

Dean (1965) developed a numerical wave theory that is somewhat similar to the approach used by Chappalear, but uses the stream function to define the flow. When wave motion is converted to steady flow by subtracting the wave celerity from the horizontal motion, the free surface and bottom become streamlines (the stream function is a constant along these two surfaces).

Using the stream function Ψ and converting to steady flow, the two-dimensional wave boundary value problem is to find a solution to the Laplace equation

$$\frac{\partial^2 \Psi}{\partial x^2} + \frac{\partial^2 \Psi}{\partial z^2} = 0 \tag{4.23}$$

for the following boundary conditions

$$\frac{\partial \Psi}{\partial x} = 0 \quad \text{at } z = -d \tag{4.24}$$

$$\frac{\partial \Psi}{\partial x} = \left(\frac{\partial \Psi}{\partial z} - C\right)\frac{\partial \eta}{\partial x} \quad \text{at } z = \eta \tag{4.25}$$

$$\frac{1}{2}\left[\left(\frac{\partial \Psi}{\partial z} - C\right)^2 + \left(\frac{\partial \Psi}{\partial x}\right)^2\right] + g\eta = Q \quad \text{at } z = \eta \tag{4.26}$$

where $u = \partial\Psi/\partial z$ and $w = -\partial\Psi/\partial x$.

The small amplitude wave theory solution in terms of the stream function is

$$\Psi = \frac{gH}{2\sigma}\frac{\sinh k(d + z)}{\cosh kd}\cos(kx - \sigma t) \tag{4.27}$$

Adding a steady uniform horizontal flow yields

$$\Psi = Cz + \frac{gH}{2\sigma}\frac{\sinh k(d + z)}{\cosh kd}\cos kx \tag{4.28}$$

Dean addressed two problems:

1. Given the wave height and period and the water depth, determine the other dependent wave characteristics.
2. Given the surface profile, determine the dependent wave characteristics.

The first is the classic wave problem which will be discussed below. The second allows one to calculate wave characteristics from a measured wave profile that may, for example, be horizontally asymmetric. Dean also allowed for a proscribed surface pressure distribution and a horizontal current that is uniform with depth. For these conditions the two surface boundary conditions (Eqs. 4.25 and 4.26) would be modified accordingly.

Considering Eq. 4.28 and the form of the Stokes higher-order equations Dean used an Nth order stream function having the following form:

$$\Psi = Cz + \sum_{n=1}^{N} X_n \sinh nk(d + z)\cos nkx \tag{4.29}$$

For the surface streamline Ψ_η this would become

$$\Psi_\eta = c\eta + \sum_{n=1}^{N} X_n \sinh nk(d + \eta)\cos nkx \tag{4.30}$$

Equation 4.30 exactly satisfies the Laplace equation, the BBC, and the KSBC (Eqs. 4.23–4.25). The basic problem is to determine the value of the coefficients X_n (to the order desired), the wave number k, and the surface value of the stream function so that they best satisfy the DSBC (Eq. 4.26). The volumes of water above and below the mean water level must also be equal. This can be accomplished with Eq. 4.30.

In basic terms this is done by evaluating the Bernoulli constant Q at a number of points along the wave surface so that the sum of the deviations of the square of the difference of each Q value from the average Q value is minimized. An iterative procedure is used to obtain the wave number, which defines the wave length, the surface stream function, which gives the surface profile, and the X_n values, which then define the stream function to the desired order. Other desired wave characteristics can be determined from the stream function.

Dean's use of the stream function is more advantageous than the approach presented by Chappalear which employs the velocity potential to define flow. Chappalear required two sets of coefficients, one in the velocity component terms and one in the surface profile term. And neither surface boundary condition is expressly satisfied. Dean's approach requires determination of only one set of coefficients employing the one surface boundary condition. This simplifies the analysis and improves the quality of results.

A set of tables is available (Dean, 1974) for determining the wave length, surface profile, particle velocity and acceleration components, dynamic pressure, group celerity, energy, and momentum flux for a specified wave height and period and water depth (and where appropriate for specified x, z coordinates in the wave). These results are tabulated for 40 conditions of H, $L_0 (= gT^2/2\pi)$, and d. These 40 conditions are for 10 values of d/L_0 spread over the shallow to deep water range of 0.002–2.0 and H_b/H = 0.25, 0.5, 0.75, and 1.0 for each d/L_0 value. Calculation of d/L_0 and H/L_0 yields the related value of H/H_b from a plot given in Dean (1974). Interpolation is required in the likely event that the d/L_0, H/L_0 condition of interest is not one of the 40 tabulated points. An interpolation procedure that allows generally adequate results for any given d/L_0 and H/H_b combination is given.

For a given wave height and period and water depth, an opposing or following current will affect the other properties of a wave. For example, a following current would increase the particle velocity under the wave crest, increase the wave length, and raise the crest elevation. The original stream function theory developed by Dean (1965) allows for a vertically uniform opposing or following current. But currents typically have a vertical velocity profile resulting in a vertical variation of vorticity. Dalrymple (1974) and Dalrymple and Cox (1976) extended the stream function wave theory to include these more complex velocity profiles.

4.6 VERIFICATION OF WAVE THEORIES

There have been many published efforts to verify the various wave theories. Typically, these efforts have considered only one or a select number of the many wave theories available. They can be placed into one of four categories (Dean and Perlin, 1986).

1. The ultimate purpose for most applications of wave theory is to calculate prototype conditions. Thus the ideal verification is to compare calculated results (surface profile, particle kinematics, etc.) with measured values. However, there is little quality field data available for this purpose. The theories are for monochromatic waves rather than the irregular sea; the field data only cover limited ranges of relative depth and wave height; and quality field measurements are difficult and expensive to obtain.
2. Most physical evaluations of wave theories have involved comparisons with laboratory measurements. Monochromatic waves of the desired height, period, and depth combination can be generated. And measurements are easier to make under smaller-scale controlled conditions. However, viscosity and surface tension scale effects occur, and wave generator/wave flume effects may diminish the value of the data. The latter include wave reflection and mass transport effects caused by a closed system.
3. The Laplace equation and the bottom boundary condition (horizontal, impervious, and frictionless bottom) are exactly satisfied by wave theories. One or both of the free surface boundary conditions are only approximately satisfied. The degree to which these surface boundary conditions are satisfied has been used to evaluate the various wave theories.
4. Some theories, based on their development, are known to be the best of those available (but not necessarily the most easy to use). Consequently, comparisons can be made with the more commonly used but perhaps less valid theories. This comparison can cover all ranges of wave conditions, including those where physical data are not available.

Some of the examples of these approaches to wave theory evaluation are discussed herein, and some of the other efforts are referenced.

Wiegel (1964) summarizes some of the early wave flume investigations that compare surface profile and particle velocity measurements with measured values. Particle velocities were measured by taking motion pictures of neutrally buoyant colored particles suspended in the wave. Results were compared to the small amplitude, Stokes second-order, and cnoidal theories. The small amplitude theory predicted particle velocities for waves of appreciable steepness for d/L larger than 0.1–0.2. As expected, Stokes second-order theory was more successful in defining the particle velocity and surface profile for steeper waves in deep and intermediate water depths. A few measurements in fairly shallow water were inconclusive in comparing the cnoidal and small amplitude theories.

LeMehaute et al. (1968) measured the horizontal water particle velocity distribution under the wave crest for eight wave conditions in intermediate and shallow water ($d/gT^2 = 0.025$ to 0.0015, see Section 2.4.3) with heights approaching the breaking height. Neutrally buoyant particles and an open camera shutter with strobe lighting were used. Results were compared to 12 theories including the small amplitude, Stokes second, third, and fifth, three cnoidal, and two solitary theories. Measurements were made before a reflected wave could return to affect wave conditions. No theory was uniformly exceptionally valid. The Stokes theories, as might be excepted for these small d/L conditions, were not very good. The small amplitude theory was "surprisingly good for the shorter waves but departs significantly from the data as the period increases." But even for the longer waves, the small amplitude theory gave reasonable results for bottom velocity. The cnoidal theory (given in Section 4.3) gave "perhaps the most generally acceptable description" of particle velocities. For easier application, one could use a solitary theory for the crest shape and particle velocities near the surface, and the small amplitude theory for velocities near the bottom.

Dean (1974) extended the comparison made by LeMehaute et al. (1968) to include his stream function theory. Given the H, T and d, the stream function theory gave consistently better results for the range of conditions evaluated than did any of the other theories considered. An example of one of the sets of velocity measurements versus the predictions of the theories presented in this chapter is shown in Figure 4.7 ($d/L = 0.065$).

Hattori (1986) made a series of wave flume measurements of surface profile and particle velocities for waves shoaling on a 1:20 slope. A laser doppler velocimeter was used to make more precise measurements of particle velocities. The small amplitude, Stokes fifth, cnoidal third and stream function fifth-order theories were compared. Results showed that the theories considered made good predictions in the regions where they would be expected to apply. They also used the stream function ninth theory to calculate particle velocities employing the measured surface profile rather than H, T and d. This yielded the best agreement over a wide range of relative depths from deep water to near breaking.

Other wave flume measurements, typically of the surface profile and particle velocities for monochromatic waves, have been made by Tsuchiya and Yamaguchi (1972), Chakraabarti (1980), and Easson et al. (1988). In each case results were compared to a selection of different wave theories with varying results.

Some field measurements have been reported in the literature. Grace and Rocheleau (1973) measured bottom velocities with a ducted propeller meter and calculated surface profiles from bottom pressure measurements. Wave

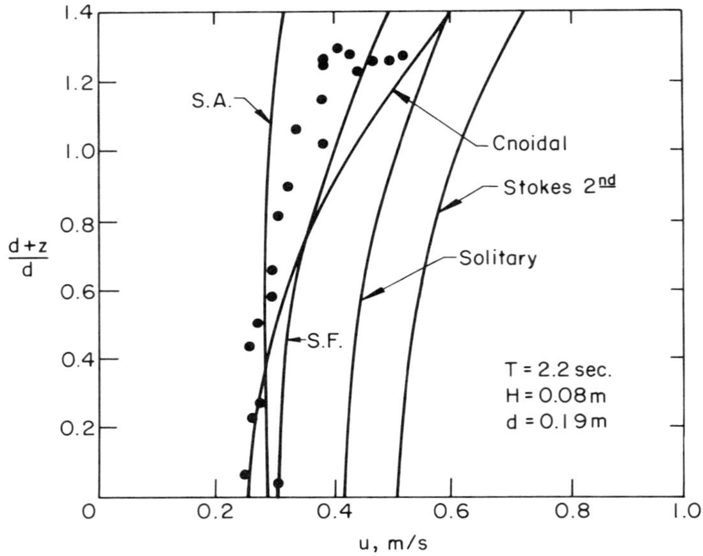

Figure 4.7 Water particle velocities under the wave crest (after Dean, 1974).

heights were less than a fourth of the breaking height. Results were compared to the small amplitude and stream function fifth-order theories. Results suggest that the small amplitude theory best predicted wave trough velocities, whereas the stream function theory best predicted crest velocities. However, velocity predictions were based on height and period values determined from the pressure measurements using a small amplitude theory analysis.

Ohmart and Gratz (1978) measured surface profiles with a wire staff gauge and particle velocities with an electromagnetic current meter. The water depth was 177 ft and wave heights were up to 60% of breaking. The linear and Stokes fifth-order theories both were generally successful in predicting particle velocities. The stream function theory employing the measured surface profile was most successful.

Dean (1970) compared the small amplitude, Stokes third and fifth, two cnoidal, two solitary, and the stream function fifth-order wave theories on the basis of the root-mean-square (rms) errors in the kinematic and dynamic surface boundary conditions. The comparison was made for the 40 wave conditions for which stream function theory solutions were made. This covers the complete deep-shallow water range for low to high steepness waves. Generally, results were more satisfactory for $d/gT^2 > 0.006$ ($d/L = 0.081$) than for smaller values. The stream function theory gave the best boundary condition fit over the entire range except for steeper waves in very shallow water. Of the analytical theories, the cnoidal theory presented herein was best in shallow water into intermediate depths, the Stokes fifth theory was best in deep water into intermediate depths, and for some in-between intermediate depths the small amplitude theory was best. This evaluation approach is limited in indicating the best theory for use. For example, Dean found that two theories had the same degree of boundary condition fit in shallow water, but when employed to calculate wave drag force (a function of particle velocity squared), gave results that differed by a factor of 4.

Chaplin (1980) considered the theory presented by Cokelet (1977) to best represent wave conditions near breaking and used some of his tabulated data for near breaking waves to evaluate the stream function theory. Evaluations were based on the calculated crest elevation, crest particle velocity, and wave length. He found that for waves having a height of up to three-fourths of the breaking height, stream function results were very good except in extremely shallow water. Interpolated results for $H = 0.9H_b$ were in error by no more than 5% in most cases. At the limiting wave height some errors were significant. For example, the crest particle velocities were underestimated by as much as 30%. For other properties the errors were not so extreme.

Finally, attempts were made to modify basic analytical wave theories so that they better fit observed data from the laboratory and field. The results may locally violate the Laplace equation or surface boundary conditions, but the modified equations provide a better description of actual wave kinematics, particularly near the surface. One interesting result is that a number of quality wave measurement programs have found that for steep deep water waves the horizontal particle velocity is higher at the trough than at the crest. Gudmestad and Connor (1986) summarize some of these attempts to correct the small amplitude and Stokes higher-order wave kinematics equations by empirical calibration and the efficacy of these modified equations to predict measured values.

4.7 RANGE OF APPLICATION OF THEORIES

It is not possible to precisely specify ranges of wave steepness and relative depth (H/gT^2 and d/gT^2) for which a specific wave theory should be used. Several factors come into play in selecting a wave theory for use. The input conditions—wave height and period for a given water depth—may be precisely specified. Or they may only be approximately known from wave hindcasts and a shoaling analysis to determine the local wave height. In the latter case, extreme sophistication in carrying out wave calculations may not be justified and small amplitude wave theory may be adequate.

The input wave height and period may be precisely specified, but calculated wave characteristics may only need to be approximately known; again small amplitude theory may suffice. However, if, for example, one wanted to determine the wave crest elevation to establish a design elevation for the underside of a pier or floating structure, an appropriate finite amplitude theory should be employed. Or, if wave particle velocities and accelerations are to be calculated from a measured wave surface profile, use of the stream function theory may be dictated because the surface profile can be specified.

The difficulty of choice of a theory to use is further complicated by the fact that a particular wave theory is better at defining certain wave characteristics than others. For example, in shallow water the small amplitude theory does a good job of predicting particle velocities, particularly near the bottom, but completely misrepresents the surface asymmetry.

Generally, we know that the small amplitude theory is adequate in deep and intermediate water depths if the wave is not too steep. The higher-order Stokes theories are good for steeper waves in deep water and for intermediate depths when the wave is not too steep. For steeper waves in shallow water and a portion of intermediate water depths cnoidal theory is generally best. For steep waves in any water depth a numerical theory may be employed.

Diagrams giving specific recommended ranges of application for the various wave theories have been given by Muir Wood (1969), LeMehaute (1969), and Komar (1976). They are based on analytical considerations (such as a limiting Ursell number for application of a certain theory), experimental evaluations of the various theories, and a good dose of personal judgment. Generally, an attempt is made to expand the range of application of the theories that are simpler to apply. Dean (1970) also presents such a diagram, based on his evaluation of the fit of the various theories to the surface boundary conditions.

Figure 4.8 is the diagram given by LeMehaute (1969) with a small modi-

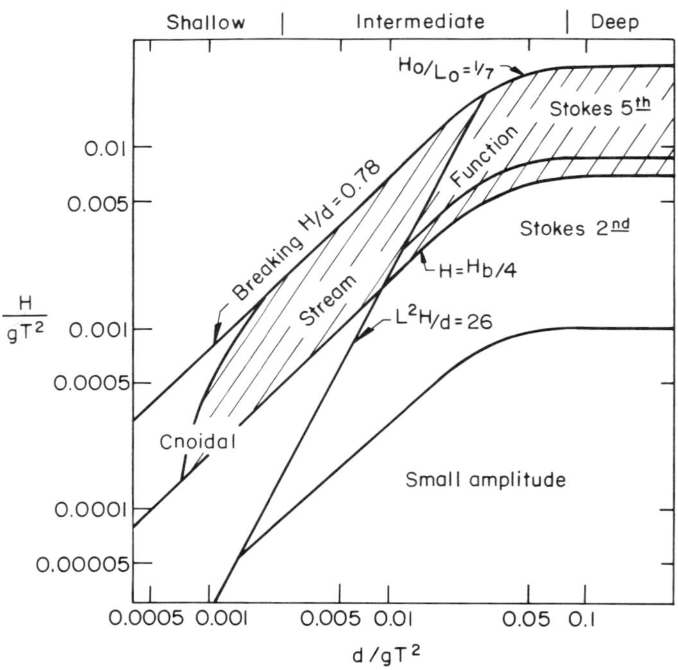

Figure 4.8 Recommended ranges for selected wave theories (after LeMehaute, 1969).

fication. The Stokes fifth theory is shown in the region where LeMehaute recommends using the Stokes third and fourth for progressively steeper waves. The solitary theory is not shown on the diagram, but it may be used for steeper waves in very shallow water. This diagram is given to demonstrate the type of diagrams available and to further suggest ranges of application for the various theories.

4.8 FINITE AMPLITUDE SHOALING CALCULATIONS

In Section 3.2 the small amplitude theory was used to develop a relationship for the change in wave height as a wave propagated from one depth to another (Eqs. 3.1 and 3.2, Table 3.1). The change in wave height from deep water to some point in intermediate or shallow water is generally of greatest interest, and for the small amplitude theory H/H_0' is a function of only the relative depth d/L_0 or d/L. If the wave steepness is small (and the bottom slope is gentle), small amplitude wave theory is adequate for calculating these changes in wave height.

For steeper waves, finite amplitude theory should be employed for shoaling calculations if an accurate prediction of wave height change is needed. (Once the shallow water height is determined, using the period and water depth the other wave properties can be calculated.) But two complexities arise. For finite amplitude waves, the change in wave height depends not only on the change in relative depth but also on the initial wave steepness. And most of the finite amplitude theories are only applicable over a portion of the range of relative depths from deep to shallow water. To extend the analysis over the entire range of depths, two theories have to be coupled at some intermediate depth.

Employing a conservation of energy flux LeMehaute and Webb (1964) and Koh and LeMehaute (1966) applied Stokes third- and fifth-order theories to the prediction of shoaling characteristics. They developed curves of H/H_0 versus d/gT^2 for selected values of H_0/gT^2 that are applicable from deep water into intermediate depths where the Ursell parameter is about 10. Results showed that H/H_0 was less than the value given by the linear theory and very slightly dependent on wave steepness for d/L_0 greater than 0.4. For relative depths less than 0.4, the Stokes wave height was higher than the small amplitude height and increasingly dependent on wave steepness. But over the range of depths considered, the difference in predicted heights between the Stokes and small amplitude theories never exceeded 5%. For relatively deep water ($d/L > 0.25$), the fifth-order theory was best, but over the entire range considered, the third-order theory was preferable.

Svendsen and Brink-Kjaer (1972) also employed conservation of energy flux, but with the cnoidal theory, to predict wave height variation as a wave shoals. The cnoidal theory was coupled with the small amplitude theory in deeper water by equating energy flux at $d/L_0 = 0.1$. This produces a discontinuity in the wave height at this point, with the cnoidal height being a few percent smaller. There is also a discontinuity in wave length, particle kinematics, pressure, and so on at the match point. (The wave period was held constant.) In shallow water, particularly for the steeper waves, the conoidal wave height was significantly greater than the height predicted by using small amplitude theory all the way in from deep water.

Svendsen and Buhr-Hansen (1977) also used cnoidal theory to evaluate wave shoaling, but matched wave heights rather than energy flux at $d/L_0 = 0.1$ (thus producing an energy flux discontinuity at the match point). Having noticed a good deal of scatter in the results of previous experiments on wave height change with shoaling, they conducted their own, very controlled experiments (paying close attention to the flume friction losses, quality of wave generated, and gentleness of slope used). Matching the wave heights produced better results than matching the energy flux.

Iwagaki (1968) investigated wave shoaling by employing a simplified cnoidal theory (called hyperbolic theory) matched to Stokes theory by the energy flux. Yamaguchi and Tsuchiya (1976) also employed cnoidal–Stokes theory matching to evaluate wave shoaling. They performed experiments to evaluate the results but the experimental data scatter was significant, making it difficult to precisely evaluate the analytical results. Stiassnie and Perigrine (1980) used Cokelet's (1977) numerical extension of the Stokes approximation and the numerical solitary wave solution of Longuet-Higgins and Fenton (1974) to investigate wave propagation from deep water to breaking. Calculations were done for 12 wave steepnesses and results of wave amplitude change were compared to available laboratory measurements. The results were "as good as can be expected without the inclusion of some dissipation in the theory." Again, the calculations were complex and only a limited number of cases was considered.

LeMehaute and Wang (1980) discuss the various approaches used to evaluate finite amplitude wave shoaling. They propose a hybrid approach employing cnoidal theory (as in Svendsen and Brink-Kjaer, 1972) to determine wave height transformation and small amplitude theory to predict changes in wave length. A computational procedure is presented to calculate wave shoaling and the point of breaking. They also demonstrate how small amplitude theory underpredicts wave breaking characteristics (see Fig. 4.9). The wave breaking height

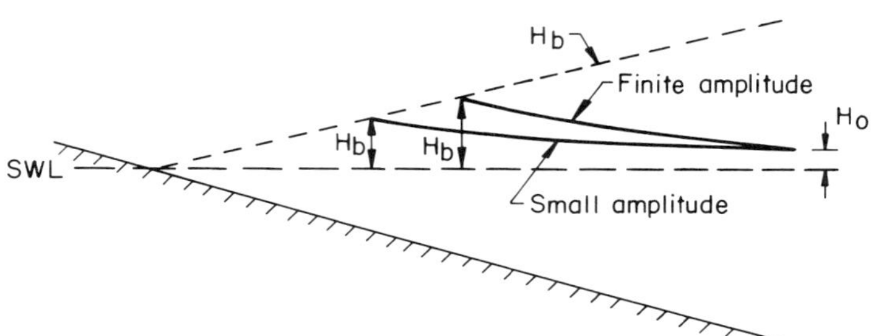

Figure 4.9 Comparison of wave heights for small and finite amplitude wave shoaling.

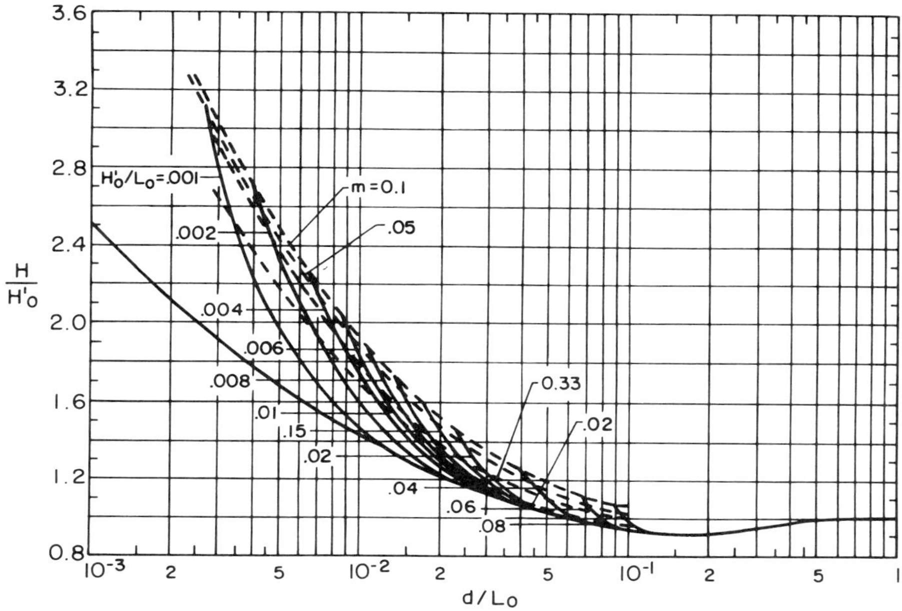

Figure 4.10 Small and finite amplitude shoaling and breaking (after Walker and Headland, 1982).

generally depends on the water depth for a given beach slope and wave period. Underprediction of the growth of the wave height would yield an underpredicted breaking wave height and depth (and related wave properties) as well as surf zone width.

The various finite amplitude shoaling analysis procedures are also surveyed by Walker and Headland (1982), with an emphasis on predicting nearshore wave breaking conditions. Employing the available theoretical and experimental shoaling curves and experimental data on wave breaking they developed the diagram given in Figure 4.10. The lowest solid line is the shoaling curve based on the small amplitude theory (i.e., a plot of columns 2 and 3 from Table 3.1). For increasing deep water wave steepnesses the appropriate higher solid line would apply. The dashed lines (which depend on beach slope) indicate breaking conditions where they intercept the solid lines. For a given deep water wave steepness one can trace the increase in wave height as the relative depth decreases and determine the final wave height and water depth at breaking. The breaker curves depend on the beach slope. The wave shoaling curves also have some dependence on beach slope, but the shoaling curves in Figure 4.10 are drawn for a representative slope $m = 0.033$.

REFERENCES

Bruun, P. (1963), "Longshore Currents and Longshore Troughs," *J. Geophys. Res.*, **68,** 1065–1078.

Chakrabarti, S. K. (1980), "Laboratory Generated Waves and Wave Theories," *J. Waterw. Port Coastal Ocean Eng. Div., Am. Soc. Civ. Eng.*, August, 349–368.

Chaplin, J. R. (1980), "Developments of Stream-Function Wave Theory," *Coastal Eng.*, **3,** February, 179–206.

Chappelear, J. E. (1961), "Direct Numerical Calculation of Wave Properties," *J. Geophys. Res.*, **66,** 501–508.

Chappelear, J. E. (1962), "Shallow Water Waves," *J. Geophys. Res.*, **67,** 4693–4704.

Cokolet, E. D. (1977), "Steep Gravity Waves in Water of Arbitrary Uniform Depth," *Philos. Trans. R. Soc., London*, Series A, 183–230.

Dailey, J. W. and Stephan, S. C. (1953), "Characteristics of the Solitary Wave," *Trans. Am. Soc. Civ. Eng.*, **118,** 575–587.

Dalrymple, R. A. (1974), "A Finite Amplitude Wave on a Linear Shear Current," *J. Geophys. Res.*, **79,** 4498–4504.

Dalrymple, R. A. and Cox, J. C. (1976), "Symmetric Finite-Amplitude Rotational Water Waves," *J. Phys. Oceanogr.*, **6,** 847–852.

Dean, R. G. (1965), "Stream Function Representation of Nonlinear Ocean Waves," *J. Geophys. Res.*, **70,** 4561–4572.

Dean, R. G. (1970), "Relative Validities of Water Wave Theories," *J. Waterw. Harbors Div., Am. Soc. Civ. Eng.*, February, 105–119.

Dean, R. G. (1974), "Evaluation and Development of Water Wave Theories for Engineering Application," Special Report No. 1, U. S. Army Coastal Engineering Research Center, Ft. Belvoir, VA (2 Vols).

Dean, R. G. and Perlin, M. (1986), "Intercomparison of Near Bottom Kinematics by Several Wave Theories and Field and Laboratory Data," *Coastal Eng.*, **9,** 399–437.

Easson, W. J., Griffiths, M. W. P. and Greated, C. A. (1988), "Kinematics of Breaking Waves in Coastal Regions," *Proceedings, 21st International Conference on Coastal Engineering,* American Society of Civil Engineers, Malaga, Spain, pp. 871–883.

Fenton, J. D. (1985), "A Fifth-Order Stokes Theory for Steady Waves," *J. Waterw. Port Coastal Ocean Eng. Div., Am. Soc. Civ. Eng.*, March, 216–234.

Galvin, C. J. (1972), "Wave Breaking in Shallow Water," *Waves on Beaches and Resulting Sediment Transport*, R. E. Myers, Ed., Academic, New York, pp. 413–451.

Grace, R. A. and Rocheleau, R. Y. (1973), "Near-bottom Velocities Under Waikiki Swell," Technical Report 31, Ocean Engineering, University of Hawaii, Honolulu.

Gudmestad, O. T. and Connor, J. J. (1986), "Engineering Approximations to Nonlinear Deepwater Waves," *Appl. Ocean Res.*, **8,** 76–88.

Hardy, T. A. and Kraus, N. C. (1987), "A Numerical Model for Shoaling and Refraction of Second Order Cnoidal Waves Over an Irregular Bottom," Miscellaneous Paper CERC 87-9, U. S. Army Waterways Experimental Station, Vicksburg, MS.

Hattori, M. (1986), "Experimental Study on the Validity Range of Various Wave Theories," *Proceedings, 20th International Conference on Coastal Engineering*, American Society of Civil Engineers, Taipei, pp. 232–246.

Ippen, A. T. (1966), *Estuary and Coastline Hydrodynamics*, McGraw-Hill, New York.

Iwagaki, Y. (1968), "Hyperbolic Waves and Their Shoaling," *Proceedings, 11th Con-*

ference on Coastal Engineering, American Society of Civil Engineers, London, pp. 124–144.

Keller, J. B. (1948), "The Solitary Wave and Periodic Waves in Shallow Water," *Commun. Pure Appl. Math.*, **1,** 323–339.

Keulegan, G. H. (1950), "Wave Motion," *Engineering Hydraulics*, H. Rouse, Ed., Wiley, New York, Chapter 11.

Keulegan, G. H. and Patterson, G. W. (1940), "Mathematical Theory of Irrotational Translation Waves," *J. Res. Nat. Bur. Stand.*, **24,** 47–101.

Koh, R. C. Y. and LeMehaute, B. (1966), "Wave Shoaling," *J. Geophys. Res.*, **71,** 2005–2012.

Komar, P. D. (1976), *Beach Processes and Sedimentation*, Prentice-Hall, Englewood Cliffs, NJ.

Korteweg, D. J. and de Vries, G. (1895), "On the Change of Form of Long Waves Advancing in a Rectangular Canal, and on a New Type of Long Stationary Waves," *Philos. Mag.*, Series 5, **39,** 422–443.

Laitone, E. V. (1960), "The Second Approximation to Cnoidal and Solitary Waves," *J. Fluid Mech.*, **9,** 430–444.

LeMehaute, B. (1969), "An Introduction to Hydrodynamics and Water Waves," Technical Report ERL 118-POL-3-2, U. S. Department of Commerce, Washington, DC.

LeMehaute, B., Divoky, D., and Lin, A. (1968), "Shallow Water Waves: A Comparison of Theories and Experiments," *Proceedings, 11th Conference on Coastal Engineering*, American Society of Civil Engineers, London, pp. 86–107.

LeMehaute, B. and Wang, J. D. (1980), "Transformation of Monochromatic Waves from Deep to Shallow Water," Technical Report 80-2, U. S. Army Coastal Engineering Research Center, Ft. Belvoir, VA.

LeMehaute, B. and Webb, L. M. (1964), "Periodic Gravity Waves over a Gentle Slope at a Third Order Approximation," *Proceedings, 9th Conference on Coastal Engineering*, American Society of Civil Engineers, Lisbon, pp. 23–40.

Longuet-Higgins, M. S. (1953), "Mass Transport in Water Waves," *Philos. Trans. R. Soc. London*, Series A, 535–581.

Longuet-Higgins, M. S. (1975), "Integral Properties of Periodic Gravity Waves of Finite Amplitude," *Proc. R. Soc. London*, Series A, 157–174.

Longuet-Higgins, M. S. and Fenton, J. D. (1974), "On the Mass, Momentum, Energy and Circulation of a Solitary Wave," *Proc. R. Soc. London*, Series A, 471–491.

Madsen, O. S. and Mei, C. C. (1969), "The Transformation of a Solitary Wave Over an Uneven Bottom," *J. Fluid Mech.*, **39,** 781–791.

McCowan, J. (1891), "On the Solitary Wave," *London, Edinburgh Dublin Mag. J. Sci.*, **32,** 45–58.

Muir Wood, A. M. (1969), *Coastal Hydraulics*, Gordon and Breach, New York.

Munk, W. H. (1949), "The Solitary Wave and its Application to Surf Problems," *Ann. NY Acad. Sci.*, **51,** 376–424.

Ohmart, R. D. and Gratz, R. L. (1978), "A Comparison of Measured and Predicted Ocean Wave Kinematics," *Proceedings, Offshore Technology Conference*, Houston, pp. 1947–1957.

Russell, R. C. H. and Osorio, J. D. C. (1958), "An Experimental Investigation of Drift Profiles in a Closed Channel," *Proceedings, 6th Conference on Coastal En-*

gineering, Council on Wave Research, Engineering Foundation, University of California, Berkeley, pp. 171–183.

Sarpkaya, T. and Isaacson, M. (1981), *Mechanics of Wave Forces on Offshore Structures*, Van Nostrand Reinhold, New York.

Schwartz, L. W. (1974), "Computer Extension and Analytical Continuation of Stokes' Expansion for Gravity Waves," *J. Fluid Mech.*, **62,** 552–578.

Skjelbreia, L. (1959), "Gravity Waves, Stokes Third Order Approximations, Tables of Functions," Council on Wave Research, Engineering Foundation, University of California, Berkeley.

Skjelbria, L. and Hendrickson, J. A. (1961), "Fifth Order Gravity Wave Theory," *Proceedings, 7th Conference on Coastal Engineering*, Council on Wave Research, Engineering Foundation, University of California, Berkeley, pp. 184–196.

Stiassnie, M. and Peregrine, D. H. (1980), "Shoaling of Finite-Amplitude Surface Waves on Water of Slowly-Varying Depth," *J. Fluid Mech.*, **97,** 783–805.

Stokes, G. G. (1847), "On the Theory of Oscillatory Waves," *Trans. Cambridge Philos. Soc.*, **8,** 441–455.

Svendsen, I. A. and Brink-Kjaer, O. (1972), "Shoaling of Cnoidal Waves," *Proceedings, 13th International Conference on Coastal Engineering*, American Society of Civil Engineers, Vancouver, pp. 365–383,

Svendsen, I. A. and Buhr-Hansen, J. (1977), "The Wave Height Variation for Regular Waves in Shoaling Water," *Coastal Eng.*, **1,** 261–284.

Tsuchiya, Y. and Yamaguchi, M. (1972), "Some Considerations on Water Particle Velocities of Finite Amplitude Wave Theories," *Coastal Eng. Jpn.*, **15,** 43–57.

U. S. Army Coastal Engineering Research Center (1984), *Shore Protection Manual*, U. S. Government Printing Office, Washington, DC.

Ursell, F. (1953), "The Long-Wave Paradox in the Theory of Gravity Waves," *Proc. Cambridge Philos. Soc.*, **49,** 685–694.

Walker, J. and Headland, J. (1982), "Engineering Approach to Nonlinear Wave Shoaling," *Proceedings, 18th International Conference on Coastal Engineering*, American Society of Civil Engineers, Cape Town, pp. 523–542.

Wiegel, R. L. (1960), "A Presentation of Cnoidal Wave Theory for Practical Application," *J. Fluid Mech.*, **7,** 273–286.

Wiegel, R. L. (1964), *Oceanographical Engineering*, Prentice-Hall, Englewood Cliffs, NJ.

Williams, J. M. (1985), *Tables of Progressive Gravity Waves*, Pitman, London.

Yamaguchi, M. and Tsuchiya, Y. (1976), "Wave Shoaling of Finite Amplitude Waves," *Proceedings, 15th International Conference on Coastal Engineering*, American Society of Civil Engineers, Honolulu, pp. 497–506.

Zabusky, N. J. and Galvin, C. J. (1971), "Shallow-Water Waves, the Kortweg–de Vries Equations and Solutions," *J. Fluid Mech.*, **47,** 811–824.

5

THREE-DIMENSIONAL WAVE TRANSFORMATIONS

Chapters 2–4 were concerned with two-dimensional wave characteristics and wave transformation in the x, z plane. Next we must consider the three-dimensional transformations that occur as long-crested monochromatic waves propagate over irregular water depths (wave refraction), are intercepted by a rigid structure and have their energy spread into the shadow zone behind the structure (wave diffraction), or obliquely reflect from a structure. We are interested in predicting the change in wave crest patterns caused by these phenomena as well as the consequent change in wave energy density and height. The local wave height that results from these three-dimensional transformations, along with the wave period and water depth, define the other wave characteristics as shown in the preceding chapters.

In intermediate and shallow water, the celerity of a wave depends on the local relative depth (d/L). If the depth varies along the crest of a wave, the portion of the wave in shallower water will have a lower celerity. This will cause the wave crest to reorient as the wave propagates forward and consequently to change its alignment toward the alignment of the bottom contours. It will also cause the wave crest length to increase or decrease, resulting in a commensurate decrease or increase in wave energy density and wave height. Thus, as a wave with a given deep water height, period, and direction of propagation travels toward the shore, refraction will cause the wave crest orientation to change and the wave height will change owing to both the effects of refraction and shoaling. Since currents affect the celerity of a wave, waves intercepted by an oblique current or a current with a lateral velocity gradient will also be refracted.

If a portion of the wave crest is intercepted by a structure, energy will flow laterally along the unintercepted portion of the wave and spread into the lee of the structure. The wave crest pattern in the structure lee essentially consists of concentric circular arcs moving out from the tip of the structure where the wave was intercepted. Wave diffraction will occur any time the wave height is not constant along the crest of a wave and the energy flow is from a point of higher to lower wave height or energy density. Wave height variations along a wave crest caused by wave refraction would be slightly diminished by wave diffraction. It is very important, for example in the design of a harbor, to be able to determine the resulting wave height and crest orientation in the lee of a protective breakwater owing to diffraction (and refraction if the depth is not constant).

When a wave encounters a partial or complete barrier, depending on the barrier characteristics, a certain portion of the wave energy is reflected. The reflected wave height is commensurately less than the incident height. Also, if the incident wave approaches the barrier with its crest oriented other than parallel to the barrier, the reflected wave orientation will be different from the incident wave orientation. In harbors with highly reflective bulkheads or quay walls, for example, it is important to be able to predict reflected wave patterns and heights.

The focus of this chapter is on the various analytical and numerical methods that may be employed to evaluate wave refraction, diffraction, and reflection—both the resulting wave crest patterns and the changes in wave height. This is done for long-crested monochromatic waves, which may represent a selected design wave height, period, and direction or be one component of a spectrum of waves. We also briefly present a related topic, the three-dimensional pattern of waves generated by a moving object.

5.1 WAVE REFRACTION

Wave refraction analyses usually consider waves that travel from deep water to some intermediate or shallow water depth. Employing the small amplitude theory, the change in wave height that occurs is obtained by rewriting Eq. 3.1 to yield

$$\frac{H}{H_0} = \sqrt{\frac{L_0}{2nL}}\sqrt{\frac{B_0}{B}} = K_s K_r \tag{5.1}$$

where the first term on the right accounts for shoaling effects (K_s) and the second term accounts for refraction effects (K_r). Recall that B is the spacing between two orthogonals (sometimes called wave rays) and that orthogonals are lines that are normal to the wave crest at all points along the orthogonal. Equation 5.1 is developed by equating the energy flux between two orthogonals in deep water and at some intermediate or shallow depth. It assumes that no energy is added or removed by the mechanisms discussed in Chapter 3 and that no energy diffracts across the orthogonal lines. If some finite amplitude theory is to be employed the first term on the right of Eq. 5.1 would have to be modified to account for the wave energy flux specification according to that theory.

There is an infinity of orthogonals in any wave refraction pattern, but usually only enough orthogonals are considered to adequately define the refraction effects. Figure 5.1 shows a simple wave refraction diagram for oblique waves traveling from deep water to the shore. Shown are the wave crests and orthogonals as well as the bottom contours that caused the refraction to occur. The wave breaking pattern is also shown. This diagram would be developed for a given wave period and deep water crest orientation. A change in either of these would change the refraction pattern. An engineering study for a particular site exposed to design waves having a range of periods and directions would require a different refraction diagram for each direction–period combination.

To evaluate the effects of refraction, the pattern of wave orthogonals that develop as a wave propagates forward must be determined. The spacing between two orthogonals in deep water B_0 is normally arbitrarily selected and the value of B is determined from the spacing of these two orthogonals at the point of interest. For the refraction pattern shown in Figure 5.1, the orthogonal spacing increases in the shoreward direction so refractive effects cause the wave height to decrease. (There may still be a net increase in wave height if the shoaling effects dominate.) The calculated change in wave height using Eq. 5.1 yields a wave height value that is the average over the orthogonal spacing B. It may be necessary to use a small orthogonal spacing to obtain an adequate evaluation of the wave height at a specific location, particularly for complex

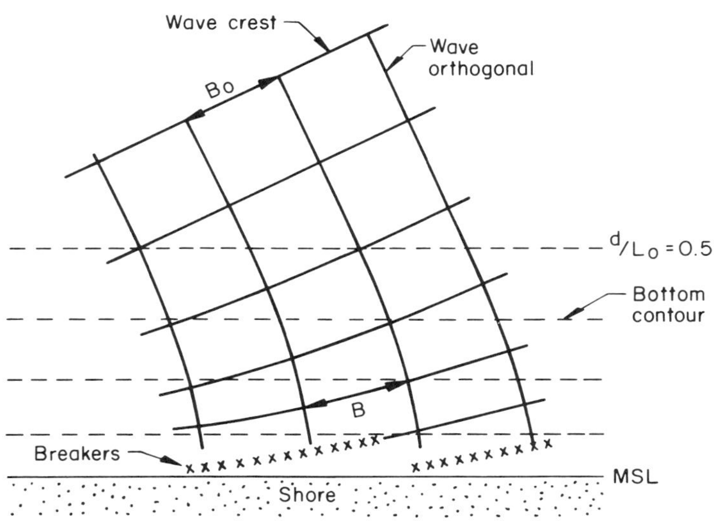

Figure 5.1 Wave crest and orthogonal pattern for shoaling waves.

bottom hydrography. Also note the change in wave direction and wave crest orientation as the wave approaches the shore.

The effects of shoaling (K_s) would be determined as discussed in Chapters 3 and 4; the focus of this section is the development of the wave refraction (orthogonal) patterns. These refraction analyses were initially done by manual construction of refraction diagrams, but now they are mostly done by numerical/computer analysis, except for situations involving one or a few wave conditions for a specific location. We will also briefly consider some special situations such as wave refraction by currents and the development of caustics.

5.1.1 Basic Wave Refraction Equations

Although it may at times be interesting to know the wave crest pattern that develops as wave refraction occurs, it is more useful to be able to predict the resulting wave orthogonal pattern. The orthogonal pattern yields the local wave propagation direction and most importantly allows us to determine the change in wave height owing to refraction. The wave crest pattern can easily be constructed from the orthogonal pattern.

Arthur et al. (1952) developed the equations for a wave orthogonal as a wave propagates through water of changing depth. Their development parallels equivalent derivations in physics texts for the refraction of light as it passes through mediums of different density. Their development is summarized herein.

Consider Figure 5.2, which shows a wave orthogonal and crest located in the horizontal x, y plane and crossing a bottom contour line. As before, θ is the angle between a tangent to the wave orthogonal and the x axis. If x and y are the coordinates of a point on the wave orthogonal and s is the distance along the orthogonal from some arbitrary point, then the equations defining the orthogonal are

$$\frac{dx}{ds} = \cos\theta \qquad (5.2a)$$

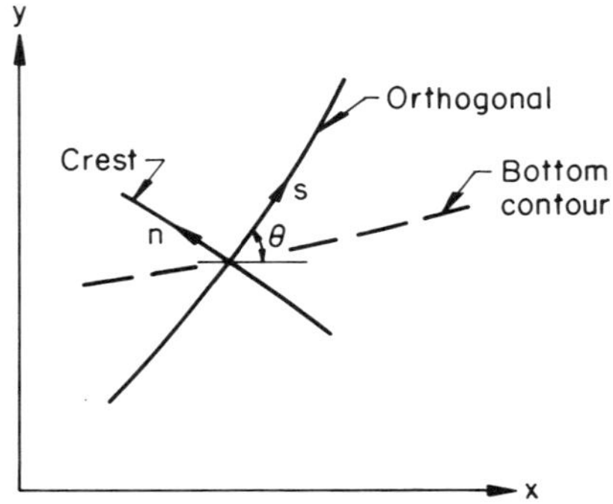

Figure 5.2 Definition diagram for wave orthogonal equation.

$$\frac{dy}{ds} = \sin\theta \tag{5.2b}$$

$$\frac{d\theta}{ds} = -\frac{1}{C}\frac{dC}{dn} \tag{5.2c}$$

In Eq. 5.2c C is the wave celerity and n is the direction normal to the wave orthogonal. Equation 5.2c states that the curvature of the orthogonal depends on the gradient of the wave celerity normal to the wave direction and that the wave orthogonal bends in the direction of the region of lower wave celerity. This is a basic description of the refraction process. It requires that C and its derivatives with respect to x and y be continuous. By the chain rule, Eq. 5.2c becomes

$$\frac{d\theta}{ds} = \frac{1}{C}\left[\sin\theta\,\frac{dC}{dx} - \cos\theta\,\frac{dC}{dy}\right] \tag{5.3}$$

Equations 5.2a and b and 5.3 are the basic wave orthogonal equations. Knowing the incoming orthogonal direction and determining the wave celerity and the gradient of the celerity in the x and y directions at a point one can determine the change in orthogonal direction at that point. Step-by-step solution of this equation numerically would determine the sequential variation of θ and the path of the orthogonal.

To evaluate the changes that occur as a wave refracts from a point in deep water to some point near the shore, a pair of orthogonals could be constructed using the orthogonal equations. From this, the change in wave height could be determined by employing Eq. 5.1 with the measured spacings between the orthogonals at the two points of interest. However, it is more advantageous to employ the concepts developed by Munk and Arthur (1951). Besides the wave orthogonal equations, they developed an equation that defines the separation of a closely spaced pair of orthogonals at any point along the orthogonal path. These two equations were then combined to yield an equation for the wave intensity at any point along the wave orthogonal. Their results are summarized below.

The development is based on

$$\beta = \frac{B}{B_0} = (K_r)^{-1/2} \tag{5.4}$$

which is called the orthogonal separation factor. From the geometry of a pair of adjacent orthogonals an equation was derived for the orthogonal separation factor in terms of the distance s along the wave orthogonal. The result is

$$\frac{d^2\beta}{ds^2} + p\,\frac{d\beta}{ds} + q\beta = 0 \tag{5.5}$$

where

$$p = \frac{\cos\theta}{C}\frac{\partial C}{\partial x} - \frac{\sin\theta}{C}\frac{\partial C}{\partial y}$$

$$q = \frac{\sin^2\theta}{C}\frac{\partial^2 C}{\partial x^2} - \frac{2\sin\theta\cos\theta}{C}\frac{\partial^2 C}{\partial x\,\partial y} + \frac{\cos^2\theta}{C}\frac{\partial^2 C}{\partial y^2}$$

For given bottom contours the wave celerity would be defined as a function of location by the dispersion equation (Eq. 2.11). Then, for a given deep water incident wave direction, Eqs. 5.2a and b and Eq. 5.5 would be solved simultaneously to yield positions along the wave orthogonal and the orthogonal separation factor at these positions. The wave height change can then be evaluated from Eqs. 5.4 and 5.1.

The developments presented so far in this section involve wave refraction by following a selected wave orthogonal as the wave propagates forward. For each step the orthogonal is extended forward to some a priori unknown point. Conditions along the orthogonal to the unknown point must typically be interpolated from known conditions at fixed grid points in the vicinity. When conditions (e.g., water depths, calculated wave celerities) are known at fixed grid positions, it is more convenient to employ the conservation of waves equation for wave refraction analysis. Development of this equation is presented in Dean and Dalrymple (1984); a summary is given herein.

For a two-dimensional (x, y) region, the conservation of waves equation equates the number of waves entering and leaving the region. The waves are long crested, may have any orientation relative to the x, y coordinate system, and have a wave period that is steady with time. If so, the resulting relationship is

$$\frac{\partial(k\sin\theta)}{\partial x} - \frac{\partial(k\cos\theta)}{\partial y} = 0 \tag{5.6}$$

where $k = 2\pi/L$. Expanding Eq. 5.6 yields

$$k\cos\theta\,\frac{\partial\theta}{\partial x} + k\sin\theta\,\frac{\partial\theta}{\partial y} = \cos\theta\,\frac{\partial k}{\partial y} - \sin\theta\,\frac{\partial k}{\partial x} \tag{5.7}$$

Equation 5.7 would be solved numerically over a coastal grid to determine θ values at all points on the grid. The input deep water values of θ would provide a boundary condition for the solution. With the known deep water θ values and calculated grid θ values, the orthogonal pattern is defined.

5.1.2 Manual Construction of Refraction Diagrams

Given the incident wave period and direction and a hydrographic chart of the area of interest, one can manually construct a refraction diagram by either the wave crest or orthogonal method. The former, which involves the direct construction of wave crest patterns, is not now used, but is of historic interest and is instructive of the wave refraction processes. The latter involves the direct construction of wave orthogonals from which wave height and direction changes can be directly determined. And it is less dependent on personal interpretation. Thus this is the approach commonly used for the graphical construction of refraction diagrams. Each is discussed below.

The wave crest method simply involves construction of a diagram (over the hydrographic chart) that shows the wave crest positions at one or a multiple of wave length intervals as a wave propagates from deep water to the shore. Given the initial wave crest position, each point along the crest can be advanced perpendicular to the crest an integer number of wave lengths by calculating the wave length(s) for the average water depth in front of the crest using the dispersion equation (Eq. 2.12). This process is continued toward the shore with the plotting of successive wave crest positions. Orthogonals can be added later by constructing lines normal to the wave crest in a manner analogous to the manual construction of flow nets. This method is discussed in more detail and templates to assist both the crest and orthogonal line construction are given in Wiegel (1964a).

The orthogonal method derives from Eq. 5.3 which, with simplifying assumptions, is applied between a pair of bottom contours. Assuming that the x axis is oriented in the mean direction of the two contours, that the depth varies uniformly between the two contours, and that the wave celerity varies uniformly across the contour interval, Eq. 5.3 becomes

$$\frac{d\theta}{ds} = -\frac{\cos\theta}{C}\frac{dC}{dy} \tag{5.8}$$

Here, dC/dy is a constant (negative when the orthogonal goes from deep to shallow water and vice versa). Combining Eqs. 5.8 and 5.2b yields an equation in terms of C and θ having the solution

$$\frac{\cos\theta}{C} = \text{constant} = \frac{\cos\theta_1}{C_1} = \frac{\cos\theta_2}{C_2}$$

where the subscripts 1 and 2 refer to the conditions at the two contour lines. Defining α as the angle between the wave crest and the mean contour direction or x axis gives $\alpha = \pi/2 - \theta$. The equations shown above then yield

$$\frac{\sin\alpha_1}{\sin\alpha_2} = \frac{C_1}{C_2} = \frac{L_1}{L_2} \tag{5.9}$$

$$\frac{d\alpha}{ds} = \frac{\sin\alpha_1}{C_1}\frac{dC}{dy} = \text{constant} \tag{5.10}$$

Equation 5.9 is the classic Snell's law for refraction at an interface. For example, if a wave crest encounters an abrupt change in water depth so that the

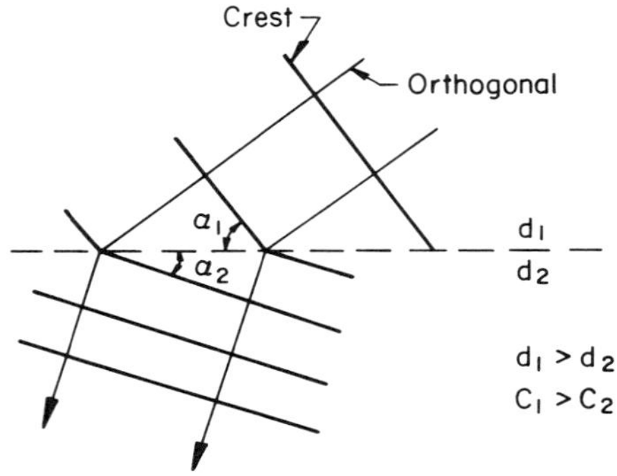

Figure 5.3 Wave crest and orthogonal pattern over a depth discontinuity.

wave celerity changes from C_1 to C_2, the incoming wave crest orientation α_1 changes to α_2 (see Fig. 5.3). Snell's law can also be easily developed from Eq. 5.6 (see Dean and Dalrymple, 1984).

If the nearshore bottom contours are essentially straight and parallel (often a reasonable assumption for many areas) as shown in Figure 5.1, the effects of refraction can be determined by a simple application of Snell's law. Considering Figure 5.4, and applying Snell's law,

$$\frac{\sin \alpha_0}{L_0} = \frac{\sin \alpha}{L} = \frac{1}{x}$$

$$\frac{B_0}{\cos \alpha_0} = x = \frac{B}{\cos \alpha}$$

if we choose B_0 and B so that the orthogonal lengths equal L_0 and L as shown. Then, from Figure 5.4,

$$K_r = \sqrt{\frac{B_0}{B}} = \sqrt{\frac{\cos \alpha_0}{\cos \alpha}} \tag{5.11}$$

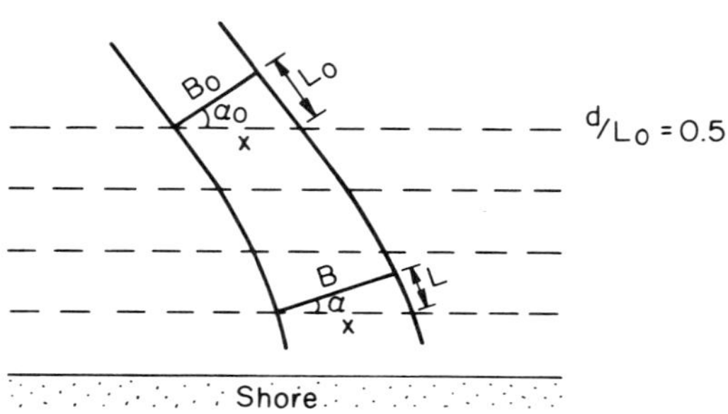

Figure 5.4 Wave refraction over straight parallel bottom contours.

where

$$\alpha = \text{arc sin}\left(\frac{C}{C_0}\sin\alpha_0\right) \tag{5.12}$$

From Eqs. 5.11 and 5.12 the nearshore refraction coefficient and wave direction can be calculated given the wave offshore direction and period and the nearshore depth (to calculate C and C_0). For example, a 7-sec wave approaching the shore at a deep water angle of 35° (α_0) would refract to a nearshore angle (α) of 18°5′ when it reaches a depth of 4 m. The refraction coefficient (K_r) at this depth would be 0.93.

Remember, Eqs. 5.11 and 5.12 apply to nearshore regions having essentially straight and parallel bottom contours. They have, however, been applied as an approximation for regions where the bottom contours are more irregular, to obtain a general indication of refractive effects.

Equation 5.10 shows that, for the assumptions made, the orthogonal is a circular arc between the two contours. It provides a formula for the calculation of this curvature. Arthur et al. (1952) presented a procedure for constructing wave orthogonals, but rather than construct the circular arcs, they just extend the incoming orthogonal to a contour located midway between the two working contours and deflect the incoming orthogonal at that point by the change in angle ($\alpha_1 - \alpha_2$) given by Eq. 5.9. To assist this graphical procedure, they developed the template in Figure 5.5, which would be reproduced on a transparent sheet for use. When the template is constructed so that the distance from the turning point to the orthogonal is about 15 cm, one can construct refraction diagrams on most hydrographic charts.

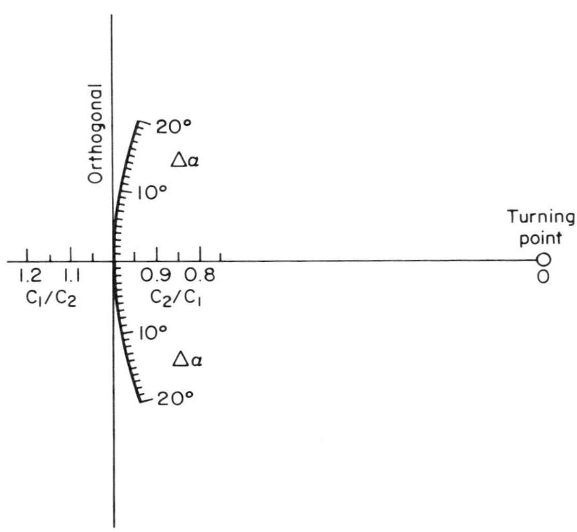

Figure 5.5 Template for constructing wave orthogonals (Arthur et al. 1952).

The procedure for constructing refraction diagrams may be summarized as follows:

1. Locate the depth contour represented by $d/L_0 = 0.5$ on the hydrographic chart. Then label each of the shallower chart contours in terms of their relative depth d/L_0. Bottom contour irregularities that are smaller than about one wave length do not appreciably affect the wave behavior and may be smoothed out.

2. For each contour and the one landward of it calculate the ratio of wave celerities C_1/C_2 where C_1 is the celerity at the deeper contour of the pair. From Eq. 2.11

$$\frac{C_1}{C_2} = \frac{\tanh(2\pi d_1/L_1)}{\tanh(2\pi d_2/L_2)}$$

where from Eq. 2.16

$$\frac{d}{L}\tanh\left(\frac{2\pi d}{L}\right) = \frac{d}{L_0}$$

so d/L can be calculated by trial to use in calculating C_1/C_2.

3. Starting at the two most seaward contours, construct a midcontour that is equidistant from the two given contours. Then extend the incoming deep-water orthogonal straight to the midcontour, and construct a line tangent to the midcontour at the intersection of the midcontour with the incoming orthogonal.

4. Lay the template (Fig. 5.5) with the line marked orthogonal over the incoming orthogonal with $C_1/C_2 = 1.0$ at the intersection of the midcontour and the orthogonal.

5. Then, with a pin in the template at the turning point, rotate the template until the calculated value of C_1/C_2 intersects the tangent to the midcontour. The line on the template labelled orthogonal now lies in the direction of the departing orthogonal. However, it is not at the correct position of the departing orthogonal.

6. With a pair of triangles, move the departing orthogonal to a parallel position such that the incoming and departing orthogonals connect and the lengths of the two orthogonals between contours are equal (thus the incoming and departing orthogonals may not meet at the midcontour).

7. Repeat the procedure for successive contour intervals to extend the orthogonal from deep water to the shoreward point of interest.

Orthogonals may be constructed from shallow to deep water using the same procedure, except C_2/C_1 values are used where C_1 is still the wave celerity at the deeper contour.

Arthur et al. (1952) recommend that the contour interval used to construct the refraction diagram be sufficiently small such that $\Delta C/C_1 < 0.2$ and $\Delta\alpha$

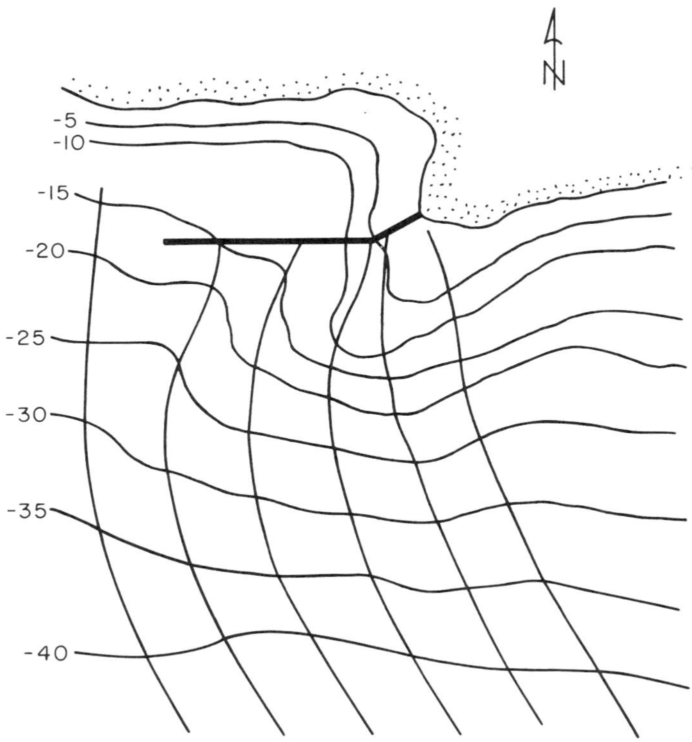

Figure 5.6 Refraction diagram for 7-sec wave from S 30°E.

$< 15°$. If the angle between the wave crest and the bottom contour exceeds 80°, the procedure described above is not sufficiently accurate and a modified procedure must be used. This modified procedure and additional detail on the outlined procedure may be found in U. S. Army Coastal Engineering Research Center (1984).

Figure 5.6 is a refraction diagram for a 7-sec wave from S 30° E approaching a small coastal harbor. For a 7-sec wave $L_0/2 = 38.25$ m so refractive effects commence inside the 40-m contour. Note how refraction concentrates wave energy near the breakwater dogleg ($B_0/B > 4$) and spreads energy at the breakwater head ($B_0/B < 1$). The orthogonal spread at the breakwater head is relatively large, so additional orthogonal lines should be constructed between the two orthogonals that bracket the head to give a more precise determination of the wave height and direction at the head.

5.1.3 Wave Refraction by Numerical Computation

In the 1960s, with the development of adequate digital computers, a number of authors published techniques for the numerical calculation of wave orthogonals and the plotting of orthogonal refraction diagrams. These were all orthogonal tracing procedures employing Eq. 5.2c or 5.3 to calculate the orthogonal curvature ($d\theta/ds$) at a point on the orthogonal and then to extrapolate the redirected orthogonal a finite distance Δs to the point where a new curvature was calculated, and so on. The computations were done on an x, y grid system of water depths. The grid system had to be fine enough to approximate the hydrography, but not so fine as to exceed the computer capacity. All the orthogonal tracing approaches require selection of representative depths at the grid points and interpolation from these grid points to points on the wave orthogonal. These efforts are time consuming and somewhat arbitrary, depending on the type of interpolation that is used.

Griswold (1963) employed the small amplitude dispersion equation (Eq. 2.11) to calculate the wave celerity at each grid point and then proceeded with orthogonal curvature calculations and extrapolations in the field of C using Eq. 5.3. A centered difference scheme was then used to calculate the refraction coefficients employing Eq. 5.5. Wilson (1966) and Jen (1969), working directly with the depth grid and using different techniques for interpolation of depths and depth gradients from the grid values, also used Eq. 5.2c to develop wave orthogonal refraction diagrams. Jen (1969) employed a constant time step Δt so that the incremental distance in each interpolation step decreased as the wave propagated in shallower water ($\Delta s = C\Delta t$), to maintain shallow water computational accuracy. He also calculated orthogonal spacings between adjacent orthogonals to determine refraction coefficients rather than use Eq. 5.5.

Keulegan and Harrison (1970) used Eq. 5.3 to construct tsunami refraction diagrams. The wave celerity, of course, is now given by Eq. 2.17, and their technique compensated for the distorted picture of the earth's surface found on a Mercator projection so results could be plotted on a Mercator map. A major purpose for developing the tsunami refraction diagrams was to determine the tsunami wave crest orientations at the location of model basin wave generators for physical model studies of tsunami effects on the coast.

Skovgaard et al. (1975) present a more sophisticated approach for numerical calculation of wave refraction using orthogonal tracing techniques. Included are development of the orthogonal separation factor equation (Eq. 5.5) in terms of time rather than distance for variable time-stepping calculations and inclusion of the effects of wave height attenuation by turbulent bottom friction (see Section 3.3.2).

The celerity and length of finite amplitude waves depends on the wave height as well as the water depth and wave period. So wave height variations will have some effect on wave refraction. The work discussed above neglected this factor by using small amplitude theory. A number of authors have combined orthogonal tracing techniques with finite amplitude wave theory for numerical computation of wave refraction. The refraction and shoaling analyses must be done iteratively, since shoaling changes the wave height which, in turn, changes the refraction pattern because of the effect of wave height on celerity.

Crowley et al. (1982) utilized vocoidal wave theory, Headland and Chu (1984) used linear theory in deep and intermediate water and cnoidal theory in shallow water and Oh and Grosch (1985) used Stokes third-order theory and energy dissipation due to bottom friction for numerical wave refraction/shoaling analyses. These finite amplitude techniques generally predict less refraction than the small amplitude technique. This might be expected because for a given water depth and wave period, finite amplitude waves have a greater celerity than small amplitude waves.

An alternate approach to numerical methods that employ orthogonal tracing was first applied by Noda et al. (1974). It involves a finite difference solution of the conservation of waves equation (Eqs. 5.6 or 5.7) to determine the wave direction directly at fixed grid positions where the water depths are known. Extrapolation to a priori unknown points along the orthogonal is not required, which leads to a less computer intensive and arbitrary effort. The result is not a set of traced orthogonal lines but an x, y grid of wave direction values. One could then sketch in wave crest and orthogonal lines to better see the wave pattern.

Perlin and Dean (1983) have presented a numerical scheme that is simpler to apply than the scheme used by Noda et al. (1974). Besides using Eq. 5.6 to calculate the wave direction at grid points they employ the conservation of energy flux equation for the x, y plane, excluding any energy input or dissipation, to calculate wave heights. This equation may be written (see Eqs. 2.35 and 2.38)

$$\frac{\partial}{\partial x}(\overline{E}\, C_g \sin\theta) + \frac{\partial}{\partial y}(\overline{E}\, C_g \cos\theta) = 0 \tag{5.13}$$

where the energy density is defined in terms of the wave height by Eq. 2.32. Small amplitude wave theory is used to determine k and C_g in applying Eqs. 5.6 and 5.13.

The conservation of waves equation and the conservation of energy flux equations have been employed with second-order cnoidal theory (Hardy and Kraus, 1987) and with third-order Stokes theory (Cialone and Kraus, 1987) for numerical refraction/shoaling analyses. Again, this requires an iterative approach since wave celerity is height dependent for finite amplitude waves.

5.1.4 Other Refraction Considerations

When a wave obliquely crosses from still water to water having a current or propagates across a current of varying velocity, the wave celerity relative to the still bottom will change, thus causing the wave to refract. In the nearshore zone wave refraction by currents may be particularly noticeable in the vicinity of tidal entrances where relatively strong current velocity gradients are found.

Current-induced wave refraction may be demonstrated by a simple case considered by Johnson (1947) and depicted in Figure 5.7. A wave propagating in still deep water crosses into a deep water region having a current velocity U. The wave crests and orthogonals patterns change as shown. From the ge-

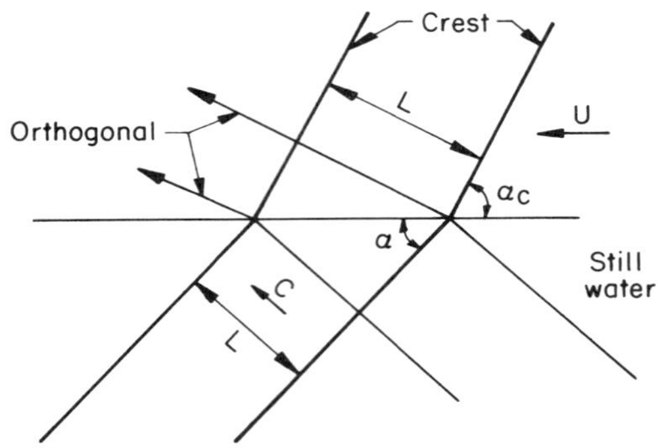

Figure 5.7 Deep water wave refraction by a current.

ometry in the figure Johnson showed that

$$\sin \alpha_c = \frac{\sin \alpha}{\left(1 - \frac{U}{C} \sin \alpha\right)^2} \tag{5.14}$$

where U/C could be positive or negative depending on the direction of the current. For the situation shown, the current refraction has two effects on the wave height. The wave length is increased, which has a diminishing effect on the wave height but the opposite effect results from a convergence of the wave orthogonals. By conservation of the energy flux between two orthogonals Johnson showed

$$\left(\frac{H_c}{L_c}\right)^2 = \left(\frac{H}{L}\right)^2 \frac{\cos \alpha}{\cos \alpha_c} \left[\frac{\left(1 - \frac{U}{C} \sin \alpha\right)^6}{\left(1 + \frac{U}{C} \sin \alpha\right)}\right] \tag{5.15}$$

where

$$L_c = L \frac{\sin \alpha_c}{\sin \alpha}$$

For intermediate or shallow water conditions, Johnson's derivation would have to be modified by using the complete dispersion equation (Eq. 2.10 rather than Eq. 2.13), resulting in more complex results. The effect of a current that gradually, rather than abruptly, increases can be pictured and evaluated by considering a series of stepped increases having the effect shown in Figure 5.7.

The full equations for the refraction of small amplitude waves propagating through a varying current field and varying water depth are presented by Jonsson and Christensen (1984). They determine resulting wave heights and directions along a wave orthogonal and include dissipation due to bottom friction. References to a number of review papers on wave–current interaction are also provided.

When the pattern of wave orthogonals is constructed for a refracted wave pattern (caused by depth or current refraction), refraction may cause wave orthogonals to cross (known as caustics). From energy flux considerations, the resulting wave height at the point of orthogonal crossing would be infinite. Of course, before the crossing point is reached wave diffraction effects may significantly decrease the wave energy between the orthogonals and wave breaking and reformation will likely occur. Nonlinear effects become important and the rate at which the orthogonals converge will affect the wave behavior prior to the point of convergence. When constructing refraction diagrams it is important to look for the occurrence of caustics and to be careful in interpreting conditions in the region beyond a caustic. Pierson (1951, 1972) provides examples and further discussion of wave behavior near caustics, and Perigrine (1981) and Perigrine and Smith (1979) discuss the nonlinear effects that might be important.

5.2 WAVE DIFFRACTION

Figure 5.8 shows a long-crested monochromatic wave propagating toward a nontransmitting semiinfinite barrier. The segment of the wave that hits the barrier will have part of its energy dissipated and part reflected. The wave segment that passes the tip of the barrier will have a portion of its energy transfer along the wave crest into the lee of the barrier. As a consequence, the wave height in the shadow region inside the dashed line will have its height

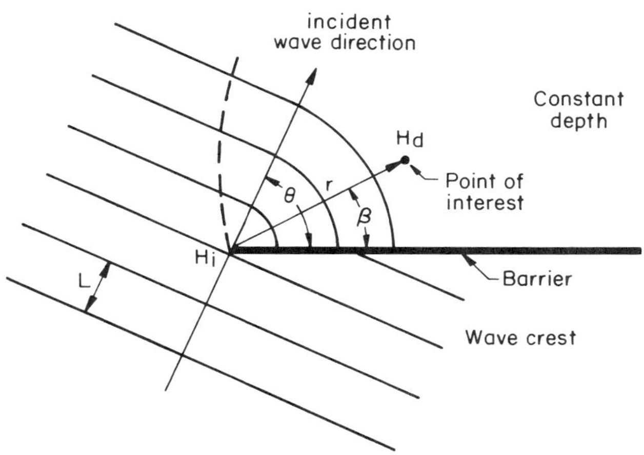

Figure 5.8 Wave diffraction behind a barrier.

reduced. The diffracted wave crest pattern in the lee of the structure will form approximate concentric circular arcs with the wave height decreasing exponentially along the crest. Note that the water depth in the structure lee in Figure 5.8 is constant; otherwise, the wave crest pattern and heights would also be affected by refraction.

If the barrier does reflect wave energy, the reflected wave crest would also diffract to form concentric circular wave crests around the tip of the barrier.

If H_i is the incident wave height at the tip of the barrier and H_d is the diffracted height at a point of interest, we define $H_d/H_i = K_d$ as the diffraction coefficient. If the point of interest in the diffraction zone is at a radial distance r from the tip and at an angle β from the barrier, then $K_d = fct(\theta, \beta, r/L)$ where θ and L are the incident wave direction and length. Then, for a given water depth and barrier geometry, K_d is a function of the incident wave period and direction. As is the case for wave refraction, the component periods and directions of a wave spectrum will be affected differently by wave diffraction.

5.2.1 Diffraction Analysis: Semiinfinite Barrier

The problem depicted in Figure 5.8 can be solved as a three-dimensional irrotational flow problem by linearizing the surface boundary conditions and assuming a velocity potential having a similar vertical dependence to that which develops in the small amplitude two-dimensional wave theory. The Laplace equation is solved with a perfectly reflecting vertical barrier as a lateral boundary condition. This solution was presented by Sommerfeld (1896) for the diffraction of light, and Penny and Price (1952) showed that the same solution applied to the surface water wave diffraction problem. The solution yields the wave crest pattern in the x, y plane and the wave height distribution throughout the affected area.

The two most common diffraction problems are the semiinfinite barrier and a barrier with a gap that allows wave passage. Penny and Price (1952) presented solutions for incident waves from different directions passing a semiinfinite barrier and for normally incident waves passing through a barrier gap. Figure 5.9 shows the results for a wave approaching normal to a semiinfinite barrier. The diagram is nondimensionalized by dividing the horizontal dimensions by the incident wave length. The only depth requirement is that the depth be constant. Note the extent of the region where wave heights are affected. Theory shows that although the wave crests are not pure circular arcs, they are well approximated by arcs concentric on the barrier tip.

Wiegel (1962) summarizes the Penny and Price (1952) solution for a semiinfinite barrier and presents extensive tabulated results (repeated in Table 5.1). The resulting equations are extremely complex to apply but, by interpolation from Table 5.1, reasonable results can be obtained. Graphic plots of the results for a semiinfinite breakwater were also presented by Wiegel (1962) for the approach angles included in Table 5.1. Plots of Wiegel's results may be found in the U. S. Army Coastal Engineering Research Center (1984). Figure 5.10

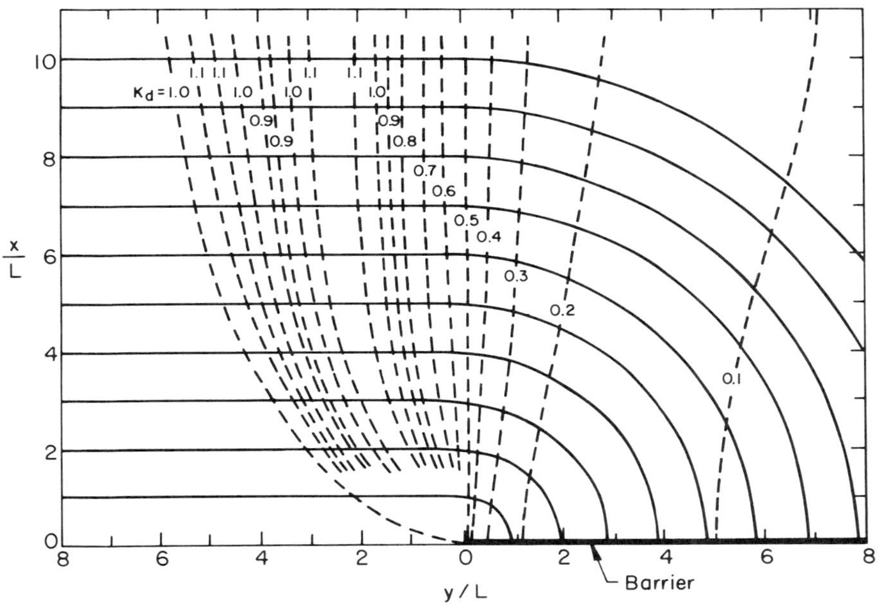

Figure 5.9 Wave crest pattern and K_d values for normally incident waves passing a barrier (after Penny and Price, 1952).

is an example from the latter source for a wave approach angle θ equal to 60°. A curious and interesting feature of the diffraction results shown in Table 5.1 is that for any wave approach angle, the value of the diffraction coefficient along a line in the lee of the barrier that extends from the breakwater tip in the direction of the approaching wave is approximately 0.5. Note in Table 5.1 and Figures 5.9 and 5.10 that there are regions outside the shadow zone where K_d is greater than unity. The values develop from the superposition of the incident wave and the diffracted portion of the wave that reflects from the barrier.

Putnam and Arthur (1948) made laboratory measurements for waves diffracting past a semiinfinite barrier. Two wave periods and six different approach directions were studied. Crest patterns were not measured, but wave diffraction coefficient values were determined by measuring wave heights along lines behind the barrier. Results generally confirm the diffraction theory. Remember that the small amplitude assumptions were employed in deriving the diffraction theory. Putnam and Arthur (1948) used relatively small amplitude waves in their experiments (average $H/L = 0.035$). For steeper waves finite amplitude effects would cause the results to differ from the small amplitude diffraction theory.

Consider, as an example, a train of 6-sec waves approaching a perfectly reflecting breakwater at an angle $\theta = 60°$. If the water depth is constant at 10 m, the wave length from Eq. 2.12 is 48.3 m. At an angle $\beta = 30°$ and a distance of 96.9 m ($r/L = 2.0$) from the breakwater tip, $K_d = 0.28$ (see Table

TABLE 5.1 Wave Diffraction Coefficients K_d as a Function of Incident Wave Direction θ and Position r/L and β[a]

	β (deg)												
r/L	0	15	30	45	60	75	90	105	120	135	150	165	180
							$\theta = 15°$						
½	0.49	0.79	0.83	0.90	0.97	1.01	1.03	1.02	1.01	0.99	0.99	1.00	1.00
1	0.38	0.73	0.83	0.95	1.04	1.04	0.99	0.98	1.01	1.01	1.00	1.00	1.00
2	0.21	0.68	0.86	1.05	1.03	0.97	1.02	0.99	1.00	1.00	1.00	1.00	1.00
5	0.13	0.63	0.99	1.04	1.03	1.02	0.99	0.99	1.00	1.01	1.00	1.00	1.00
10	0.35	0.58	1.10	1.05	0.98	0.99	1.01	1.00	1.00	1.00	1.00	1.00	1.00
							$\theta = 30°$						
½	0.61	0.63	0.68	0.76	0.87	0.97	1.03	1.05	1.03	1.01	0.99	0.95	1.00
1	0.50	0.53	0.63	0.78	0.95	1.06	1.05	0.98	0.98	1.01	1.01	0.97	1.00
2	0.40	0.44	0.59	0.84	1.07	1.03	0.96	1.02	0.98	1.01	0.99	0.95	1.00
5	0.27	0.32	0.55	1.00	1.04	1.04	1.02	0.99	0.99	1.00	1.01	0.97	1.00
10	0.20	0.24	0.54	1.12	1.06	0.97	0.99	1.01	1.00	1.00	1.00	0.98	1.00
							$\theta = 45°$						
½	0.49	0.50	0.55	0.63	0.73	0.85	0.96	1.04	1.06	1.04	1.00	0.99	1.00
1	0.38	0.40	0.47	0.59	0.76	0.95	1.07	1.06	0.98	0.97	1.01	1.01	1.00
2	0.29	0.31	0.39	0.56	0.83	1.08	1.04	0.96	1.03	0.98	1.01	1.00	1.00
5	0.18	0.20	0.29	0.54	1.01	1.04	1.05	1.03	1.00	0.99	1.01	1.00	1.00
10	0.13	0.15	0.22	0.53	1.13	1.07	0.96	0.98	1.02	0.99	1.00	1.00	1.00

						$\theta = 60°$							
½	0.40	0.41	0.45	0.52	0.60	0.72	0.85	1.13	1.04	1.06	1.03	1.01	1.00
1	0.31	0.32	0.36	0.44	0.57	0.75	0.96	1.08	1.06	0.98	0.98	1.01	1.00
2	0.22	0.23	0.28	0.37	0.55	0.83	1.08	1.04	0.96	1.03	0.98	1.01	1.00
5	0.14	0.15	0.18	0.28	0.53	1.01	1.04	1.05	1.03	0.99	0.99	1.00	1.00
10	0.10	0.11	0.13	0.21	0.52	1.14	1.07	0.96	0.98	1.01	1.00	1.00	1.00
						$\theta = 75°$							
½	0.34	0.35	0.38	0.42	0.50	0.59	0.71	0.85	0.97	1.04	1.05	1.02	1.00
1	0.25	0.26	0.29	0.34	0.43	0.56	0.75	0.95	1.02	1.06	0.98	0.98	1.00
2	0.18	0.19	0.22	0.26	0.36	0.54	0.83	1.09	1.04	0.96	1.03	0.99	1.00
5	0.12	0.12	0.13	0.17	0.27	0.52	1.01	1.04	1.05	1.03	0.99	0.99	1.00
10	0.08	0.08	0.10	0.13	0.20	0.52	1.14	1.07	0.96	0.98	1.01	1.00	1.00
						$\theta = 90°$							
½	0.31	0.31	0.33	0.36	0.41	0.49	0.59	0.71	0.85	0.96	1.03	1.03	1.00
1	0.22	0.23	0.24	0.28	0.33	0.42	0.56	0.75	0.96	1.07	1.05	0.99	1.00
2	0.16	0.16	0.18	0.20	0.26	0.35	0.54	0.69	1.08	1.04	0.96	1.02	1.00
5	0.10	0.10	0.11	0.13	0.16	0.27	0.53	1.01	1.04	1.05	1.02	0.99	1.00
10	0.07	0.07	0.08	0.09	0.13	0.20	0.52	1.14	1.07	0.96	0.99	1.01	1.00
						$\theta = 105°$							
½	0.28	0.28	0.29	0.32	0.35	0.41	0.49	0.59	0.72	0.85	0.97	1.01	1.00
1	0.20	0.20	0.24	0.23	0.27	0.33	0.42	0.56	0.75	0.95	1.06	1.04	1.00
2	0.14	0.14	0.13	0.17	0.20	0.25	0.35	0.54	0.83	1.08	1.03	0.97	1.00
5	0.09	0.09	0.10	0.11	0.13	0.17	0.27	0.52	1.02	1.04	1.04	1.02	1.00
10	0.07	0.06	0.08	0.08	0.09	0.12	0.20	0.52	1.14	1.07	0.97	0.99	1.00

TABLE 5.1 (*Continued*)

	β (deg)												
r/L	0	15	30	45	60	75	90	105	120	135	150	165	180
							$\theta = 120°$						
½	0.25	0.26	0.27	0.28	0.31	0.35	0.41	0.50	0.60	0.73	0.87	0.97	1.00
1	0.18	0.19	0.19	0.21	0.23	0.27	0.33	0.43	0.57	0.76	0.95	1.04	1.00
2	0.13	0.13	0.14	0.14	0.17	0.20	0.26	0.16	0.55	0.83	1.07	1.03	1.00
5	0.08	0.08	0.08	0.09	0.11	0.13	0.16	0.27	0.53	1.01	1.04	1.03	1.00
10	0.06	0.06	0.06	0.07	0.07	0.09	0.13	0.20	0.52	1.13	1.06	0.98	1.00
							$\theta = 135°$						
½	0.24	0.24	0.25	0.26	0.28	0.32	0.36	0.42	0.52	0.63	0.76	0.90	1.00
1	0.18	0.17	0.18	0.19	0.21	0.23	0.28	0.34	0.44	0.59	0.78	0.95	1.00
2	0.12	0.12	0.13	0.14	0.14	0.17	0.20	0.26	0.37	0.56	0.84	1.05	1.00
5	0.08	0.07	0.08	0.08	0.09	0.11	0.13	0.17	0.28	0.54	1.00	1.04	1.00
10	0.05	0.06	0.06	0.06	0.07	0.08	0.09	0.13	0.21	0.53	1.12	1.05	1.00
							$\theta = 150°$						
½	0.23	0.23	0.24	0.25	0.27	0.29	0.33	0.38	0.45	0.55	0.68	0.83	1.00
1	0.16	0.17	0.17	0.18	0.19	0.22	0.24	0.29	0.36	0.47	0.63	0.83	1.00
2	0.12	0.12	0.12	0.13	0.14	0.15	0.18	0.22	0.28	0.39	0.59	0.86	1.00
5	0.07	0.07	0.08	0.08	0.08	0.10	0.11	0.13	0.18	0.29	0.55	0.99	1.00
10	0.05	0.05	0.05	0.06	0.06	0.07	0.08	0.10	0.13	0.22	0.54	1.10	1.00

						θ = 165°							
½	0.23	0.23	0.23	0.24	0.26	0.28	0.31	0.35	0.41	0.50	0.63	0.79	1.00
1	0.16	0.16	0.17	0.17	0.19	0.20	0.23	0.26	0.32	0.40	0.53	0.73	1.00
2	0.11	0.11	0.12	0.12	0.13	0.14	0.16	0.19	0.23	0.31	0.44	0.68	1.00
5	0.07	0.07	0.07	0.07	0.08	0.09	0.10	0.12	0.15	0.20	0.32	0.63	1.00
10	0.05	0.05	0.05	0.06	0.06	0.06	0.07	0.08	0.11	0.11	0.21	0.58	1.00
						θ = 180°							
½	0.20	0.25	0.23	0.24	0.25	0.28	0.31	0.34	0.40	0.49	0.61	0.78	1.00
1	0.10	0.17	0.16	0.18	0.18	0.23	0.22	0.25	0.31	0.38	0.50	0.70	1.00
2	0.02	0.09	0.12	0.12	0.13	0.18	0.16	0.18	0.22	0.29	0.40	0.60	1.00
5	0.02	0.06	0.07	0.07	0.07	0.08	0.10	0.12	0.14	0.18	0.27	0.46	1.00
10	0.10	0.05	0.05	0.04	0.06	0.07	0.07	0.08	0.10	0.13	0.20	0.36	1.00

[a] Data from Wiegel (1962).

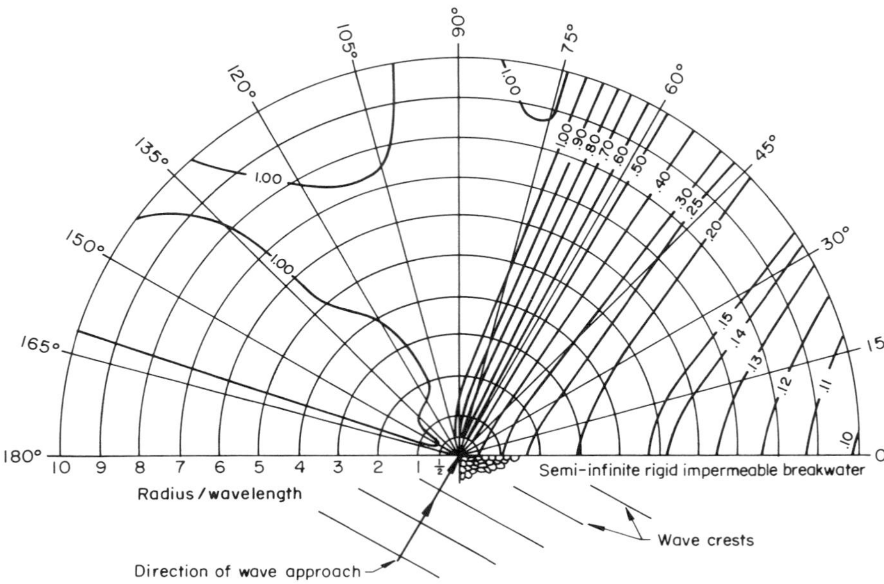

Figure 5.10 Values of K_d versus position for a wave approach angle of 60° (U. S. Army Coastal Engineering Research Center, 1984).

5.1). Thus, a 1-m high incident wave at the breakwater tip would be 0.28 m high at this point and would be propagating in the direction β. If the breakwater has a reflection coefficient that is less than 1, the result given above would not be very different because the diffraction of the reflected wave would be so great that effects would be negligible at the point of interest.

Note, from Table 5.1, that for a given location in the lee of a breakwater a spectrum of waves all coming from the same direction will generally experience a greater percentage decrease in wave energy density for successively smaller wave periods of the spectrum. That is, for given θ and β values, decreasing wave periods mean shorter wave lengths or larger values of r/L. Thus, the energy density concentration will shift toward the higher wave periods in the spectrum.

Waves approaching a barrier of finite length will diffract at each end and meet in the lee of the barrier. The resulting pattern can be developed by combining the results for semiinfinite barrier diffraction at each end. Figure 5.11 shows a resulting wave crest pattern for a 45° wave approach angle to a barrier that is four wave lengths long. (The diffracted crests that have propagated past the barrier tip would curve around to the front of the barrier, but this is not shown on the figure to make the diagram simpler to understand.) Note that this figure just shows the pattern for one instant as the waves propagate forward. The highest wave amplitudes will occur along the lines of intersection of the wave crests at the instant that the crests meet. One of these lines of intersection

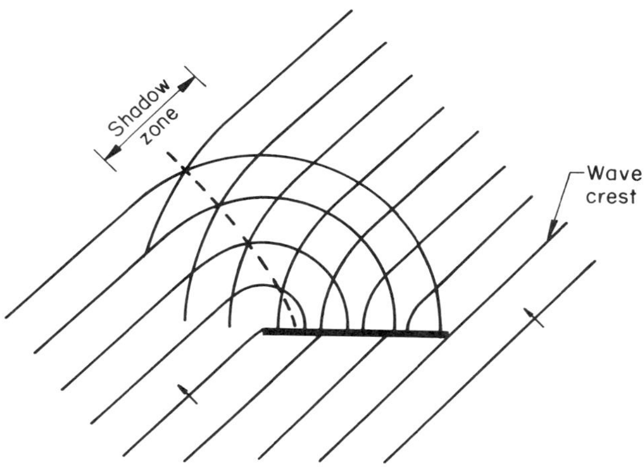

Figure 5.11 Diffraction pattern for waves approaching a short barrier.

is shown by a dashed line. Assuming small amplitude waves, the wave height at the point of intersection would be the sum of the heights of the two component waves from the individual diffraction diagrams. This problem has been investigated analytically by Penny and Price (1952), Montefusco (1968), and Goda et al. (1971). Harms (1979) has developed a numerical computer analysis of the problem and conducted laboratory experiments to evaluate his results.

5.2.2 Diffraction Analysis: Barrier Gap

Penny and Price (1952) also applied the Sommerfeld solution to normally incident wave diffraction through a barrier gap. They essentially superimposed the solutions for two mirror image semiinfinite barriers, one with a reflection coefficient of unity and one with a zero reflection coefficient. Blue and Johnson (1949) conducted laboratory studies on diffraction through a gap and verified the solution for gap widths as small as $1.41L$. Johnson (1952) presented plots of diffraction coefficient for normal wave incidence to gaps having widths between one and five wave lengths. Figure 5.12 is an example of one of these plots for a gap width of $2.5L$. Beyond two or three wave lengths from the gap, wave crests are essentially circular arcs concentric about the midpoint of the gap. (The K_d plots for other gap widths can be found in U. S. Army Coastal Engineering Research Center, 1984). If the gap width exceeds five wave lengths, the zones of influence of the diffraction zones at the two barrier tips generally will not overlap so the diffraction pattern and K_d values may be determined from the separate solutions for two semiinfinite barriers.

The angle of wave incidence to a breakwater may be other than 90°. Johnson (1952) showed that the results given above for normally incident waves can be used by employing an imaginary gap width as defined in Figure 5.13.

Carr and Stelzriede (1952) employed a different analytical approach than

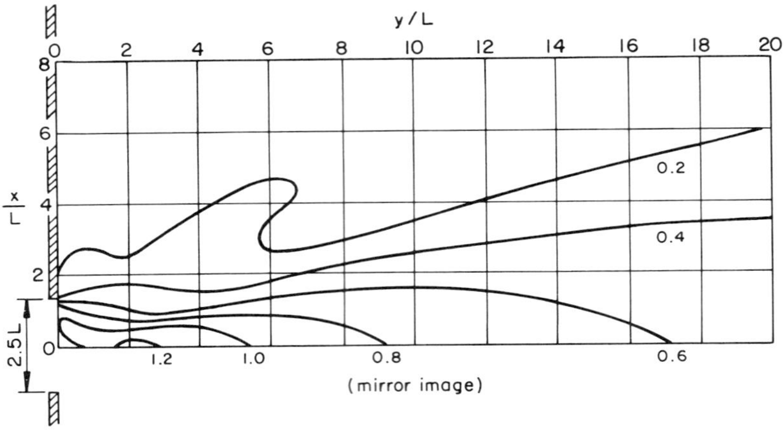

Figure 5.12 Values of K_d for normally incident waves entering a barrier gap 2.5 wave lengths wide.

Sommerfeld to develop diffraction patterns for barrier gaps that are small compared to the wave length. Johnson (1952) employed their approach to also develop diffraction coefficient pattern plots for a range of wave approach angles and a gap width equal to one wave length. These plots are also presented in U. S. Army Coastal Engineering Research Center (1984).

5.2.3 Practical Application

Often, barrier geometries that are not identical to the specific geometries presented above will be encountered. However, approximate but useful results may still be achieved by employing some ingenuity in applying the gap and semiinfinite breakwater geometries to bracket the encountered geometry. Or recourse may be made to physical model tests (see Chapter 10) or numerical computer model analyses such as those for combined refraction–diffraction (which is discussed below).

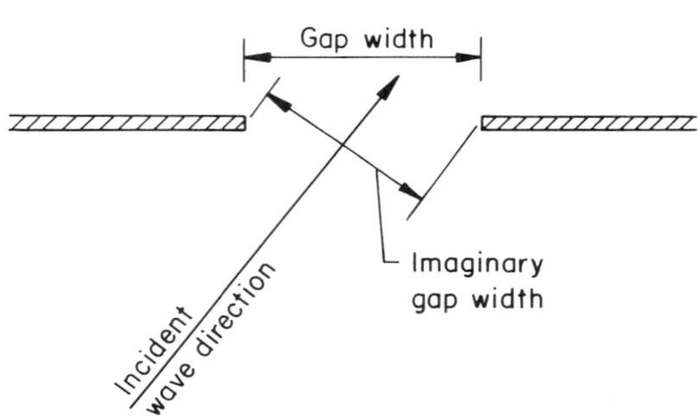

Figure 5.13 Definition of imaginary gap width.

5.3 COMBINED REFRACTION AND DIFFRACTION

We have considered refraction and diffraction separately. Only in uncommon coastal circumstances does pure refraction or diffraction occur. But in many situations, one of these processes is so dominant that an effective analysis can be carried out by considering that process alone.

Consider the coastal section and incoming wave pattern depicted in Figure 5.6. Besides knowing the wave heights along the breakwater, a designer would need to know the wave height and direction at certain points in the harbor. Wave refraction and shoaling would be dominant up to the breakwater. Determination of K_s from the change in water depth and K_r and the wave direction from a refraction analysis would yield the wave height (Eq. 5.1) and direction at the breakwater head. Immediately beyond the breakwater head diffraction would dominate. A pure diffraction analysis using the incident wave height and direction determined at the breakwater head would provide an adequate indication of wave heights in the lee of the breakwater. In the breakwater lee, the depth is relatively constant (as is the case in most harbors) and the bottom contours approximately parallel the diffracting wave crest pattern. The diffraction analysis would be carried to points where the rate of change of bottom contours is significant and the wave crests are sufficiently oblique to the bottom contours for refraction to again dominate. Given the wave height and direction at this transition point as determined by the diffraction analysis, the refraction analysis can be continued to the point of interest or the point of wave breaking.

This example demonstrates a situation where an analysis, alternately using refraction and diffraction alone, may be adequate. Of course there are many situations where this approach would be inadequate. Besides a harbor where the depth contours are very irregular, waves approaching the shore might be propagating over a large shallow offshore shoal. Wave energy would diffract in the lee of the shoal and caustics may occur on the shoal. Or waves propagating toward the shore over a deep and wide navigation channel may undergo significant combined refraction–diffraction effects in the vicinity of the channel. Such problems can be studied by conducting physical model tests in a wave basin or by numerical computer analyses using techniques that have been developed relatively recently.

For an analytical solution of a combined refraction–diffraction problem we basically need to solve the three-dimensional Laplace equation with appropriate surface, bottom, and lateral boundary conditions over an area with irregularly varying depths. A significant step toward the solution of such problems was the development and application of the mild slope equation by Berkhoff (1972). This is a two-dimensional elliptical partial differential equation that describes the complete transformation of small amplitude waves including refraction and diffraction. The mild slope equation is solved as a boundary value problem which requires a priori knowledge of all the lateral boundary conditions. Analytical solutions have been developed for a few special cases but, most effectively, the solutions for real situations require a numerical finite element or finite differences solution. See Behrendt and Johnson (1984) and Massel (1989) for further discussion on the development of the mild slope equation. Berkhoff et al. (1982), Booij (1983), Ebersole (1985), Houston (1981), and Tsay and Liu (1983) are some accessible publications describing numerical applications of the mild slope equation to refraction–diffraction problems.

To develop the mild slope equation Berkhoff (1972) employed the three-dimensional Laplace equation with the usual linearized free-surface boundary condition and a bottom boundary condition that allowed a mild sloping bottom rather than a horizontal bottom. He assumed that the velocity potential has a cosh $k(d + z)$ variation and the Laplace equation was integrated over depth to yield a two-dimensional (x, y) equation having the form

$$\frac{\partial}{\partial x}\left(CC_g \frac{\partial \phi_0}{\partial x}\right) + \frac{\partial}{\partial y}\left(CC_g \frac{\partial \phi_0}{\partial y}\right) + \sigma^2 \frac{C_g}{C} \phi_0 = 0 \tag{5.16}$$

which is the mild slope equation. In Eq. 5.16 ϕ_0 is a two-dimensional complex velocity potential.

For coastal situations involving short waves, the mild-slope equation becomes computationally difficult when the lateral extent of the area being investigated is greater than about 10 wave lengths (Massel, 1989). For larger areas, up to 100 wave lengths, a parabolic approximation to Eq. 5.16 has been developed (Dalrymple et al., 1984; Kirby, 1986; Kirby and Dalrymple, 1983; Lozano and Liu, 1980; Radder, 1979). This approach, which modifies the orthogonal refraction theory to allow energy to flow in a direction normal to the direction of wave propagation, requires only the specification of initial conditions rather than all of the lateral boundary conditions.

5.4 WAVE REFLECTION

Two-dimensional wave reflection was briefly discussed in Sections 2.5.3 and 3.7 where the reflection coefficient $C_r = H_r/H_i$ was defined. When a wave obliquely approaches a reflecting barrier, the wave will have a reflected crest (and orthogonal) angle equal to the incident crest (and orthogonal) angle as shown in Figure 5.14. The reflected wave height will be dependent on the

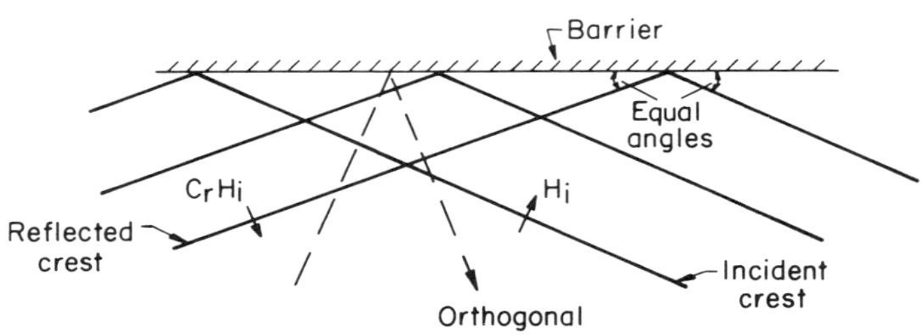

Figure 5.14 Incident and reflected wave crest patterns, constant water depth.

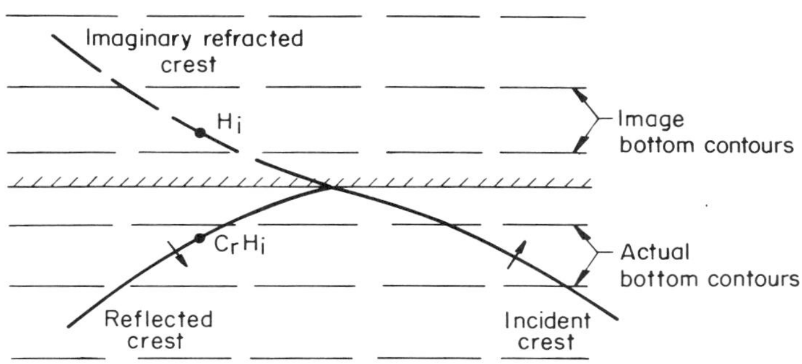

Figure 5.15 Reflection of refracted wave crest.

reflection coefficient, as shown. The diamond-shaped pattern created by the incident and reflected wave crests can create rather complex particle orbit, velocity, and dynamic pressure patterns. The reader is referred to Fuchs (1952) and Silvester (1974) for a detailed discussion of these patterns. Silvester (1986) discusses the effect the resulting bottom water particle motions have on sediment transport near the barrier.

Figure 5.14 depicts the simple case of a straight-crested wave approaching a straight barrier in water of constant depth. Figure 5.15 shows a refracting wave crest approaching a barrier. The reflected wave crest pattern can be constructed by constructing imaginary mirror image hydrography on the other side of the barrier, then constructing the wave crest pattern that would develop over this imaginary hydrography, and then drawing the mirror image of the imaginary wave crest pattern as the real reflected wave. This is demonstrated in Figure 5.15 along with the reflected wave height.

Another example is depicted in Figure 5.16, which shows a long-crested

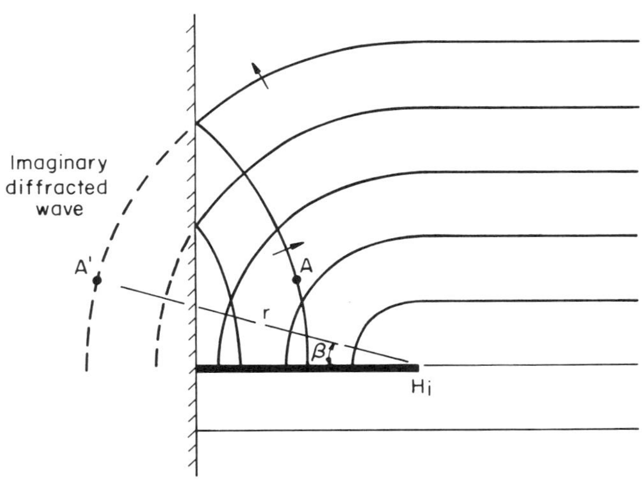

Figure 5.16 Reflection of diffracted wave crest.

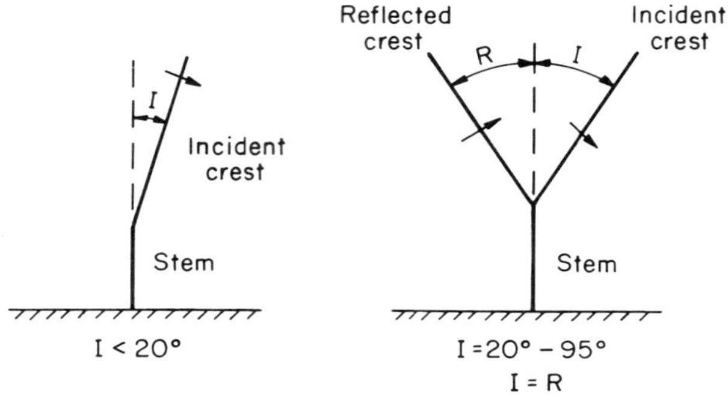

Figure 5.17 Mach-stem reflection (after Wiegel, 1964b).

wave diffracting past a barrier and then reflecting off a second barrier. The water depth is constant. The imaginary diffracted wave is carried past the reflecting barrier and its mirror image is constructed to depict the reflected wave pattern. The wave height at point A would equal the height the diffracted wave would have at A′ times the reflection coefficient (i.e., $H_A = H_i(K_d)_{A'}C_r$). By applying the concepts demonstrated in Figures 5.14–5.16 one can develop the reflection patterns and resulting wave heights for relatively complex situations.

When the angle between the incident wave crest and the reflecting barrier is 45° or less, normal reflection as shown in Figure 5.14 occurs. For larger incidence angles, Wiegel (1964b) has demonstrated that the incident and reflected waves develop a mach stem pattern similar to that which develops in acoustic waves (see Fig. 5.17). When angle I is less than 20° (i.e., the angle between incident wave crest and barrier is greater than 70°), there is no reflected wave and a stem develops. The stem grows as the incident wave propagates along the barrier. When angle I is between 20° and 45° the stem also develops and grows, but the incident and reflected crest segments are separate, as shown. See Wiegel (1964a, b) for a detailed discussion of the mach stem effect.

5.5 WAVES GENERATED BY A MOVING OBJECT

A moving object that penetrates the water surface, or water flowing past a still object that penetrates the surface, will generate a pattern of surface waves that is steady with respect to the object (if the object or water speed is constant). We consider the case of a moving object (a vessel), but the wave pattern is the same for the still object in flowing water.

As a vessel travels in originally still water, water flows back past the vessel at a relative speed equal to the vessel speed plus the absolute speed of the water required to fill the evacuated space behind the vessel. This flow causes the pressure to rise at the vessel bow, to fall below the free stream pressure over most of the side of the hull, and to rise again at the vessel stern. The water surface profile along the hull responds to this pressure distribution, causing a rapid rise and fall of the water surface at the bow and, to a lesser extent (owing to flow separation), at the stern. Inertial effects cause the water surface to "overshoot" its equilibrium position, establish a surface oscillation, and generate sets of waves that propagate out from the bow and stern of the vessel. The pressure distribution and resulting height of the generated waves at the bow depend on the vessel speed, the bow geometry, and the clearance between the bow and the channel bottom and sides. The periods and propagation directions of the generated waves depend only on the ship speed and the water depth.

Thompson (1887) and Havelock (1908) did the original theoretical work on the pattern and heights of waves generated by a moving disturbance; Sorensen (1973) gives a review of vessel-generated waves and recent analytical and experimental work. An overview of the pattern of vessel waves and their characteristics is given herein.

Figure 5.18 shows the pattern of wave crests generated by the bow of a ship that is moving over deep water. It consists of symmetrical sets of diverging waves that move obliquely out from the sailing line and a single set of transverse waves that move in the direction of the vessel. The transverse and diverging waves meet to form cusps located along lines that are 19° 28′ out from the sailing line. The largest wave amplitudes are found at the cusp points. If the vessel speed is increased, this pattern retains the same form but expands in size as the individual wave lengths increase. A similar pattern of waves, typically with lower amplitudes, would be generated at the vessel's stern.

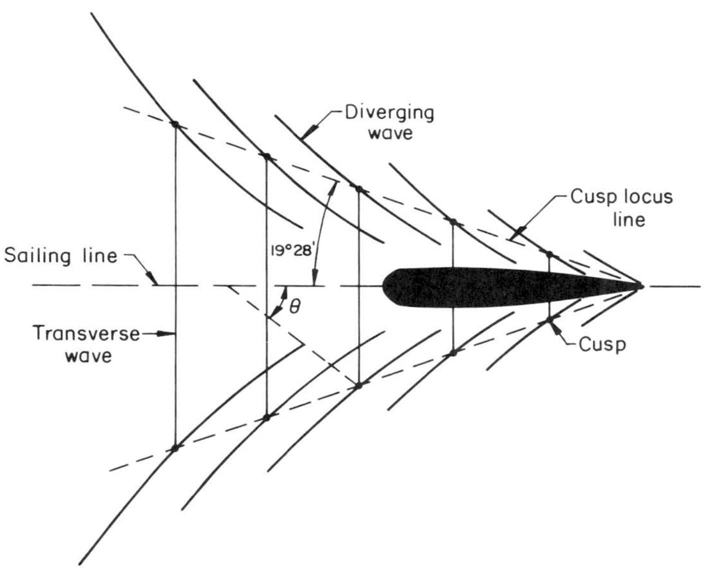

Figure 5.18 Deep water wave crest pattern generated at the bow of a moving vessel.

Since the wave system remains steady with respect to the vessel, the waves that form the pattern must have a celerity given by

$$C = V \cos \theta \tag{5.17}$$

where V is the vessel speed and θ is the angle between the sailing line and the direction of wave propagation (see Fig. 5.18). For diverging waves in deep water the theoretical value of θ is 35° 16′.

Successive transverse and diverging waves aft of the vessel bow have increasing crest lengths (and commensurate lower energy densities and amplitudes) owing to wave diffraction as they propagate forward. Havelock (1908) demonstrated analytically that the wave heights at the cusp points should decrease at a rate that is inversely proportional to the cube root of the distance from the bow, while the transverse wave heights along the sailing line decrease at a rate that is inversely proportional to the square root of the distance from the bow. Thus, at greater distances from the vessel, the diverging waves become relatively higher than the transverse waves.

At first look, it is somewhat puzzling that the diverging waves propagate forward at an angle to the sailing line but the pattern remains steady with respect to the vessel. This phenomenon is best explained by considering the group celerity (see Section 2.4.7) for deep water waves. As the diverging waves propagate forward one wave length, half of their energy is left behind. The outer end of the diverging wave receives no energy from the wave in front of it, but the inner portion does. So, as the diverging waves propagate forward obliquely to the sailing line they diminish on the outer end and grow on the inner end, allowing them to remain stationary relative to the vessel. By the same process, the wave system adds one wave for each wave length that the vessel advances, with the waves at the bow maintaining a constant amplitude, but the waves farthest from the bow continually diminishing in amplitude.

Wave amplitudes increase exponentially as the vessel speed increases in deep water. Of course this causes an exponential increase in the vessel wave drag which, along with the increased friction drag, will limit the vessel speed at some point (owing to the limit in vessel motive power). In shallower water, when the vessel speed increases to a point where the waves are long enough to "feel" the bottom ($d/L < 0.5$), the wave crest pattern starts to change. This occurs at a Froude number F in excess of approximately 0.7 where

$$F = V/\sqrt{gd} \tag{5.18}$$

As the Froude number increases from 0.7 to 1.0, the transverse wave heights increase at a faster rate than do the diverging wave heights, so they become more prominent toward a Froude number of unity. The cusp locus angle increases from the deep water value of 19° 28′ to 90° at a Froude number of 1. At $F = 1$, with a cusp locus angle of 90°, the transverse and diverging waves have coalesced with their crests oriented perpendicular to the sailing line. Also, most of the energy in the wave system is concentrated in the first wave at the bow.

Beyond a Froude number of unity, no transverse waves can exist since the vessel speed exceeds the speed of a shallow water wave. Diverging waves extend back from the vessel at an angle equal to arcsin (F^{-1}) similar to the mach angle in aerodynamics. But, self-propelled vessels in channels cannot exceed a Froude number of unity and most vessels operate at speeds such that their Froude number is 0.9 or less (see Schofield, 1974).

Very light vessels, moving at increasing speeds, develop sufficient hydrodynamic lift to cause them to plane. After the onset of planing there is usually no significant increase in generated wave heights for further increases in vessel speed.

There have been extensive field and laboratory measurements of vessel-generated waves made during the last three decades. For the common range of vessel speeds between 5 and 15 knots the maximum wave heights generated by a vessel and measured near the vessel typically range between 0.2 and 0.9 m. Common wave periods range from 1 to 2.5 sec. For summaries of these data and the techniques that have been developed for the prediction of the height, period, and direction of vessel-generated waves the reader is referred to Sorensen (1989), Sorensen and Weggel (1984), and Weggel and Sorensen (1986).

REFERENCES

Arthur, R. S., Munk, W. H., and Isaacs, J. D. (1952), "The Direct Construction of Wave Rays," *Trans. Am. Geophys. Union*, **33,** 855–865.

Behrendt, L. and Johnson, I. G. (1984), "The Physical Basis for the Mild-Slope Equation and an Engineering Application," *Proceedings, 19th International Conference on Coastal Engineering*, American Society of Civil Engineers, Houston, pp. 941–954.

Berkhoff, J. C. W. (1972), "Computation of Combined Refraction–Diffraction," *Proceedings, 13th International Conference on Coastal Engineering*, American Society of Civil Engineers, Vancouver, pp. 471–490.

Berkhoff, J. C. W., Booij, N., and Radder, A. C. (1982), "Verification of Numerical Wave Propagation Models for Simple Harmonic Linear Water Waves," *Coastal Eng.*, **6,** 255–279.

Blue, F. L. and Johnson, J. W. (1949), "Diffraction of Water Waves Passing Through a Breakwater Gap," *Trans. Am. Geophys. Union*, **33,** 705–718.

Booij, N. (1983), "A Note on the Accuracy of the Mild-Slope Equation," *Coastal Eng.*, **7,** 191–203.

Carr, J. H. and Stelzriede, M. E. (1952), "Diffraction of Water Waves by Breakwaters," *Gravity Waves*, Circular 521, National Bureau of Standards, Washington, DC, pp. 109–125.

Cialone, M. A. and Kraus, N. C. (1987), "A Numerical Model for Shoaling and Refraction of Third-Order Stokes Waves Over An Irregular Bottom," Miscellaneous Paper CERC 87-10, U. S. Army Waterways Experiment Station, Vicksburg, MS.

Crowley, J. B., Fleming, C. A., and Cooper, C. K. (1982), "Computer Model for the Refraction of Nonlinear Waves," *Proceedings, 18th International Conference on Coastal Engineering*, American Society of Civil Engineers, Cape Town, pp. 384–403.

Dalyrymple, R. A., Kirby, J. T., and Hwang, P. A. (1984), "Wave Diffraction Due to Areas of Energy Dissipation," *J. Waterw. Port Coastal Ocean Eng. Div., Am. Soc. Civ. Eng.*, February, 67–79.

Dean, R. G. and Dalrymple, R. A. (1984), *Water Wave Mechanics for Engineers and Scientists*, Prentice-Hall, Englewood Cliffs, NJ.

Ebersole, B. A. (1985), "Refraction–Diffraction Model for Linear Water Waves," *J. Waterw. Port Coastal Ocean Eng. Div., Am. Soc. Civ. Eng.*, November, pp. 939–953.

Fuchs, R. A. (1952), "On the Theory of Short-Crested Oscillatory Waves," *Gravity Waves*, Circular 521, National Bureau of Standards, Washington, DC, pp. 187–200.

Goda, Y., Yoshimura, T., and Ito, M. (1971), "Reflection and Diffraction of Water Waves by an Insular Breakwater," Report, Port and Harbor Research Institute, vol. 10, June, pp. 3–52.

Griswold, G. M. (1963), "Numerical Calculation of Wave Refraction," *J. Geophys. Res.*, **68,** 1715–1723.

Hardy, T. A. and Kraus, N. C. (1987), "A Numerical Model for Shoaling and Refraction of Second Order Cnoidal Waves Over an Irregular Bottom," Miscellaneous Paper CERC 87-9, U. S. Army Waterways Experiment Station, Vicksburg, MS.

Harms, V. W. (1979), "Diffraction of Water Waves by Isolated Structures," *J. Waterw. Port Coastal Ocean Eng. Div., Am. Soc. Civ. Eng.*, May, 131–147.

Havelock, T. H. (1908), "The Propagation of Groups of Waves in Dispersive Media, with Application to Waves on Water Produced by a Travelling Disturbance," *Proc. R. Soc. London*, Series A, 398–430.

Headland, J. R. and Chu, H. -L. (1984), "A Numerical Model for Refraction of Linear and Cnoidal Waves," *Proceedings, 19th International Conference on Coastal Engineering*, American Society of Civil Engineers, Houston, pp. 1118–1131.

Houston, J. R. (1981), "Combined Refraction and Diffraction of Short Waves Using the Finite Element Method," *Appl. Ocean Res.*, **3,** 163–170.

Jen, Y. (1969), "Wave Refraction near San Pedro Bay, California," *J. Waterw. Harbors Div., Am. Soc. Civ. Eng.*, August, pp. 379–393.

Johnson, J. W. (1947), "The Refraction of Surface Waves by Currents," *Trans. Am. Geophys. Union*, **28**, 867–874.

Johnson, J. W. (1952), "Generalized Wave Diffraction Diagrams," *Proceedings, 2nd Conference on Coastal Engineering*, Council on Wave Research, Berkeley, pp. 6–23.

Jonsson, I. G. and Christensen, J. B. (1984), "Current Depth Refraction of Regular Waves," *Proceedings, 19th International Conference on Coastal Engineering*, American Society of Civil Engineers, Houston, pp. 1103–1117.

Keulegan, G. H. and Harrison, J. (1970), "Tsunami Refraction Diagrams by Digital Computer," *J. Waterw. Harbors Div., Am. Soc. Civ. Eng.*, May, 219–233.

Kirby, J. T. (1986), "Higher Order Approximation in the Parabolic Equation Method for Water Waves," *J. Geophys. Res.*, **91,** 933–952.

Kirby, J. T. and Dalrymple, R. A. (1983), "A Parabolic Equation for the Combined Refraction–Diffraction of Stokes Waves by Mildly Varying Topography," *J. Fluid Mech.*, **136,** 453–466.

Lozano, C. J. and Liu, P. L. -F. (1980), "Refraction–Diffraction Model for Linear Surface Water Waves," *J. Fluid Mech.*, **101,** 705–720.

Massel, S. R. (1989), *Hydrodynamics of Coastal Zones*, Oceanography Series, vol. 48, Elsevier, Amsterdam.

Montefusco, L. (1968), "The Diffraction of a Plane Wave by an Isolated Breakwater," *Meccanica*, **3,** 156–166.

Munk, W. H. and Arthur, R. S. (1951), "Wave Intensity Along a Refracted Ray," *Gravity Waves*, Circular 521, National Bureau of Standards, Washington, DC, pp. 95–108.

Noda, E. K., Sonu, C. J., Rupert, V. C., and Collins, J. I. (1974), "Nearshore Circulation Under Sea Breeze Conditions and Wave Current Interaction in the Surf Zone," Tetra Technical Report P-72-149-4, Pasadena, CA.

Oh, I. S. and Grosch, C. E. (1985), "Numerical Study of Finite Amplitude Wave Refraction," *J. Waterw. Port Coastal Ocean Eng. Div., Am. Soc. Civ. Eng.*, January, 78–95.

Penny, W. G. and Price, A. T. (1952), "The Diffraction Theory of Sea Waves by Breakwaters, and the Shelter Afforded by Breakwaters," *Philos. Trans. R. Soc. London, Series A*, **244,** 236–253.

Peregrine, D. H. (1981), "Refraction of Finite Amplitude Water Waves; Deep-Water Waves Approaching Circular Caustics," *J. Fluid Mech.*, **109,** 63–74.

Peregrine, D. H. and Smith, R. (1979), "Nonlinear Effects Upon Waves Near Caustics," *Philos. Trans. R. Soc. London, Series A*, 341–370.

Perlin, M. and Dean, R. G. (1983), "An Efficient Algorithm for Wave Refraction/Shoaling Problems," *Proceedings, Coastal Structures '83 Conference*, American Society of Civil Engineers, Arlington, VA, pp. 988–999.

Pierson, W. J. (1951), "The Interpretation of Crossed Orthogonals in Wave Refraction Phenomena," U. S. Army Beach Erosion Board, Technical Report 21, Washington DC.

Pierson, W. J. (1972), "Wave Behavior Near Caustics in Models and in Nature," *Waves on Beaches*, Academic, New York, pp. 163–180.

Putnam, J. A. and Arthur, R. S. (1948), "Diffraction of Water Waves by Breakwaters," *Trans. Am. Geophys. Union*, **29,** 481–491.

Radder, A. C. (1979), "On the Parabolic Equation Method for Water-Wave Propagation," *J. Fluid Mech.*, **95**, 159–176.

Schofield, R. B. (1974), "Speed of Ships in Restricted Navigation Channels," *J. Waterw. Harbors Coastal Eng. Div., Am. Soc. Civ. Eng.*, May, pp. 133–150.

Silvester, R. (1974), *Coastal Engineering I*, Elsevier, Amsterdam.

Silvester, R. (1986), "The Influence of Oblique Reflection of Breakwaters," *Proceedings, 20th International Conference on Coastal Engineering*, American Society of Civil Engineers, Taipei, pp. 2253–2267.

Skovgaard, O., Jonsson, I. G., and Bertelsen, J. A. (1975), "Computation of Wave Heights due to Refraction and Friction," *J. Waterw. Harbors Coastal Eng. Div., Am. Soc. Civ. Eng.*, February, pp. 15–32.

Sommerfeld, A. (1896), "Mathematische Theory of Diffraction," *Math. Ann.*, **47,** 317–374.

Sorensen, R. M. (1973), "Ship-Generated Waves," *Advances in Hydroscience*, Academic Press, New York, vol. 9, pp. 49–83.

Sorensen, R. M. (1989), "Port and Channel Bank Protection from Ship Waves," *Proceedings, Ports '89 Conference*, American Society of Civil Engineers, Boston, pp. 393–401.

Sorensen, R. M. and Weggel, J. R. (1984), "Development of Ship Wave Design Information," *Proceedings, 19th International Conference on Coastal Engineering*, American Society of Civil Engineers, Houston, pp. 3227–3243.

Thompson, W. (Lord Kelvin) (1987), "On Ship Waves," *Transactions*, Institution of Mechanical Engineers, London, pp. 409–433.

Tsay, T. -K., and Liu, P. L. -F. (1983) "A Finite Element Model for Refraction and Diffraction," *Appl. Ocean Res.*, **5,** 30–37.

U. S. Army Coastal Engineering Research Center (1984), *Shore Protection Manual*, U. S. Government Printing Office, Washington, DC.

Weggel, J. R. and Sorensen, R. M. (1986), "Ship Wave Prediction for Port and Channel Design," *Proceedings, Ports '86 Conference*, American Society of Civil Engineers, Oakland.

Wiegel, R. L. (1962), "Diffraction of Waves By a Semiinfinite Breakwater," *J. Hydraul. Div.*, Am. Soc. Civ. Eng., Jan., pp. 27–44.

Wiegel, R. L. (1964a), *Oceanographical Engineering*, Prentice-Hall, Englewood Cliffs, NJ.

Wiegel, R. L. (1964b), "Water Wave Equivalent of Mach-Reflection," *Proceedings, 9th Conference on Coastal Engineering*, American Society of Civil Engineers, pp. 82–102.

Wilson, W. S. (1966), "A Method for Calculating and Plotting Surface Wave Rays," Technical Memorandum 17, U.S. Army Coastal Engineering Research Center, Washington, DC.

6

WIND-GENERATED WAVES

Up to this point we have only considered two- and three-dimensional long-crested monochromatic waves. This wave motion was irrotational (except for the brief discussion of wave attenuation in Section 3.3) because surface wind stress and bottom stress have been ignored. We now shift our focus to the real waves at sea—both the waves actively being generated by the wind and the swell that propagate out from the generating area.

Waves that are actively being generated by the wind have a very complex surface form and the wind stress causes motion near the surface to be rotational and turbulent. Wave crests are short and poorly defined, and the waves travel in a range of directions around the dominant direction of the wind. The energy transferred to the waves from the wind generates a range of wave periods or frequencies. The waves grow in average height and period as they propagate through the generating area. After they leave the generating area they much more approximate the irrotational waves discussed in previous chapters. Wave crests become longer and more easily discerned. Water particle motions are nominally irrotational. Waves continue to loose some energy owing to surface and internal friction; but angular spreading causes a more significant reduction in energy density. And frequency dispersion occurs owing to the larger celerity of the longer waves. That is, the low-frequency longer waves propagate out ahead of the main body of waves and the shorter waves lag behind.

First we present a brief physical description of the wave generating process as far as it is understood. This includes a schematic depiction of wave growth and decay, and the concepts of fetch and duration limited waves. It also includes a discussion of typical sea wave surface records and the basic approaches commonly employed to analyze these wave records.

One approach to wave record analysis is to identify individual waves in the record, and then to statistically analyze the height and period distributions of these individual waves. Joint wave height/period correlations are also considered, as are the effects of shallow water on wave height distributions.

The other common approach to wave record analysis is to determine the wave spectrum—a plot of the component energy as a function of wave period or frequency. Common spectral models and their characteristics are presented for both one-dimensional and directional spectra. Defining parameters for spectral shape and their significance, as well as the key descriptors of basic wave characteristics that can be derived from a wave spectrum are presented, as are the related phenomenon of the appearance of waves in groups and the definition and characteristics of these groups.

An important tool for the coastal and ocean engineer and marine scientist is to be able to predict resulting wave characteristics—either representative wave heights and periods or wave spectra—for a particular wind condition. This can be done by simple empirical procedures or by more sophisticated numerical models that employ the spectral energy balance equation. Other factors that affect wave generation, such as shallow water and lateral restrictions on the wind generating area, are also considered.

Finally, the transformations that occur as wind-generated waves propagate toward shore are discussed. Specifically these include the effects of shoaling, refraction and diffraction, and nearshore breaking.

6.1 WIND WAVE CHARACTERISTICS

6.1.1 Wave Generation

The sea surface and the immediately overlying wind field exhibit very complex and changing patterns as energy and momentum are transferred from the wind to the waves to cause wave growth. While short waves are being formed, longer and higher waves are growing, and some of the steepest waves are breaking—each of these phenomena involving somewhat different air–sea and wave–wave interaction mechanisms. The wind wave generation process is not fully understood but some of the basic mechanisms involved in this process are. A good relatively recent analytical and descriptive review of wind wave generation is presented by LeBlond and Mysak (1978) and an in-depth discussion of the important original analytical work done on this subject is given by Kinsman (1965). Herein, we just present a brief physical description of the basic mechanisms of wind wave generation.

As a starting point consider a wind field with a fully developed turbulent boundary layer that is blowing over a flat water surface. Wind stress at the air–sea boundary will develop a shear layer just below the water surface as a wind-generated surface current develops. The question is, how are the initial wave undulations developed? A resonance model developed by Phillips (1957, 1960) appears to best explain the initiation and beginning stages of wave generation.

The turbulent wind boundary layer contains random three-dimensional pressure fluctuations (eddies) that vary in size, frequency, and duration of life. They move forward with the air flow at varying speeds depending on their size and consequent position in the boundary layer (larger eddies move faster). These pressure fluctuations exert a normal force on the water surface, causing surface undulations to develop. But, in themselves, the pressure fluctuations are not sufficient to develop and maintain surface waves. The key to Phillips' mechanism is the resonant interaction that occurs between the forward advected pressure fluctuations and the growing free waves that travel at the same speed. This resonance mechanism causes the rapid and continuous early growth of the waves. Phillips assumed irrotational waves having a linearized surface boundary condition (small amplitude). As nonlinear effects become important, his analysis is insufficient to account for the observed wave growth. Also, physically a momentum transfer mechanism is required to explain further wave growth when the wave amplitude becomes large enough to significantly affect the air flow pattern over the wave surface. A shear flow wave generation mechanism proposed by Miles (1957) provides a useful description of this momentum transfer mechanism.

When air flows over a fixed wave profile the vertical convergence and subsequent separation of streamlines would cause a pressure distribution that is in phase with the wave surface profile (i.e., a lower pressure over the wave crest and a higher pressure over the wave trough). For a moving wave the air flow pattern is more complex. The vertical velocity profile is continuous from the air into the water across the air–sea interface. So, relative to the wave profile celerity, there is a layer of reverse flow just above the wave. The boundary point between the relative forward and reverse flows (known as the critical level) is at a height of about a tenth of the wave length above the wave. And this flow pattern causes a relative flow circulation in the vertical plane around a point at the critical level. This more complex flow pattern causes the pressure distribution along the surface to be out of phase with the surface displacement. This wind pressure–water surface profile phase difference causes the input of wind energy into the wave.

Miles (1957) developed the equations to define this shear flow pattern and the resulting wave growth for different components of the wave spectrum. In a later paper (Miles, 1960) he combined the resonance and shear flow mechanisms to develop a more general wave generation theory. Later numerical calculations (see LeBlond and Mysak, 1978 for a discussion) have refined the understanding of the shear flow pattern and resulting energy transfer mechanism.

Other factors besides the resonance and shear flow mechanisms, which both function through pressure forces, are also involved. A shear force is exerted by the wind on the water surface. This contributes to wave growth and profile modification, but its contribution is less than that from applied pressures. Wind flow separation from the wave surface profile (downwind of the wave crest) only occurs when a wave is breaking. But the resulting flow separation drag caused by the pressure differences on the wave form should further contribute to the input of wind energy to the wave spectrum. And nonlinear wave–wave interactions will cause energy transfer from short to longer waves under certain conditions. This mechanism would assist the continued growth of a wind wave spectrum.

A purely analytical prediction of the wind wave spectrum caused by a given wind field is not presently possible. But the development of the mechanisms discussed above has significantly increased both our basic understanding of the wave generation process and, with empirical calibrations, our ability to make wave predictions.

6.1.2 Typical Wind Wave Record

A common surface-piercing wave gage located in a region where active wave generation is occurring might produce a record similar to that shown in Figure 6.1. This is a plot of the surface elevation as a function of time at a single point. It should not be confused with a plot of the surface elevation as a function of the horizontal distance in the direction of wave propagation. (The two would be the same for a monochromatic wave, but not for irregular wind waves.)

At a short distance away from the gage in the direction of wave propagation, the record would be significantly different. This recorded surface time history includes waves traveling in various directions which would produce a different mix of waves at the point a short distance away. Also, it consists of individual waves having different periods and celerities so components that travel in the same direction will still have a different mix a short distance away. And this difference is compounded by the group phenomenon so that none of the individual components have a permanent form. But, over a short time or space interval, the statistical properties of this wave record should not change.

An interesting feature of the wave record shown in Figure 6.1 is the groups of higher waves that occur. This phenomenon is quite common, to a lesser or greater extent than shown in the figure, in most of the wave records measured at sea. Grouping of wave heights, rather than their random distribution in the record, is important for many practical concerns. For example, these concerns

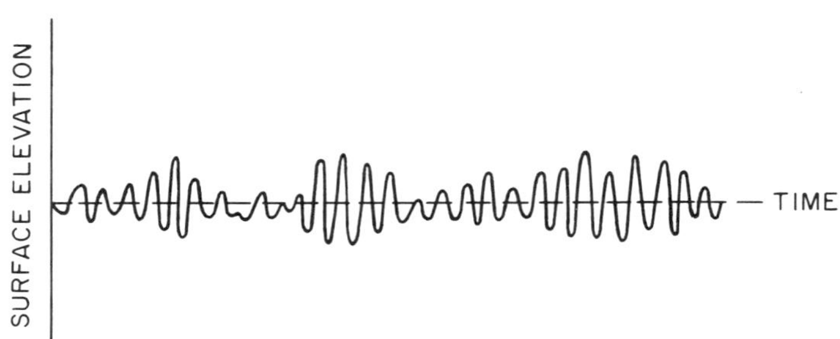

Figure 6.1 Typical wave record.

include the stability of stone mound structures, wave-induced nearshore setup/setdown and related wave-induced alongshore currents in the surf zone, and strong resonant responses that develop in structures and water bodies having resonant periods similar to the group periods.

A fundamental practical question is how to analyze a wave record like that shown in Figure 6.1. The answer largely depends on what type of information we want from the record. Do we want just a representative wave height and period or more detailed knowledge of the distribution of wave energy over the range of frequencies in the record? And perhaps we may want an indication of the degree of wave grouping and the typical group characteristics. The answer also depends on how much effort we are willing to put into an analysis of each of many possible wave records. In earlier days when wave records were commonly a line drawn by a pen on a chart, detailed analyses of wave records were quite time consuming. Now, when wave records are commonly digital time series on tape that can be directly analyzed by a computer, detailed analyses can more routinely be carried out.

There are two basic approaches to analyzing a surface wave record. One is to identify individual wave heights and periods in the wave record and then to do a statistical analysis of these heights and periods. For design purposes primary emphasis is commonly placed on the distribution of wave heights in the record and the consequent extreme heights that occur. The most often used representative wave height from this distribution of wave heights is commonly called the significant height (H_s). This representative statistical parameter was introduced by Sverdrup and Munk (1947), who did pioneering wave forecasting work during the Second World War. The significant height is defined as the average height of the highest one-third of the waves in the record. The average period of these highest third of the waves is denoted as the significant period (T_s). The significant height is approximately the height an experienced observer will report when visually estimating the height of waves at sea.

The other basic approach to wave record analysis is to conduct a Fourier analysis of the record to develop the wave spectrum. The recorded surface profile is assumed to consist of small amplitude sinusoidal components of varying frequency and phase position that, when added linearly, will recombine to produce the wave record. Figure 6.2 is a typical resulting wave spectrum presented as a plot of the energy density as a function of wave frequency. This is a one-dimensional spectrum, based on a point measurement of the water surface time history. If measurements are made at more than one point in a confined area, wave directionality can be included in the spectrum. The result is a two-dimensional or directional spectrum.

The wave-by-wave and spectral approaches to wave record analysis are both important. To quote Goda (1974), "The two methods reflect dual features of sea waves: i.e., nonlinearity and irregularity. They compliment each other and neither one alone is not (sic) sufficient for successful application of wave data for engineering problems. For example, linear phenomena such as wave diffraction, refraction, and inertia forces on structures are well analyzed with the

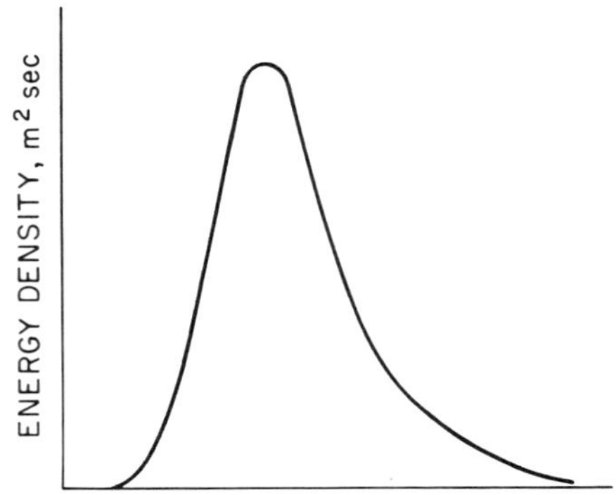

Figure 6.2 Typical wind wave spectrum.

calculation of spectral functions. On the other hand, nonlinear phenomena of random wave breaking, wave overtopping, drag forces on pilings and others are better treated by the wave-by-wave analysis with the knowledge of regular wave action of large amplitude.''

Both the individual wave-by-wave analysis and the spectral approaches are discussed in much more detail in subsequent sections. Note that neither approach retains individual component phase information, so other analyses of the wave record are required if wave grouping is to be evaluated.

6.1.3 Schematic Depiction of Wave Growth and Decay

The resulting wave characteristics at the downwind edge of a storm primarily depend on the wind velocity U, the fetch length F (i.e., the horizontal distance over which the wind blows), and the wind duration t_d. Other factors, which in certain circumstances may be quite important, include the fetch width measured normal to the wind direction, the water depth and bottom conditions if a portion of the spectrum of waves are intermediate or shallow water waves, the air–water temperature difference, and the spatial and temporal variations in the wind field during wave generation.

Waves are generated with propagation directions that vary up to about 90° to either side of the dominant wind direction. Average wave periods are longest for waves traveling in the direction of the wind and average periods decrease with increasing obliqueness to the wind direction. Thus the greater the fetch width, the smaller the percentage of wave energy that is lost owing to waves leaving the fetch at the sides, and, consequently the higher the average wave height for a given wind speed and distance along the fetch.

The water depth (for intermediate and shallow water waves) affects the surface profile form and water particle kinematics and thus the transfer mechanism for wind energy to the waves. Shallow water depths limit the nonbreaking wave heights. And bottom friction, percolation, and bed movement (see Section 3.3) all dissipate wave energy, which limits the rate of growth and ultimate height of the waves that are generated. As a consequence, for a given wind speed, fetch and duration waves of greater height and period will be generated in deep than in intermediate or shallow water.

The ratio of air to water temperature affects atmospheric stability and, consequently, the air velocity profile in the boundary layer. This, in turn, has an affect on the wave-generating mechanisms as discussed above.

No storm has a spatially or temporally constant wind velocity during the period of wave generation. Wind fields can be quite irregular and complex or they can have the more defined irregularity found in the circular pattern of winds in a hurricane. However, for a conceptual discussion of wind generation and for certain simpler wave prediction procedures, it is commonly assumed that the wind has a constant velocity over a fixed fetch length for a given duration.

Figure 6.3 is a schematic depiction of the growth of waves in the direction of the wind (x) for a constant wind speed having a given duration. The wave characteristics are defined by their significant height and period. If the wind duration exceeds the time required for the waves being generated to travel the entire fetch length, the waves will grow to OAB along the fetch and their characteristics at the end of the fetch will depend only on the fetch length and wind velocity. For the wind duration to exceed the wave travel time over the fetch requires that $t_d > F/C_g$. (Note that C_g will increase along the fetch as the wave period grows and will be different for the various components of the spectrum.) This is the "fetch limited" condition.

If the wind duration is less than F/C_g wave growth stops short of the conditions at point B. At the end of the wind duration, the wave conditions along the fetch would be defined by OAC. Short of point A conditions are controlled by the generation distance; beyond point A conditions are controlled by the wind duration. This is the "duration limited" condition.

If both the fetch and duration are sufficiently large, the curve OAB becomes

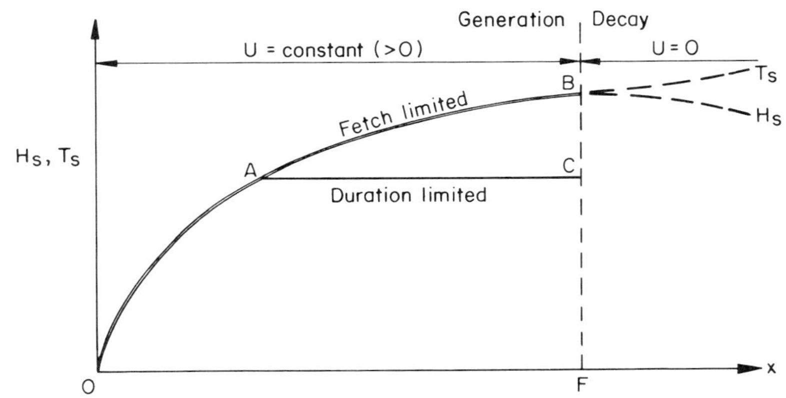

Figure 6.3 Schematic depiction of wave growth and decay.

essentially horizontal at the downwind edge and a "fully developed sea" has been generated for that particular wind velocity. This requires an extremely large fetch and duration that are not often encountered at sea, particularly if the downwind end of the storm fetch is intercepted by a land boundary.

In the region beyond the end of the fetch (sometimes called the decay region), where the waves are propagating as swell, the significant height will decrease and the significant period will increase. Lateral spreading of the waves in the decay region will decrease the wave height and since this effect is greater for the shorter period waves, the significant wave period will increase. The same effects, but to a lesser degree, occur because of wave dissipation owing to internal and surface resistance, which has a greater impact on the shorter and steeper waves.

Further insight into the nature of wave growth along a fetch is given by Figure 6.4 which shows typical wave spectra at successive points along the fetch. Note how the peak period in the spectrum, which is related to the significant period, grows. So does the total area under the spectrum, which indicates the total energy density and is related to the significant wave height. The higher frequency (shorter period) waves grow to an energy level limited by wave breaking and the transfer of energy to lower frequencies by wave–wave interaction. Since wave energy is proportional to H^2L, longer (lower frequency) waves require a greater energy input for a given increase in wave height. Thus the resulting decrease in the energy density observed at frequencies below the peak frequency.

If wave spectra were evaluated for wave records taken at a fixed point in the wave generating area at successive times, the spectra would exhibit the same growth pattern as depicted in Figure 6.4 for increasing distances along the fetch.

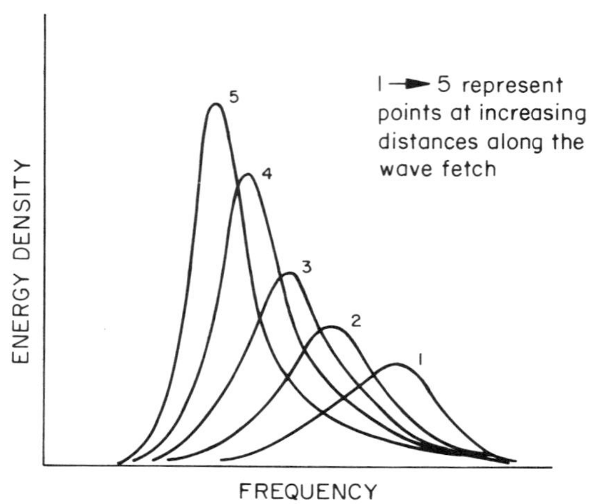

Figure 6.4 Growth of wave spectra along a fetch.

6.1.4 Spectral Energy Balance Equation

The growth of a wind-generated wave field can be represented by the spectral energy balance equation (The SWAMP Group, 1985).

$$S_{\text{in}} + S_{\text{nl}} + S_{\text{ds}} = \frac{\partial S(f, \theta)}{\partial t} + C_g \cdot \nabla S(f, \theta) \tag{6.1}$$

where

$$\nabla = i\frac{\partial}{\partial x} + j\frac{\partial}{\partial y}$$

The terms on the left of Eq. 6.1 represent the energy input from the wind S_{in}, the nonlinear energy transfer from one frequency to another by wave–wave interactions S_{nl}, and the energy dissipation S_{ds}. The term on the right is the propagation operator which defines spectral wave growth as a function of time and space. On the right-hand side, C_g is the wave group celerity and $S(f, \theta)$ is the directional wave spectrum. For a fully developed sea there is no further wave growth so $S_{\text{in}} + S_{\text{nl}} + S_{\text{ds}} = 0$.

Equation 6.1, employing either the one-dimensional $S(f)$ or directional $S(f, \theta)$ spectral representations, is the basis for the development of numerical wave prediction models. The energy input functions are typically based on the Miles and Phillips wave generation mechanisms and the energy dissipation function represents internal dissipation and wave breaking, as well as bottom friction and percolation for intermediate/shallow water wave generation.

6.2 WAVE RECORD ANALYSIS FOR HEIGHT AND PERIOD DISTRIBUTIONS

6.2.1 Wave Height Distribution

A typical point wind wave record will look like the surface elevation versus time plot given in Figure 6.1. We need a formalized repeatable procedure, that is statistically meaningful, for defining individual waves from this record. A question arises as to which surface undulations should be considered as waves. And what are the height and period of these undulations? There is no absolutely correct procedure for this analysis. Common practice is to utilize the zero-upcrossing or zero-downcrossing method for analyzing a wave record; the first method is most frequently used (see Pierson, 1954).

Figure 6.5 demonstrates the zero-upcrossing procedure. This figure shows a typical measured water surface elevation as a function of elapsed time. A mean water surface elevation is determined from the record and then each point

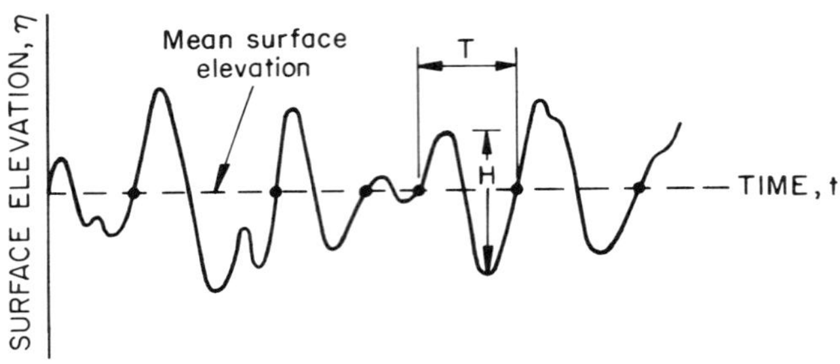

Figure 6.5 Water surface time history record.

where the water surface crosses the mean water level in the upward direction is marked. The time interval between each mark is the wave period, and the maximum vertical distance between a crest and trough is the wave height. This procedure is somewhat arbitrary as, for example, surface undulations that don't cross the mean water level would be discounted while smaller amplitude undulations that do would be counted. Consequently, there is some filtering of higher frequency components because the wave undulations that do not cross the mean water level are likely to be of higher frequency. Given this, the analysis is repeatable and statistically meaningful. A digital or analog wave record on tape can easily be edited and analyzed for zero-upcrossing heights and periods by computer.

A field wave record from a storm, containing say a hundred waves or more, would be analyzed to determine the individual wave heights and the results might be plotted as a height-frequency distribution. This would typically yield the distribution shown in Figure 6.6 [where $p(H)$ is the frequency or probability of the occurrence of wave height H]. Also shown on Figure 6.6 is the significant wave height H_s, the average height of the shaded upper third of the wave heights.

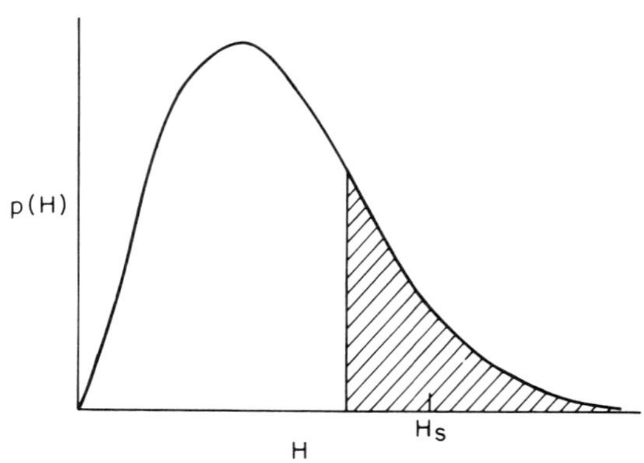

Figure 6.6 Wave height–frequency distribution.

Other representative wave heights are often used. The root-mean-square height is

$$H_{rms} = \sqrt{\frac{\Sigma H_i^2}{N}} \tag{6.2}$$

where H_i are the individual wave heights in a record containing N waves. And, in a manner similar to the significant wave height, we can define a height H_n, which represents the average of the highest n percent of waves. By this designation the significant height would be H_{33} and the average height would be H_{100}. Another important wave in a record is H_{max}, the highest wave found in the record. The H_{max} is different from H_n in that its value depends on the number of waves in the record and generally increases as the number of waves increases.

For research and design purposes it would be extremely valuable to have a model for the distribution shown in Figure 6.6 that applies to storm-generated waves. With this, and knowing a representative wave height such as H_s or H_{rms}, one could estimate any H_n and the percentage of waves that would exceed this H_n value. Longuet-Higgins (1952) demonstrated that a Rayleigh distribution best defines the distribution of wave heights in a storm.

Specifically, Longuet-Higgins assumed that the waves have a narrow band of frequencies and that the phases of the individual waves are randomly distributed. As a consequence, the water surface elevations would have a Gaussian distribution, which is realistic for small amplitude waves, but less so for steep or breaking waves and waves in shallow water which would have a skewed distribution of surface amplitudes.

For our purposes the Rayleigh distribution may be written

$$p(H) = \frac{2H}{(H_{rms})^2} e^{-(H/H_{rms})^2} \tag{6.3}$$

where H_{rms} is used to give a base magnitude to the distribution. The cumulative probability distribution $P(H)$ (i.e., the percentage of waves having a height equal to or less than H) is given by

$$P(H) = \int_0^H p(H)\, dH = 1 - e^{-(H/H_{rms})^2}$$

Of greater interest to us is the percentage of waves having a height greater than a given height or

$$1 - P(H) = e^{-(H/H_{rms})^2} \tag{6.4}$$

TABLE 6.1 Ratios of H_n/H_s for Given Values of n from the Rayleigh Distribution

n	H_n/H_s
1	1.67
2	1.56
5	1.40
10	1.27
20	1.12
50	0.89
100	0.63

Thus, $1 - P(H)$ represents the area under the $p(H)$ versus H curve to the right of the given H value (see Fig. 6.6). Employing the Rayleigh distribution yields $H_s = 1.416\ H_{rms}$ and the ratios given in Table 6.1.

Thus, for example, if the significant height for waves in a storm is 5 m, the average height of the highest 1% of the waves (H_1) would be 8.35 m and the average height of all of the waves would be 3.15 m. And, from Eq. 6.4 it can be shown that $1 - P(H_s) = 0.135$ or 13.5% of the waves exceed the significant height (i.e., less than a half of 33% owing to the skewed distribution of wave heights, as shown by Figure 6.6).

Figure 6.7, adapted from the U. S. Army Coastal Engineering Research

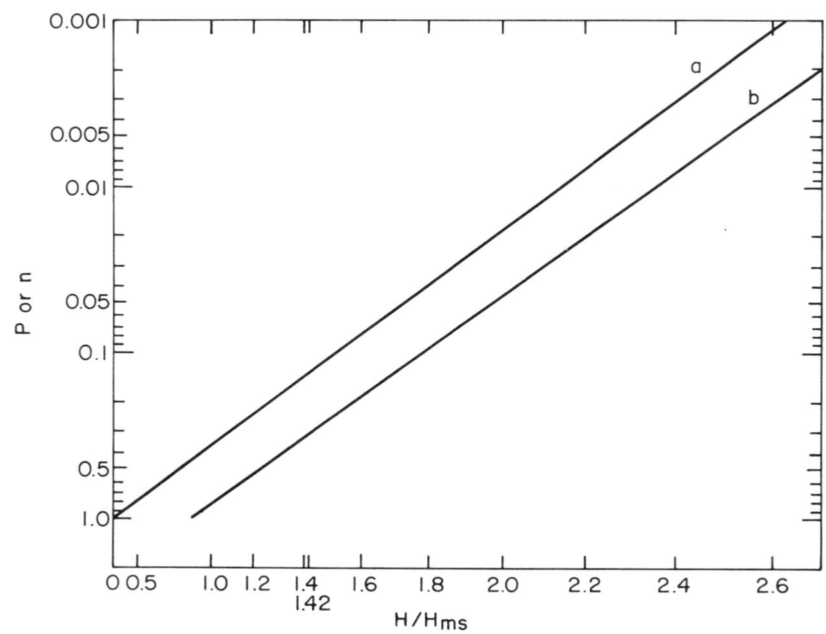

Figure 6.7 Rayleigh distribution for wave heights (adapted from U.S. Army Coastal Engineering Research Center, 1984).

Center (1984), is useful for applications of the Rayleigh distribution. The upper line (*a*) indicates the probability P that any wave height (H/H_{rms}) is greater than the value indicated by the line. For example, for H_s/H_{rms} which equals 1.42, the figure yields $P = 0.135$ or 13.5% (as noted above). The lower line (*b*) gives the average height of the n highest percent of waves. Again, for H_s/H_{rms} this gives $n = 0.33$ or 33%.

Several authors (e.g., Chakrabarti and Cooley, 1971; Collins, 1967; Earle, 1975; Goda, 1974; and Goodnight and Russel, 1963) have published comparisons of the Rayleigh distribution with field measurements of wave height and have generally concluded that the Rayleigh distribution yields acceptable results. Forrestal (1978) analyzed hurricane wave data from the Gulf of Mexico and concluded that the Rayleigh distribution overpredicted the heights of the highest waves. He proposed an empirically derived variation of the Rayleigh distribution based on these data. But Ochi (1982) observed that the locations in the Gulf, where the data were collected, did not appear sufficiently deep in comparison with the wave lengths so shallow water effects may account for the overpredicted wave heights. Hence, for waves generated by a single storm and in deep water, the Rayleigh distribution is generally accepted as an adequate definition of the expected wave height distribution, except for the extreme wave heights.

6.2.2 Maximum Wave Height

The Rayleigh distribution has no upper bound; as n decreases toward zero, H_n increases toward infinity. In a storm, the limit on n would be dictated by the number of waves in the storm. Longuet-Higgins (1952) derived the mean or expected value of H_{max} to be given by

$$H_{max} = 0.707\, H_s \left[\sqrt{\ln N} + \gamma/2\, \sqrt{\ln N}\right] \tag{6.5}$$

where N is the number of waves and γ is Euler's constant (0.5772). For a large number of waves, Eq. 6.5 reduces to

$$H_{max} = 0.707\, H_s \sqrt{\ln N} \tag{6.6}$$

For example, for a storm having a 3-h duration of high waves that have an average period of 8 sec, $N = 1350$ and $H_{max} = 1.90H_s$. The duration of high waves during a storm is difficult to quantify, but it is important to select an H_{max} value for the design of certain types of structures. Goda (1985) suggests the use of $H_{max} = 2.0H_s$ (or higher) for the design of offshore structures and $H_{max} = 1.8H_s$ for the design of vertical (caisson-type) breakwaters. It should be remembered that the Rayleigh distribution tends to slightly overpredict the heights of higher waves (owing to wave nonlinearity) which adds a bit of conservatism in the selection of H_{max}.

6.2.3 Nearshore Wave Height Distribution

The distribution of wave heights in the nearshore region is important, for example, to a better understanding of surf zone characteristics and resulting processes such as longshore sediment transport, and to the design of coastal structures. As waves shoal, the surface profile vertical asymmetry increases so the surface elevation distribution becomes increasingly non-Gaussian. At some point the higher waves begin to break and breaking increases as the waves further shoal. Breaking causes a truncation of the height distribution.

The ideal wave height frequency distribution might be equivalent to the Rayleigh distribution in deep water, develop a skew in intermediate and shallow water depths to allow for developing wave profile asymmetries, and at the commencement of breaking have higher wave heights appropriately truncated. Several authors have developed wave height distributions that modify the Rayleigh distribution to account for wave shoaling and breaking.

Collins (1970), employing shoaling and refraction analysis plus depth-related wave breaking criteria to truncate the height distribution in the surf zone, derived probability distributions for breaking waves in terms of an assumed deep water Rayleigh distribution. Battjes (1972) also employed a depth-related breaking criteria to truncate the Rayleigh distribution for a study of wave-generated setup in the surf zone. In both of these studies empirical relationships between the breaker height/water depth ratio and factors such as beach slopes and wave steepness were utilized.

Ibrageemov (1973) presented a wave height distribution for in and near the surf zone that is a Rayleigh distribution modified by an empirical factor that is dependent on the ratio of the water depth/(average wave period)^2. His distribution reduces to the Rayleigh distribution in deep water.

Kuo and Kuo (1974) and Goda (1975) presented truncated wave height distributions that distribute the energy from the broken waves back over the smaller wave heights in the distribution to account for the portion of energy retained by the broken waves. Kuo and Kuo truncated the Rayleigh distribution at the breaking wave height, but Goda assumed that breaking occurs over a small range of wave heights owing to the variation of individual wave periods and other factors.

More recently, Hughes and Borgman (1987), employing a high-quality field data set, have presented a beta-Rayleigh wave height distribution. The distribution is given in terms of a depth limited breaking wave height, and the root mean square and root mean quad wave heights (which are related by the Rayleigh distribution). These three parameters are, in turn, empirically related to the significant wave height, the peak spectral wave period, and the water depth by employing the field data set. This allows for practical application. This distribution reduces to the standard Rayleigh distribution in deep water.

The various nearshore wave height distributions discussed above are generally more complex than the basic Rayleigh formulation. Thus, the reader should consult the various references for more detail.

6.2.4 Distribution of Wave Height and Period

Longuet-Higgins (1975) presents an analytical formulation of the probability distribution of wave periods for a narrow banded spectrum with Gaussian surface elevations. The distribution is a symmetrical bell-shaped (non-Gaussian) curve having a peak at the average period of the spectrum. That is, most of the zero-upcrossing periods (i.e., waves) in a wave spectrum will be clustered around the average period.

Of greater interest is the joint probability distribution of wave heights and periods. Employing different approximations of varying complexity, a few authors (Longuet-Higgins, 1975, 1983; Cavanie et al., 1976; and Lindgren and Rychlik, 1982) have presented analytical formulations of joint wave height–period probability distributions. The resulting shape of the distribution depends on the width of the spectrum, but the general shape is depicted schematically in Figure 6.8. This is a plot of the wave height versus period where each is nondimensionalized by dividing by the average height and period respectively. The solid lines are lines of equal probability of occurrence of a height–period combination.

Note that for smaller (but not smallest) wave heights the distribution of wave periods is rather broad, but the period distribution narrows as wave height increases. The wave energy clusters around the average period and diminishes toward the extremes of lower and higher period. The highest waves and greatest energy are close to, but not necessarily at, the average period. The significant period (average period of the highest third of the waves) is slightly less than

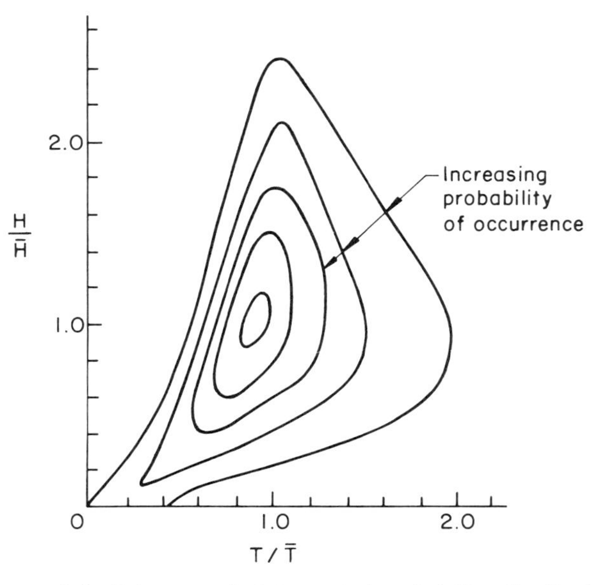

Figure 6.8 Joint wave height–period probability distribution.

the average period of all of the waves. Because the average period is considered to be less statistically stable, and because the higher waves are most important, it is best to use the significant period (T_s) or the spectral peak period (T_p) as a representative wave period for a spectrum of waves. Based on empirical data the U. S. Army Coastal Engineering Research Center (1984) recommends the relationship $T_s = 0.95\ T_p$.

6.3 WIND WAVE SPECTRA

6.3.1 Wave Spectra Characteristics

A water surface time history measured at a point can be reconstructed by linearly adding a large number of component sine waves having different amplitudes, periods, phase positions, and directions. A directional wave spectrum is a plot of the energy in these component waves plotted as a function of wave frequency f and direction θ. If the wave energy is plotted as a function of only frequency, without considering wave direction, we have a one-dimensional frequency spectrum. The wave energy density at a particular frequency may be denoted $S(f)$. A less common form, but one sometimes used in engineering literature, is the one-dimensional period spectrum where wave energy density $S(T)$ is plotted against wave period.

The energy density of a wave is $\rho g H^2/8$ (see Eq. 2.32). It is common to leave out the product of fluid density and the acceleration of gravity, yielding the following expression for the directional wave spectrum:

$$S(f, \theta)\, df\, d\theta = \sum_{f}^{f+df} \sum_{\theta}^{\theta+d\theta} \frac{H^2}{8} \tag{6.7}$$

where H represents the height of the individual component waves. For a one-dimensional frequency spectrum, we can integrate Eq. 6.7 to yield

$$S(f)\, df = \int_{-\pi}^{\pi} S(f, \theta)\, df\, d\theta = \sum_{f}^{f+df} \frac{H^2}{8} \tag{6.8}$$

From Eq. 6.8, the dimensions for $S(f)$ would be length squared times time (e.g., $m^2 \cdot sec$). Consequently, the dimensions for $S(T)$ would be length squared divided by time (e.g., m^2/sec).

Figure 6.9 shows a schematic frequency spectrum and the equivalent period spectrum. The shaded areas in each spectrum represent equivalent wave energies. That is,

$$S(f)\, df = -S(T)\, dT$$

where the minus sign is needed because dT is negative when df is positive.

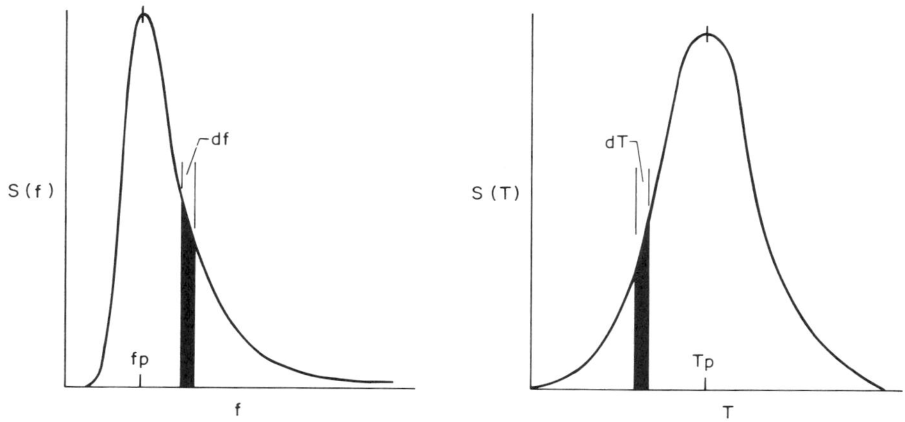

Figure 6.9 Equivalent frequency and period spectra.

Since

$$f = 1/T, \qquad df = -dT/T^2$$

so

$$S(f)\,df = S(T)\,T^2\,df$$

or

$$S(f) = S(T)\,T^2 \tag{6.9}$$

Equation 6.9 gives the relationship between the equivalent spectral energy components for the two equivalent spectra. Also noted on Figure 6.8 are the peak spectral frequency f_p and equivalent peak spectral period T_p. As noted above, this is one of the preferred frequencies or periods to represent a wave spectrum.

The exact shape and scale of a wave spectrum will, of course, vary depending on the wind speed, position within the fetch, and other factors. However, the general shape of a wave spectrum has certain consistent features. For a given spectrum, the tail of the spectrum having frequencies greater than the peak frequency (the shorter waves) should reach an upper limit or saturated condition. Continuing input of wave energy to this portion of the spectrum is balanced by energy loss caused by wave breaking (waves reach limiting steepness) and energy transfer to other frequencies by wave–wave interactions. Phillips (1958) carried out a dimensional analysis for this equilibrium range and found that for deep water waves it should have the form given by

$$S(f) = \frac{\alpha g^2 f^{-5}}{(2\pi)^4} \tag{6.10}$$

where α is a universal constant. From measured wind wave spectra, Phillips found the value of α to be 0.0074. Subsequent studies (see below) have demonstrated that α is not a constant but depends on the wind speed and fetch. Toba (1978) has suggested that the equilibrium range should follow an f^{-4} profile in deep water. And Kitaigorodskii et al. (1975) showed that shallow water effects should transform the equilibrium range of the spectrum from an f^{-5} profile to an f^{-3} profile as the spectrum of waves propagates into shallow water.

The low-frequency (longer wave length) portion of the wave spectrum has not grown to the saturated or equilibrium condition. Continued growth of the spectrum causes the saturated portion to expand into the unsaturated portion and a progressive decrease of the peak frequency. A general spectral form may be devised by modifying the Phillips formulation for the equilibrium range (Eq. 6.10) to account for the increasing wave energy density in the unsaturated region. This yields an equation of the form where A and B adjust the scale of the spectrum and commonly depend

$$S(f) = \frac{A}{f^5} e^{-B/f^4} \tag{6.11}$$

on a representative wave height and frequency (e.g., H_s and f_p). And, from our simple forecasting relationships, the representative height and frequency in turn depend on the wind speed, fetch, and duration.

The analysis of a water surface time history to develop the frequency spectrum is an exacting task, the details of which are beyond the scope of this text. Summaries of the theory of spectral analysis are given by Sarpkaya and Isaacson (1981), Ochi (1982), Goda (1985), and Tucker (1991). A good listing of the basic references on spectral analysis is given in the paper by Wilson et al. (1974) which also evaluates several factors affecting the spectral analysis of wave records including the number of and time interval between surface elevation data points, as well as the type of smoothing and filtering routines used in the analysis.

6.3.2 Spectral Moments

The moments of a wave spectrum are important in characterizing the spectrum and useful in relating the spectral description of waves to the wave-by-wave analysis discussed above. With a computer analysis of digital wave records to develop wave spectra, it becomes a relatively easy matter to determine the various spectral moments. We can define the nth moment of a wave spectrum as

$$m_n = \int_0^\infty S(f) f^n \, df \tag{6.12}$$

The zeroth moment m_0 is just the area under the spectral curve which is equal to the total energy density of the spectrum (divided by the product of fluid density and the acceleration of gravity). And, an average frequency of the spectrum can be determined from

$$\bar{f} = \frac{m_1}{m_0} \tag{6.13a}$$

which is just the first moment divided by the area which equals the centroidal value. For a zero-upcrossing analysis of a wave record, an alternative average frequency (Longuet-Higgins, 1975, 1983) is given by

$$\bar{f} = \left(\frac{m_2}{m_0}\right)^{1/2} \tag{6.13b}$$

Thus, we can easily determine representative frequencies from a spectrum such as the average (Eq. 6.13) or peak (Fig. 6.8) frequency. We also need a way to determine a representative wave height, such as the significant height.

To do so, consider the sequence of surface elevation values η from a wave record. For small amplitude waves, the total energy density is twice the potential energy density, which can be written in terms of the surface elevation as

$$\overline{E} = 2\overline{E}_P = \frac{1}{T_*}\int_0^{T_*} \rho g \eta \left(\frac{\eta}{2}\right) dt$$

or

$$\overline{E} = \rho g \overline{\eta}^2 = \frac{\rho g \Sigma \eta^2}{N_*} \tag{6.14a}$$

where T_* is the wave record length containing N_* values of η. From our definition of H_{rms} we have

$$\overline{E} = \frac{\rho g H_{rms}^2}{8} = \frac{\rho g H_s^2}{16} \tag{6.14b}$$

since $H_s = 1.416 H_{rms}$ from the Rayleigh distribution (Section 6.2.1). Thus, with m_0 for the wave spectrum being equal to the energy density (divided by ρg), we have from Eq. 6.14a and b

$$\overline{E} = \rho g m_0 = \rho g \frac{\Sigma \eta^2}{N_*} = \frac{\rho g H_s^2}{16} \tag{6.15}$$

or

$$H_s = 4\sqrt{m_0} \tag{6.16}$$

To denote the different origin of the significant height given by Eq. 6.16, it is usually denoted as H_{m0}. That is, H_s is based on a wave-by-wave analysis whereas H_{m0} is related to the total energy density as given by the zeroth moment of the wave spectrum. Their equivalence assumes the steps employed above, including the assumption that wave heights have a Rayleigh distribution and waves are of small amplitude with a sinusoidal surface profile.

Also, note from Eq. 6.15 that m_0 is equal to the mean squared displacement of the water surface elevation. Thus m_0 and consequently H_s can be evaluated directly from the N_* digitized values of the water surface displacement.

Experience shows that in deep water and for waves that are not too steep, $H_s = H_{m0}$ effectively. However, as the wave steepness increases and as waves propagate into shallow water, H_s becomes significantly larger than H_{m0} for the same set of irregular waves. This is well demonstrated by Figure 6.10, modified from Thompson and Vincent (1985) and based on field and laboratory data. This shows the ratio of H_s and H_{m0} as a function of the relative water depth. The solid line is the average of the data and the upper dashed line is an upper envelope of the data.

The basic reason for the deviation of H_s and H_{m0} is the wave profile vertical asymmetry that develops as waves become steeper in deep water and when shoaling. Considering the surface profile, one can see that a steep wave has

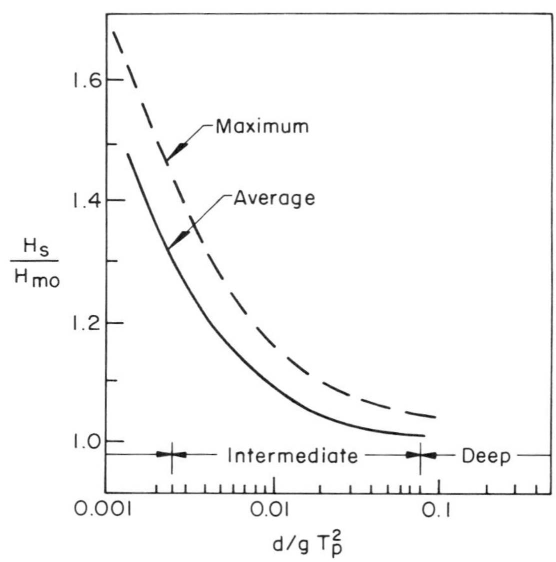

Figure 6.10 H_s/H_{m0} versus relative depth (after Thompson and Vincent, 1985).

less potential energy than a sine wave having the same height and length (steepness). Thus the representative wave based on individual surface measurements (H_{m0}) would be lower than the representative wave (H_s) based on wave-by-wave analysis assuming linear (sine) waves.

As wave record analysis by computer becomes more common, the Eq. 6.16 definition of significant height is most commonly used. Care should be taken to note the difference from the significant height, defined as the average of the one third highest waves from a zero-upcrossing analysis. When H_s and H_{m0} differ significantly, it would not be appropriate to apply the Rayleigh distribution to a value of H_{m0} to determine other H_n values. Use Figure 6.10 to convert H_{m0} to H_s first.

Some additional comments on Figure 6.10 are needed. After significant breaking of waves in the spectrum occurs, the ratio H_s/H_{m0} tends to decrease and thus deviate from the results shown in the figure. See Thompson and Vincent (1985) for a further discussion of this effect and how to account for it. Some of the scatter of the experimental data is due to the relative broadness or peakedness of the various spectra. More peaked spectra, having the same value of H_{m0} in the same relative depth, would have a slightly larger ratio of H_s/H_{m0}.

Whether a wave spectrum is broad or narrow banded is another important characteristic of the spectrum. The best description of a spectrum of waves would involve three parameters: H_{m0}, f_p, or T_p, and a parameter that quantifies the spectral shape. A number of suggestions of dimensionless parameters for the quantification of spectral shape have been offered. Cartwright and Longuett-Higgins (1956) proposed a spectral width parameter defined as

$$\epsilon = \left[1 - \frac{m_2^2}{m_0 m_4}\right]^{1/2} \tag{6.17}$$

This parameter varies from 0 to 1 for a very narrow to very broad spectrum, respectively. Later, Longuett-Higgins (1975) defined another parameter to define the width of a spectrum given as

$$\nu = \left[\frac{m_2 m_0}{(m_1)^2} - 1\right]^{1/2} \tag{6.18}$$

which tends to have a smaller range of values than ϵ. Goda (1970) defined a parameter called the peakedness parameter, Q_p, given by

$$Q_p = \frac{2}{m_0^2} \int_0^\infty f\, S^2(f)\, df \tag{6.19}$$

Q_p varies from 1 for white noise to around 2 for wind waves and values greater than 2 for swell.

Rye (1977) evaluated the parameters given above (Eqs. 6.17–6.19) by analyzing their performance with a known (JONSWAP, see Section 6.33) spectral equation. He recommends the use of Q_p to define spectral shape since it best distinguished between broad and very sharp-peaked spectra and because it was the only parameter whose numerical value was not very dependent on the high-frequency cut-off value used in evaluating the parameter. Figure 6.11 shows a pair of wave spectra measured in the Atlantic Ocean (U. S. Army Corps of Engineers, 1989) that have essentially the same peak period (10.5 sec) and significant height (3.3 m) but different values of Q_p (3.1 versus 1.9). Note the consequent differences in the spectral shapes.

6.3.3 One-Dimensional Wave Spectra

A number of one-dimensional wave spectra models have been proposed in the literature. Generally they have the form of Eq. 6.11 and are based on empirical fits to wave spectra measured under a reasonably well-defined set of conditions. As mentioned above, the parameters A and B in Eq. 6.11 are typically a function of some representative wave height and frequency or period which must be known to define the wave spectrum. Or, employing the spectra model as a wave forecasting tool, the parameters A and B are given as a function of the wind speed, fetch, and duration.

Four of these wave spectra models (Bretschneider, Pierson–Moskowitz, JONSWAP, and TMA) are presented. They are given in either the frequency or period form that they most commonly appear; Eq. 6.9 can be used to convert to the alternate form. The first three are for deep water conditions and the fourth (TMA) is adjusted for the affects of water depth. These four spectra are presented because they are of historic interest, they represent a variety of

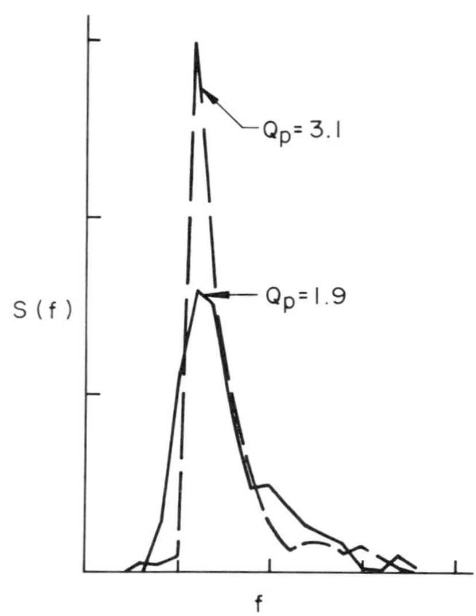

Figure 6.11 Two ocean wave spectra having different Q_p values (U.S. Army Corps of Engineers, 1989).

conditions, and they see the most use in practice. A number of other spectra models are referenced.

Bretschneider Spectrum (Bretschneider, 1959). This spectrum was originally given in the form

$$S(T) = \frac{\alpha g^2}{(2\pi)^4} T^3 e^{-0.675(gT/2\pi U F_2)^4} \tag{6.20}$$

where U is the wind speed, and the dimensionless coefficients α and F_2 are given by

$$\alpha = 3.437 \frac{F_1^2}{F_2^2}$$

$$F_1 = \frac{g\overline{H}}{U^2}; \qquad F_2 = \frac{g\overline{T}}{2\pi U}$$

In the equations given above, $\overline{H}$ and $\overline{T}$ are the average wave height and period respectively. Insertion of α, F_1, and F_2 into Eq. 6.20 yields

$$S(T) = \frac{3.44T^3(\overline{H})^2}{(\overline{T})^4} e^{-0.675(T/\overline{T})^4} \tag{6.21}$$

So, if the average wave height and period are known, the spectrum can be calculated from Eq. 6.21. If the significant wave height and the spectral peak period are known, the average wave height can be determined from the Rayleigh distribution and the average period from the relationship $\overline{T} = 0.77T_p$ (Ochi, 1982). Then the spectrum can be calculated from Eq. 6.21.

The parameters F_1 and F_2 are really a dimensionless wave height and period respectively. Bretschneider (1959) empirically related these to a dimensionless fetch (Fg/U^2) and duration (gt_d/U) as a wave forecasting relationship. Given the wind speed, fetch, and duration, one can then forecast the height, period, and spectrum for waves that are either fetch or duration limited. This is discussed in more detail in Section 6.5.2.

Note that when Eq. 6.21 is converted to a frequency spectrum employing Eq. 6.9, the high-frequency portion of the spectrum has an f^{-5} dependency.

Pierson–Moskowitz Spectrum (Pierson and Moskowitz, 1964). Employing spectra derived from wave records taken by weather ships in the north Atlantic Ocean, the following wave spectrum was developed

$$S(f) = \frac{\alpha g^2}{(2\pi)^4 f^5} e^{-0.74(g/2\pi U f)^4} \tag{6.22}$$

where $\alpha = 8.1 \times 10^{-3}$ and U is the wind speed measured at an elevation of 19.5 m rather than the standard elevation of 10 m. (This was done to avoid ship geometry effects on wind measurements.) Wind speeds at the 19.5-m elevation are typically 5–10% higher than at the 10-m elevation (Silvester, 1974), so an appropriate correction employing the standard logarithmic velocity profile (Pierson, 1964) should be made. This is discussed further in Section 6.5.1 where Eq. 6.42 can be used to make this correction.

This spectrum is independent of the fetch and duration of the wind and is thus only for fully developed seas. The wind speeds (20–40 knots) covered by the data set were sufficiently low for this to be realistic. At significantly higher wind speeds it is likely that conditions will be fetch and/or duration limited. For fully developed seas there will be a distinct dependence of the significant height and peak period on the wind speed. These are (Ochi, 1982).

$$H_{m0} = \frac{0.21U^2}{g} \tag{6.23}$$

and

$$f_p = \frac{0.87g}{2\pi U} \tag{6.24}$$

JONSWAP Spectrum (Hasselman et al., 1973). In 1968–1969 an extensive set of wave measurements was conducted in the North Sea by laboratories from four countries. This program was called the Joint North Sea Wave Program (JONSWAP). Thirteen stations were operated along a line extending about 100 miles WNW from the German isle of Sylt. The results from this experiment were used to develop a deep water, fetch limited wave spectrum. The spectrum is basically a Pierson–Moskowitz spectrum with a variable fetch dependent value for the coefficient α and a factor γ that significantly enhances the spectral peak.

If we combine Eqs. 6.24 and 6.22 by eliminating U we can write the Pierson–Moskowitz spectrum as follows:

$$S(f) = \frac{\alpha g^2}{(2\pi)^4 f^5}\, e^{-1.25(f_p/f)^4} \tag{6.25}$$

The JONSWAP spectrum is

$$S(f) = \frac{\alpha g^2}{(2\pi)^4 f^5}\, e^{-1.25(f_p/f)^4}\, \gamma^{a} \tag{6.26}$$

where

$$a = e^{-[(f-f_p)^2/2\sigma^2 f_p^2]} \tag{6.27}$$

In Eqs. 6.26 and 6.27

$$\sigma = 0.07 \quad \text{when } f < f_p$$

$$\sigma = 0.09 \quad \text{when } f \geq f_p$$

$$\alpha = 0.076 \left(\frac{gF}{U^2}\right)^{-0.22} \tag{6.28}$$

$$f_p = \frac{3.5g}{U} \left(\frac{gF}{U^2}\right)^{-0.33} \tag{6.29}$$

and γ, known as the spectral peak shape factor, had values that ranged from about 1.6 to 6 with an average recommended value of 3.3.

Note that Eqs. 6.28 and 6.29 give the growth of α and f_p as a function of increasing fetch for a given wind speed. These are indicative of the growth of wave energy ($S(f)$ is proportional to α) and of the wave length of the spectral peak (see Figure 6.3).

Letting $f = f_p$ in the JONSWAP spectrum shows that γ is the ratio of the magnitude of the JONSWAP spectral peak to the Pierson–Moskowitz spectral peak (i.e., Eq. 6.25 to Eq. 6.26 and 6.27 with $f = f_p$). This is demonstrated in Figure 6.12, which shows superimposed JONSWAP and Pierson–Moskowitz

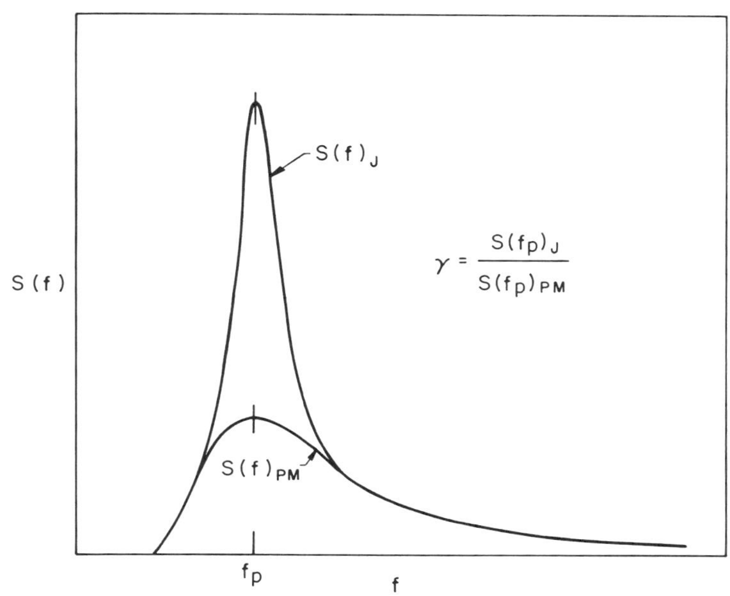

Figure 6.12 Comparison of Pierson–Moskowitz and JONSWAP spectra.

spectra. Thus one would expect a distinct relationship between γ and the peakedness parameter Q_p. Rye (1977) found the following:

$$\gamma = 1.0 \qquad 3.3 \qquad 7.0$$
$$Q_p = 2.0 \qquad 3.15 \qquad 4.65$$

The JONSWAP spectrum is very commonly used for design practice and for laboratory investigations employing irregular waves. The original data used to develop the spectrum was collected for generally light wind conditions, but later data collected by a number of authors for more severe storm conditions has produced results that compare reasonably well with the JONSWAP spectrum (Rye, 1977). Mitsuyasu et al. (1980) also found good agreement between the JONSWAP spectrum and spectra derived from ocean wave records taken during 1971 and 1976 near Japan, but he found that γ varied with the dimensionless fetch according to

$$\gamma = 7\left(\frac{gF}{U^2}\right)^{-0.143} \tag{6.30}$$

which might be used for design in place of the average value of 3.3.

TMA Spectrum (Kitaigorodskii et al., 1975; Bouws et al., 1985). The three preceding spectra were developed for deep water conditions. As deep water waves propagate into finite water depths the spectral form will change. It is desirable, for example, for the design of nearshore structures, to have a wave spectrum that accounts for these changes. The TMA spectrum (named after the three data sets used in its development) is one such spectrum in common use.

The TMA spectrum is a modified JONSWAP spectrum where the JONSWAP spectrum is multiplied by a function that is depth and frequency dependent. That is, $S(f)_{TMA} = S(f)_J\ \Phi(f, d)$ where $\Phi(f, d)$ is a complex function of wave frequency and water depth that is adequately approximated (less than 4% difference) by

$$\begin{aligned} \phi &= 2\pi f^2\, d/g \quad \text{for } f < (2\pi d/g)^{-0.5} \\ &= 1 - 0.5\,[2 - 2\pi f(d/g)^{0.5}]^2 \quad \text{for } f > (2\pi d/g)^{-0.5} \end{aligned} \tag{6.31}$$

The function $\Phi(f, d)$, which is plotted versus $2\pi f(d/g)^{0.5}$ in Figure 6.13, decreases from near unity in deep water toward zero as the water depth decreases. Also (see Hughes, 1984), the factors α and γ in the JONSWAP spectrum can be determined from the following for the TMA spectrum:

$$\alpha = 0.0078\left(\frac{2\pi U^2}{gL_p}\right)^{0.49} \tag{6.32}$$

$$\gamma = 2.47\left(\frac{2\pi U^2}{gL_p}\right)^{0.39} \tag{6.33}$$

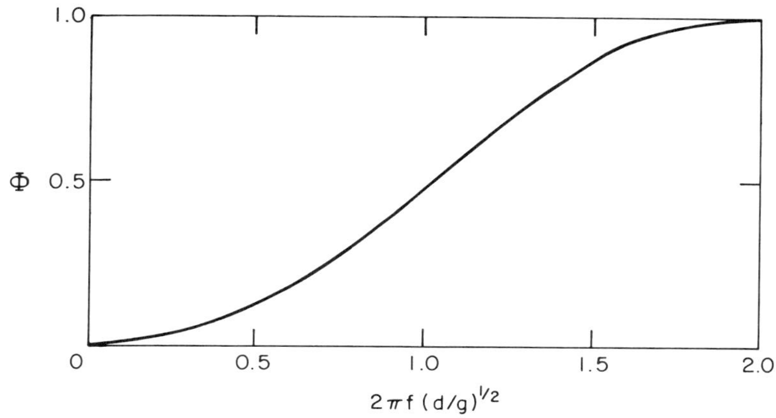

Figure 6.13 TMA spectrum function.

where L_p is the wave length for the wave peak frequency at the water depth of interest. Thus, for a given wind speed over deep water, fetch and finite water depth the nearshore wave spectrum can be determined.

Other One-Dimensional Spectra. Several other basic one-dimensional spectra have been proposed. These include Darbyshire (1952), Neumann (1953), Scott (1965), and Mitsuyasu (1972). And Ochi and Hubble (1976) have proposed a six-parameter spectrum that exhibits two spectral peaks. One peak is the dominant wind waves and the second accounts for background swell. The six parameters are the significant height, peak frequency, and a spectral shape parameter for each of the two components. These six parameters are determined empirically.

6.3.4 Directional Wave Spectra

As noted above, the sea surface in a storm consists of superimposed directional components propagating in various directions that combine (Eq. 6.7) to give the full directional wave spectrum. However, a point measurement of the water surface time history loses the directionality of the various components. Most field data are collected in this fashion and recourse is made to the one-dimensional spectrum of Eq. 6.8.

It is difficult to make reliable field measurements that yield information on the directionality of the various spectral components. But directional wave gages do exist (see Chapter 7) and, from the data collected by these gages, a few directional wave spectra formulations have been developed. The directional spread of the wave energy is frequency dependent, so these formulations are commonly presented as

$$S(f, \theta) = S(f)G(f, \theta) \tag{6.34}$$

where $S(f)$ is the one-dimensional spectrum and $G(f, \theta)$ is a dimensionless directional spreading function. Thus $S(f)$ gives the absolute value of the wave energy at a given frequency and $G(f, \theta)$ modifies this value for direction. The angle θ is usually measured clockwise from zero degrees at the principal direction. In general, the higher frequency components of the wave spectrum have a wider angular spread of energy. The energy is more focused in the dominant direction around the spectral peak frequency.

Thus $G(f, \theta)$ must satisfy

$$\int_{-\pi}^{\pi} G(f, \theta)\, d\theta = 1 \tag{6.35}$$

although practical limits for θ would be $-\pi/2$ and $+\pi/2$. The total wave energy density is

$$m_0 = \int_0^{\infty} \int_{-\pi}^{\pi} S(f, \theta)\, d\theta\, df \tag{6.36}$$

as before.

An early form of the directional spreading function (St. Denis and Pierson, 1953) was a simple "cosine squared" function independent of wave frequency. That is,

$$G(f, \theta) = G(\theta) = \frac{2}{\pi} \cos^2\theta \tag{6.37}$$

where θ varies between $+\pi/2$ and $-\pi/2$.

A more commonly used directional spreading function is that given by Eq. 6.38 from Mitsuyasu et al. (1975), which is based on their field measurements with a clover-leaf type directional wave guage.

$$G(f, \theta) = G(s) \cos^{2s}\left(\frac{\theta}{2}\right) \tag{6.38}$$

In Eq. 6.38, $G(s)$ must satisfy Eq. 6.35 or

$$G(s) = \frac{2^{2s-1}}{\pi} \frac{\Gamma^2 (s + 1)}{\Gamma (2s + 1)} \tag{6.39}$$

where s is a function of wave frequency and Γ is the gamma function given in books that tabulate mathematical functions. The variation of s is such that it causes a direction-wise concentration of spectral energy near the peak frequency. Away from the peak frequency there is a greater directional spread of wave energy. Mitsuyasu et al. (1975) employed their spreading function with a JONSWAP one-dimensional spectrum.

Mitsuyasu et al. (1975) originally wrote the parameter s as a function of wave frequency, wave peak frequency, and wind speed. Goda and Suzuki (1975) (see Goda, 1985) simplified the definition of s by writing it in terms of a maximum value s_{max}, the frequency and the peak frequency as

$$\begin{aligned} s &= s_{max}\,(f/f_p)^5 \quad &\text{when } f < f_p \\ &= s_{max}\,(f/f_p)^{-2.5} \quad &\text{when } f > f_p \end{aligned} \tag{6.40}$$

where s_{max} is empirically determined. Mitsuyasu et al. (1975) give a relationship for s_{max} as follows:

$$s_{max} = 11.5\left(\frac{2\pi f_p U}{g}\right)^{-2.5} \tag{6.41}$$

The term s_{max} varies inversely with wave steepness. Wave steepness increases (thus s_{max} decreases) for increasing wind fetches and speeds (e.g., see Eq. 6.41). As waves propagate as swell the wave steepness then decreases (s_{max} increases). For design purposes Goda (1985) recommends the following representative values.

Wind waves	$s_{max} = 10$
Swell with short decay distance	$s_{max} = 25$
Swell with long decay distance	$s_{max} = 75$

Thus, given a value of s_{max}, the directional spreading function can be evaluated as a function of wave frequency and direction by employing Eqs. 6.38–6.40. And, a one-dimensional frequency spectrum can be used to determine the directional spectrum by Eq. 6.34. Figure 6.14 is a schematic plot (for a given value of s_{max}) of $G(f, \theta)$ as a function of θ for given wave frequencies in the spectrum. The various spectral frequencies are normalized in terms of the peak frequency.

It should be noted that the discussion of directional spectra applies to deep water. Other deep water directional spreading functions have been presented by Cote et al. (1960) and Hasselman et al. (1980). As a wave spectrum approaches the shore, shoaling and refraction will reduce the spreading of wave directions for the various frequencies in the spectrum. In effect, the value of s_{max} would increase as the relative depth decreases and the effect would be greater for stronger wave refraction (see Goda, 1985).

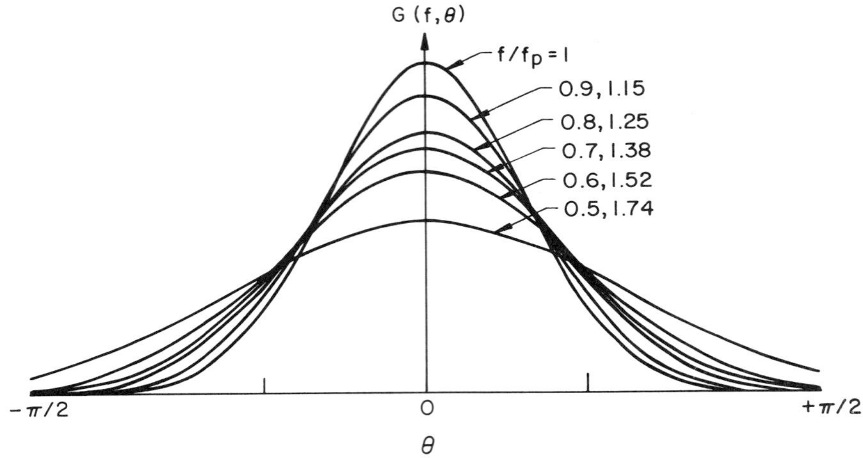

Figure 6.14 Directional spreading function versus direction and normalized frequency for given values of S_{max}.

6.4 WAVE GROUPING

Review of a number of ocean wave records will show the existence of alternate groups of high and low waves in many of the records (e.g., see Fig. 6.1). An observer standing in the surf zone will also notice that the higher breakers often come in groups. This phenomenon has been long recognized, but only recently has there been extensive investigation of the characteristics of wave grouping and analytical means for describing this grouping.

The occurrence of wave groups will have a significant impact in several ways. In the surf zone wave grouping will generate surf beat, a longer period (than the incident wave period) rise and fall of the mean water level owing to wave-induced setup and setdown. Related wave-induced alongshore currents in the surf zone will develop a pulsating character when driven by strongly grouped waves. The elevation of wave runup on beaches and structures is closely related to wave height variations and wave-induced surf beat. Harbor resonance and wave loading on moored vessels and floating breakwaters are affected by wave grouping. And, it has been observed that the stability of rubble mound structures exposed to wave spectra having the same H_{m0} and f_p can be markedly different if the component waves in the spectra have different levels of wave grouping. Strongly grouped higher waves diminish the stability of stone mound structures.

In basic terms, a wave group can be defined as a succession of two or more waves having a height in excess of some selected height such as H_s. This succession of waves is commonly known as a wave run and the number of waves is the run length. The number of waves between the first wave in a wave run and the first wave in the following wave run is the total run of waves.

Goda (1976) provides some information on group characteristics from an analysis of 171 deep water ocean wave records containing from 55 to 198 waves per record. As would be expected, as the run length increased, the number of groups in the 171 records decreased. Using the significant height as the cutoff value, the following number of run lengths was found:

Run Length	Number of Runs
2	374
3	122
4	37
5	9
6	2
7	1

The group containing the highest wave in the record tended to have a much longer than average run length. Rarely was the highest wave in a record an isolated wave.

Generally, there is a correlation between the narrowness or peakedness of wave spectrum and the groupiness of a wave record as indicated, for example, by the mean wave run length and the mean total run of waves. Field data (see Goda, 1985) indicate an approximate linear relationship between the mean run length and the peakedness parameter (Q_p, Eq. 6.19) of a wave record (i.e., increasing run length with increasing Q_p).

Some authors (see Burcharth, 1980) have considered other characteristics of the sequencing of individual waves in a wave record. Burcharth defined a wave jump as a sequence where a wave of low height is immediately followed by a high wave. A higher relative frequency of wave jumps correlated with increased wave runup and damage to rubble mound structures. Runs of specific wave period ranges have also been evaluated.

The details of wave grouping analysis are relatively complex. The reader is referred to Goda (1976) and Medina and Hudspeth (1990) for reviews of wave grouping parameters and analysis methodologies in common use.

6.5 WAVE PREDICTION

6.5.1 Wind Conditions

For deep water conditions, wave prediction requires a knowledge of the near surface wind velocity including its spatial and temporal variations. For the simpler empirical wave prediction techniques that will be considered first, it is sufficient to employ representative average values of the wind speed, fetch, and duration. Often the selection of a fetch is simplified by the existence of land boundaries that define the limits of the fetch. And the fetch may be sufficiently short so that most wave conditions of interest will be fetch limited.

Otherwise, it will be difficult to select average U, F and t_d values that are representative of the more complex reality.

The preferred source of wind data is actual wind speed/direction records for the site. These would be measurements made over the water for a sufficient length of time to do a frequency or return period analysis that can then be extrapolated to the desired return period for prediction of design waves. Typical sources for local wind data such as airports, Coast Guard stations, and meteorological institutes are on land and may be somewhat removed from the coast. Some offshore data such as the U. S. Naval Weather Service Command Summary of Synoptic Meteorological Observations (SSMO) or observations at light ships may be useful. These data need to be analyzed to develop wind speed versus return period or probability of exceedence plots for the compass directions that approach the site. Often, such plots or wind roses giving the percent occurrence for selected speed ranges and directions have already been developed as a result of a previous Corps of Engineers or other design efforts in the general vicinity.

A second source of wind data is wind record compilations presented as wind speed contours (isotachs) on a map for a given return period (e.g., see American National Standards Institute, 1972 and Thom, 1960). For example, Thom (1960) presents 2-, 50-, and 100-yr return period isotachs for the continental United States. For a selected location, these three wind speed–return period values can be plotted and the wind speed for the desired return period can be determined. However, these data do not consider wind direction. The return period for a given wind speed and compass direction would be significantly longer than for that speed irrespective of direction. If a wind rose is available for the location of interest, an approximate correction for direction can be made by assuming that the wind speeds at any return period are distributed according to the distribution of the higher speeds in the wind rose (e.g., see U. S. Army Coastal Engineering Research Center, 1984, Chapter 8).

A third source of wind data is to make predictions from weather maps that show upper level pressure contours. First, the upper atmosphere geostrophic wind speed pattern is determined from the pressure gradients. Then the surface wind speed and direction are determined from the geostrophic wind pattern. A general discussion of this is presented in the *Shore Protection Manual* (U. S. Army Coastal Engineering Research Center, 1984). However, reliable predictions could only be made by experienced forecasters and development of sufficient information to develop a wind return periods analysis would be an overwhelming task. This approach would not typically be used for the simpler empirical wave prediction techniques, but is used for the more sophisticated large-scale numerical wave prediction models.

The resulting individual wind speed values may require one or more adjustments before being employed to make wave predictions. Field measurements of wind speed may not have been made at the standard elevation of 10 m above ground level. If not, the wind speed values should be corrected to the 10-m elevation. A power law relationship is adequate for measurement elevations not greater than about 20 m (U. S. Army Coastal Engineering Research Center, 1984). This would be given by

$$U_{10} = \left(\frac{10}{z}\right)^{1/7} U_z \tag{6.42}$$

where U_z is the wind speed measured at elevation z, and U_{10} is the desired wind speed at the reference 10-m elevation. For example, from Eq. 6.42 $U_{19.5}/U_{10} = 1.10$.

Wind velocities are usually quite irregular with time. Reported speeds may be values averaged over a relatively short period of 1 min or the fastest mile (the time it takes the air to travel a mile), or they may be for longer 10- or 15-min averages. As the length of time over which the wind speed is averaged increases, the average value decreases. The wind speed value used for wave prediction should be the value averaged over the time required for waves to travel the fetch (F/C_g, see Section 6.1.3). Adjustments can be made using the following equations (U. S. Army Coastal Engineering Research Center, 1984):

$$\frac{U_t}{U_{3600}} = 1.277 + 0.296 \tanh\left[0.9 \log \frac{45}{t}\right] \quad \text{for } 1s < t < 3600s \tag{6.43}$$

$$\frac{U_t}{U_{3600}} = 1.533 - 0.15 \log t \quad \text{for } 3600s < t < 36{,}000s \tag{6.44}$$

where t is the averaging time in seconds (U_{3600} is the 1-h average wind speed). For example, a 1-min average wind speed of 30 m/s would reduce to 24.4 m/s when averaged over 30 min and 23.1 m/s when averaged over 2 h.

If wind measurements from a land station are being used, the related wind speeds over water will be different because the atmospheric boundary layer over land does not immediately adjust to the frictional characteristics of the water surface. So the required adjustment for this boundary layer change will depend on both the wind speed and the fetch length. The boundary layer stability over water is dependent on the air–water temperature difference and this can impact the wave-generating capability of the wind. A discussion and some empirical guidance on the required land–water and temperature difference adjustments are given in Resio and Vincent (1977).

When wave predictions are carried out for lakes and reservoirs in mountainous terrain, strong funneling effects can occur to cause unusually high wind speeds over water. Since wind directions can also be quite variable over relatively short distances, care must be taken in making wave predictions under these conditions.

6.5.2 Early Wave Prediction Methods

During the 19th and early 20th centuries simple empirical wave prediction formulas were developed from rough observations of wave height versus wind speed and fetch (e.g., Stevensen, 1886 and Molitor, 1934). During the Second World War Sverdrup and Munk (1947), employing wave energy growth concepts, developed a wave prediction theory that was approximately verified by the small amount of data that existed at that time. This was revised a few times by Bretschneider (1952a, 1958) based on additional sets of wind and wave data; so that this wave prediction method is now referred to the Sverdrup–Munk–Bretschneider (SMB) method.

The SMB method can most easily be presented by considering a dimensional analysis of the basic deep water wave generation relationship

$$H_s, T_s = fct(U, F, t_d, g)$$

which yields

$$\frac{gH_s}{U^2} \quad \text{and} \quad \frac{gT_s}{2\pi U}, = f\left(\frac{gF}{U^2}, \frac{gt_d}{U}\right) \tag{6.45}$$

The 2π is included because deep water wave celerity $C = gT/2\pi$ and the ratio C/U, known as the wave age, is an important parameter in defining wave growth. Equation 6.45 is the dimensionless form of the basic wave growth description from Section 6.1.3. It relates the dimensionless significant wave height and period to the dimensionless wind fetch and duration.

The relationship defined by Eq. 6.45 was presented in the form of graphs (Sverdrup and Munk, 1947; Bretschneider, 1958) employing the dimensionless parameters. It has also been presented in the form of empirical equations and dimensional plots in the *Shore Protection Manual* (see U. S. Army Coastal Engineering Research Center, 1977). Figure 6.15 is the graph of Eq. 6.45. (It should be noted that these curves are based on a lot of field data, but that the data show a large amount of scatter, as would be expected when average wind speed and duration values are used to represent the more complex reality.) Given the dimensionless fetch, the dimensionless significant height and period (solid lines) can be determined. The same can be done for the dimensionless duration (dashed lines). The smaller value of the two would be the resulting wave height and period. For example, consider an adjusted wind speed of 30 m/s blowing over a deep lake with a fetch of 20 km for a duration of 2 h. The wind speed and fetch yield $H_s = 3.1$ m and $T_s = 6.6$ sec, while the wind speed and duration yield $H_s = 3.9$ m and $T_s = 7.4$ sec. So wave generation is fetch limited and the lower values would control.

In the example just cited, if the wind duration were less than 1.5 h (the duration required to give $H_s = 3.1$ m and $T_s = 6.6$ sec), wave generation would have been duration limited. For a shorter duration, the average wind speed would have likely been somewhat higher. This would yield higher wave height and period values. The point is, if enough is known about the temporal variation of the wind speed, more than one set of wind speed and duration values should be evaluated to see which yields the highest wave height and period values.

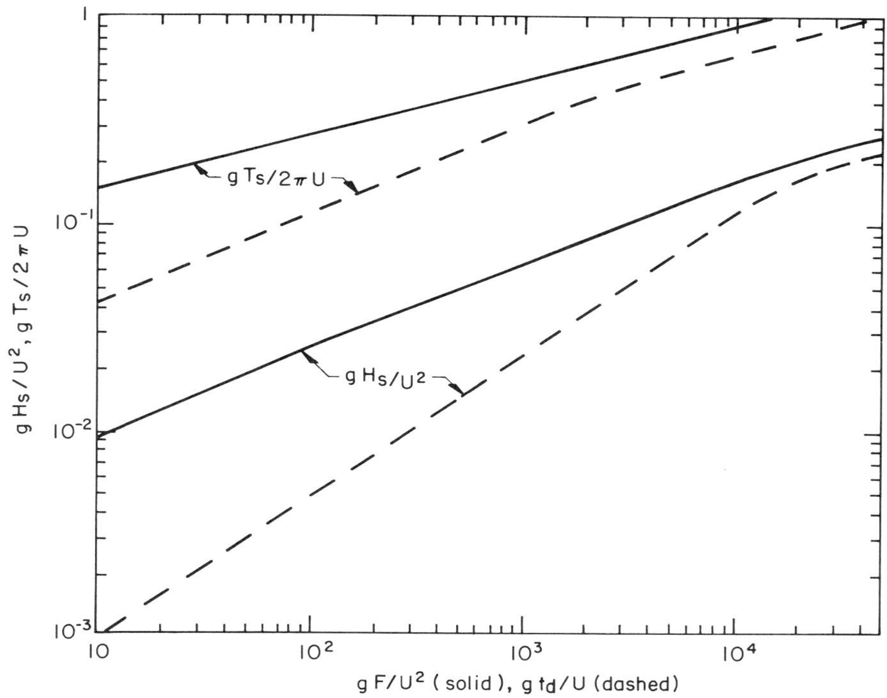

Figure 6.15 SMB wave prediction curves.

6.5.3 Wave Prediction Using Spectra Models

As previously noted, some of the one-dimensional deep water wave spectra models (Section 6.3.3) can be used for wave prediction. The Bretschneider spectrum (Eq. 6.21) is written in terms of the average wave height and period. These could be determined from the SMB wave prediction method, which yields H_s and T_s. Then, from the Rayleigh distribution, $H_{100}/H_s = 0.63$ and the average period is approximately equal to the significant period.

The Pierson–Moskowitz spectrum is only for a fully developed sea. The spectrum can be calculated from Eq. 6.22, given the wind speed at 19.5 m elevation. This speed can be determined from Eq. 6.42, given the speed at 10 m elevation. And H_{m0} and f_p can be determined directly from Eqs. 6.23 and 6.24. But, if conditions are at all fetch or duration limited, quite incorrect results can be obtained. This can be seen, for example, by employing Eqs. 6.23 and 6.24 to determine H_{m0} and f_p for the fetch limited example presented in the previous section and comparing these values to those obtained by the SMB method.

The JONSWAP spectrum was developed for fetch limited conditions. Equation 6.29 yields f_p directly and then the spectrum can be plotted from Eqs. 6.26–6.28. Figure 6.16 is a plot of the JONSWAP spectrum for the example where $U = 30$ m/s, $F = 20$ km, and γ is taken as 3.3. The spectral peak

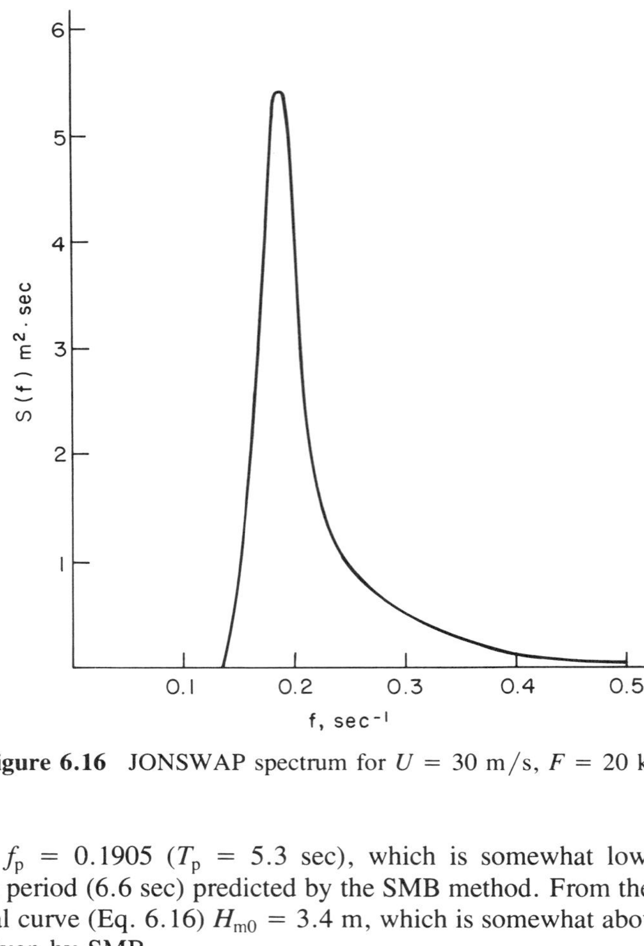

Figure 6.16 JONSWAP spectrum for U = 30 m/s, F = 20 km.

frequency $f_p = 0.1905$ ($T_p = 5.3$ sec), which is somewhat lower than the significant period (6.6 sec) predicted by the SMB method. From the area under the spectral curve (Eq. 6.16) $H_{m0} = 3.4$ m, which is somewhat above the value (3.1 m) given by SMB.

The 1984 edition of the *Shore Protection Manual* (U. S. Army Coastal Engineering Research Center, 1984) recommends that deep water wave prediction be done using a parametric model based on JONSWAP (Hasselmann et al., 1976) rather than the SMB method recommended in previous versions on the manual. The procedure is applicable to fetch or duration limited conditions and is presented by sets of both dimensional and dimensionless plots as well as by the following equations.

In these equations, U_A is a wind stress factor (called the adjusted speed)

$$\frac{gH_{m0}}{U_A^2} = 0.0016 \left(\frac{gF}{U_A^2}\right)^{1/2} \tag{6.46}$$

$$\frac{gT_p}{U_A} = 0.286 \left(\frac{gF}{U_A^2}\right)^{1/3} \tag{6.47}$$

$$\frac{gt_d}{U_A} = 68.8 \left(\frac{gF}{U_A^2}\right)^{2/3} \tag{6.48}$$

given by

$$U_A = 0.71\ U_{10}^{1.23} \tag{6.49}$$

where U_A and U_{10} are given in meters per second. They also give the relationship $T_s = 0.95T_p$ for conversion to the significant period.

To apply this procedure, Eqs. 6.46 and 6.47 would be used to calculate H_{m0} and T_p respectively. This would employ only the wind speed and fetch, and be the appropriate values for the fetch limited condition. Then the limiting duration should be calculated from Eq. 6.48. If the actual duration is less than the limiting duration, wave generation is duration limited, a new fetch is calculated from Eq. 6.48 using the actual duration, and new values of H_{m0} and T_p are calculated from Eqs. 6.46 and 6.47.

Equations 6.46–6.48 are only valid up to the fully developed sea condition given by

$$\frac{gH_{m0}}{U_A^2} = 0.243 \tag{6.50}$$

$$\frac{gT_p}{U_A} = 8.13 \tag{6.51}$$

$$\frac{gt_d}{U_A} = 7.15 \times 10^4 \tag{6.52}$$

Thus, the final values should be checked against Eqs. 6.50–6.52 to see that the fully developed conditions are not exceeded. It is most likely that they are not, but if they are, the fully developed conditions would control.

Weesakul and Charulakana (1990) recently published a comparison of the two *Shore Protection Manual* wave prediction procedures (SMB and SPM, 1984) with measured wave and wind conditions at an offshore platform in the Gulf of Thailand. As typically happens, comparisons of the measured and predicted values of H_s and T_s showed extensive scatter. The two methods were of generally comparable accuracy except for significant heights less than 1 m when the SPM (1984) method had a lower root mean square error. Considering the accuracy with which wind speed, duration, and fetch will typically be known, it is likely in most instances that either method is as good as the other.

6.5.4 Limited Fetch Width

While employing the SMB method to predict waves in inland reservoirs for comparison to measured wave conditions, Saville (1954) observed that when the width of the fetch was small compared to its length, predicted waves were much higher than those measured. For a fetch length he had used the greatest straight distance over open water in the direction of the wind. Consequently, to eliminate the discrepancy between measured and predicted values, he developed an "effective fetch" (Saville et al., 1962) for use with the SMB method for wave prediction at narrow and irregular fetches, as found in reservoirs, lakes, and coastal embayments.

The effective fetch concept is based on two primary assumptions: (1) waves are generated over a range of 45° to either side of the wind direction and the energy transfer from the wind to the waves is proportional to the cosine of the angle between the wind and waves and (2) wave growth is proportional to fetch length. To calculate the effective fetch (F_{eff}), 15 radials are drawn out at 6° intervals over 45° to either side of the wind direction and the length of each of the 15 fetches (F_i) is measured. Then,

$$F_{eff} = \frac{\Sigma F_i \cos^2\theta_i}{\Sigma \cos \theta_i} \tag{6.53}$$

where θ_i is the angle between the wind direction and F_i.

The effective fetch is easy to determine and has been extensively used. It is argued that the effective fetch concept works for the SMB method because this method overpredicts wave heights for small fetches as found in reservoirs and bays (U. S. Army Coastal Engineering Research Center, 1984). With the advent of wave prediction models based on the JONSWAP formulae (Eqs. 6.46–6.48), other fetch definitions are required for restricted fetches.

Seymour (1977) proposed a more complicated method to calculate the fetch length for a restricted fetch. It is based on two assumptions: (1) wave energy is distributed according to a $\cos^2 \theta$ function over a 180° arc and (2) the weighted average of individual fetch direction components is based on the energy along each direction as given by the JONSWAP formulae. The Seymour approach is better grounded in wave generation mechanics than the effective fetch approach, but it is very tedious to apply. Seymour (1977) compared the two methods with measured wave data at four bay sites. Generally, when the fetch width and length were approximately equal, the two approaches gave similar results but, when the fetch widths were appreciably smaller than the fetch lengths, Seymour's method produced wave predictions that better agreed with measured data.

The *Shore Protection Manual* (U. S. Army Coastal Engineering Research Center, 1984) recommends a simpler fetch determination procedure that is also based on a narrower spread of energy in the wave spectrum. This simple fetch is found by taking the average of fetch lengths measured at 3° intervals over a range of 12° to either side of the wind direction. It appears that this approach should be used for easy analyses employing the JONSWAP formulae, the effective fetch should be used with the SMB method, and the Seymour approach might be used with the JONSWAP formulae if the excess effort is justified.

Based on data collected on the Great Lakes, Donelan (1980) developed wave prediction formulae that have a general form similar to the JONSWAP formulae but include the angle θ between the direction of the wind and the resulting waves. These are

$$H_s = 0.00366 g^{-0.62} F^{0.38} (U \cos \theta)^{1.24} \tag{6.54}$$

$$T_p = 0.54 g^{-0.77} F^{0.23} (U \cos \theta)^{0.54} \tag{6.55}$$

$$t_d = 30.1 g^{-0.23} F^{0.77} (U \cos \theta)^{-0.54} \tag{6.56}$$

These equations emphasize the fact that for irregular water bodies, a longer fetch that is situated at an angle to the wind may produce higher waves than the shorter fetch oriented in the direction of the wind. Consequently, the dominant wave energy being generated (as manifest by the frequencies near the spectral peak) would travel at "off-wind" angles. For wave prediction, Donelan maximizes the product $(\cos \theta)^{0.54} F^{0.23}$ to establish the dominant wind/wave direction. Smith (1991) has developed a numerical model, based on an improved formulation of the Donelan equations, for deep water wave prediction on restricted fetches.

6.5.5 Rapidly Moving Storms

When a storm over open water has a wind field that is rapidly varying in time and space, it becomes difficult to employ prediction techniques based on an approximate constant wind speed, fetch, and duration. Wilson developed a procedure for wave prediction that is applicable to a fixed wind direction over which the wind speed can vary both in space and time. He presents both a graphical procedure (Wilson, 1955) and a computer-applicable stepwise numerical integration procedure (Wilson, 1963) for applying his method.

Wilson's method is based on an empirical formulation of Eq. 6.45 written as

$$\frac{C_s}{U} = f_1 \left(\frac{gF}{U^2} \right) \tag{6.57}$$

and

$$\frac{gH_s}{U^2} = f_2 \left(\frac{gF}{U^2} \right) \tag{6.58}$$

where C_s is the celerity of the significant waves and the functions f_1 and f_2 differ somewhat from the SMB relationships. (The SMB or the JONSWAP formulae could be employed with the Wilson approach.) The wave celerity and height in Eqs. 6.57 and 6.58 are differentiated with respect to distance along the fetch to yield

$$\frac{dC_s}{dx} = f_1'\left(\frac{C_s}{U}\right) \tag{6.59}$$

and

$$\frac{dH_s}{dx} = f_2'\left(\frac{gH_s}{U^2}\right) \tag{6.60}$$

where the wind speed is now a function of distance along the fetch. Then, Eqs. 6.59 and 6.60 are integrated numerically as wave energy travels along the fetch at the deep water group celerity ($C_s/2$). This yields the distribution of significant height and period as a function of time at the down wind point of interest. Examples of the application of this approach are given in Patterson (1972), Bea (1974), and Ward et al. (1977).

In the earlier paper Wilson shows how Eqs. 6.57 and 6.58 can be used to construct sets of wave prediction curves relating wave height and period to distance along the fetch and time elapsed for a set of wind speed values. These are then used graphically to predict the wave height and period generated by a moving wind field in which the wind varies with time and position. The required input for a given storm would be a plot of the wind velocity variation along a line oriented toward the site of interest versus time. Sample applications of his method are also shown in Ippen (1966) and Horikawa (1978).

6.5.6 Hurricane Wave Prediction

The controlling wind field for many coastal and offshore design situations is that produced by a hurricane. This consists of a forward moving, inward spiraling wind pattern with velocities that increase to a maximum near the center and then rapidly decay at the center. A discussion of the radial pressure and velocity field in a hurricane is given in the *Shore Protection Manual* (U. S. Army Coastal Engineering Research Center, 1984). The key parameters in defining the hurricane wind field are the atmospheric pressure at the center, the radius to maximum wind velocity, and the forward speed of the hurricane.

Good sources of wind field information for design are the Standard Project Hurricane and the Probable Maximum Hurricane established by the National Oceanic and Atmospheric Administration (U. S. Department of Commerce, 1979). These are hypothetical hurricanes developed for the U. S. Gulf and Atlantic coasts based on a return period analysis of key hurricane parameters measured in the field. Figure 6.17 shows a Probable Maximum Hurricane wind field for the Atlantic Ocean off South Carolina. Wind speeds in excess of 40 mph extend over a distance of about 300 mi. The forward speed of advance of the hurricane would be added to the sea level velocities shown in the figure.

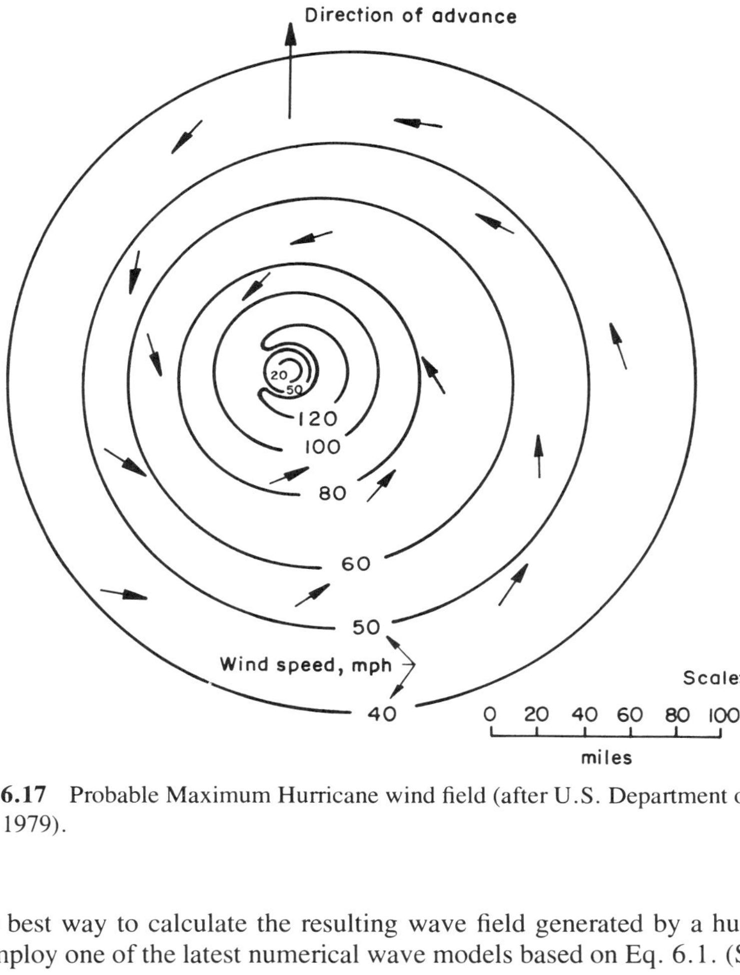

Figure 6.17 Probable Maximum Hurricane wind field (after U.S. Department of Commerce, 1979).

The best way to calculate the resulting wave field generated by a hurricane is to employ one of the latest numerical wave models based on Eq. 6.1. (Section 6.5.9). However, for less effort one can employ the graphical or numerical integration procedures developed by Wilson. One would just work with the component of the wind velocity that acts along a line through the hurricane extending forward to the site of interest. The forward speed of the hurricane would define the time variation of the wind field.

A rough but easy way to obtain a low-order estimate of the peak significant wave height and period generated by a hurricane is to employ the simple equations developed by Bretschneider (1957). Based on an analysis of 13 hurricanes in the Atlantic Ocean off the U. S. east coast, the following equations were proposed:

$$H_s = 16.5e^{0.01R\Delta P}\left[1 + \frac{208\alpha V_F}{\sqrt{U_R}}\right] \tag{6.61}$$

$$T_s = 8.6e^{0.005R\Delta P}\left[1 + \frac{104\alpha V_f}{\sqrt{U_R}}\right] \tag{6.62}$$

where R = distance from the center out to the point of maximum wind velocity (nautical miles), ΔP = pressure difference from the center to the periphery of the hurricane (inches of mercury), V_F = forward speed of the hurricane (knots), U_R = maximum wind speed at R (knots), and α is a correction factor based on the hurricane speed which may be taken as unity for a slow-moving hurricane. The calculated significant height is in feet and the significant period is in seconds. The calculated peak significant wave height develops in the vicinity of the point of maximum wind velocity (see Fig. 6.17). If the hurricane would deviate from its forward path on a straight line at a constant speed, these peak waves, of course, would move away from the point of highest winds. The *Shore Protection Manual* (U. S. Army Coastal Engineering Research Center, 1984) provides a diagram that shows how the significant height varies throughout the hurricane. A simple related equation is given to estimate the significant period distribution.

Recently, Young (1988) has developed a different parametric model for calculating the significant height distribution in a hurricane. An equivalent hurricane fetch, that is a function of the hurricane forward velocity and the maximum wind velocity, was first developed. This was then used in the fetch-limited JONSWAP formulae (Eqs. 6.46 and 6.47) to calculate the significant height and period. This parametric model was empirically developed from the results of hurricane characteristics calculations for 43 hurricanes made using a numerical model developed by the author. His numerical model, in turn, was validated by comparison with field data from several hurricanes off the northwest coast of Australia.

6.5.7 Wave Prediction in Shallow Water

Typically, when wave generation commences, the average wave period is sufficiently small so that the waves being generated are deep water waves. But for some relatively shallow bays and estuaries the wave lengths quickly increase to where the dominant waves feel bottom and continued wave growth is affected by bottom conditions. These conditions—bottom friction, bottom percolation, and bottom movement as well as finite depth affects on the wave surface profile and water particle kinematics—were discussed in Chapter 3.

Thus, for a water body where wave growth dominantly occurs under intermediate/shallow water conditions (i.e., when $d/gT^2 < 0.08$, see Section 2.4.3), Eq. 6.45 should be modified to

$$\frac{gH_s}{U^2} \text{ and } \frac{gT_s}{2\pi U} = f\left(\frac{gF}{U^2}, \frac{gt_d}{U}, \frac{gd}{U^2}\right) \tag{6.63}$$

where d is the water depth. The primary effect of the bottom is to remove energy from the wave system so, for a given wind speed, fetch, and duration the resulting significant wave height and period should be progressively less for decreasing water depths.

The commonly used method for prediction of depth dependent waves is that first presented by Bretschneider (1954, 1958). In essence, Bretschneider combined the SMB deep water wave forecasting relationship to determine energy input to the waves and the Bretschneider and Reid (1954) relationship for energy dissipation by bottom friction to develop a wave forecasting relationship. These two effects were combined by a numerical method of successive approximation. Results were calibrated by comparison with wind-generated shallow water wave growth data collected by the U. S. Army Corps of Engineers (1955) at Lake Okeechobee, Florida. This resulted in the use of a bottom friction factor equal to 0.01 in the Bretschneider and Reid (1954) relationship for bottom dissipation. Bretschneider (1958) calls this a "calibrated friction factor" since it takes into account dissipation effects other than just bottom friction.

The depth-dependent wave prediction relationships can be stated by the following equations:

$$\frac{gH_{m0}}{U_A^2} = 0.283 \tanh\left[0.53\left(\frac{gd}{U_A^2}\right)^{3/4}\right] \tanh\left[\frac{0.00565\left(\dfrac{gF}{U_A^2}\right)^{1/2}}{\tanh\left[0.53\left(\dfrac{gd}{U_A^2}\right)^{3/8}\right]}\right] \tag{6.64}$$

$$\frac{gT_s}{U_A} = 7.54 \tanh\left[0.833\left(\frac{gd}{U_A^2}\right)^{3/8}\right] \tanh\left[\frac{0.0379\left(\dfrac{gF}{U_A^2}\right)^{1/2}}{\tanh\left[0.833\left(\dfrac{gd}{U_A^2}\right)^{3/8}\right]}\right] \tag{6.65}$$

$$\frac{gt_d}{U_A} = 537\left(\frac{gT_s}{U_A}\right)^{7/3} \tag{6.66}$$

which would be applied in the same way as Eqs. 6.46–6.48 for deep water. Equations 6.64–6.66 are presented in dimensional form as a series of plots for various water depths in the *Shore Protection Manual* (U. S. Army Coastal Engineering Research Center, 1984).

It should be emphasized that the wave prediction relationships presented by Eqs. 6.64–6.66 are based on a limited data set, and should be used with caution. Vincent and Hughes (1985), employing a previously developed method for estimating the upper bound of energy in a depth limited wind wave field, compared their estimated wave heights to those predicted by Eq. 6.64 taken to the shallow water limit. The results showed excellent agreement. However, a comparison of the deep water wave prediction formulae (Eqs. 6.46 and 6.47) with the depth dependent formulae (Eqs. 6.64 and 6.65) by Hurdle and Stive (1989) indicated that these formulae do not adequately match at the transition points. They also questioned the value of Eq. 6.66 for the limiting duration. Modified forms of these equations, based on forcing them to provide a better asymptotic match rather than on "any theory of wave development or decay" were presented.

To get an indication of depth effects on wave generation we can continue the example presented in Section 6.5.3. For deep water, a wind speed of 30 m/sec over a fetch of 20 km yielded a significant wave height of 3.4 m and a significant period of 5.3 sec. If the average depth over this fetch is 5 m, the resulting significant height and period would be 1.8 m and 4.8 secs, respectively.

6.5.8 Numerical Wave Prediction Models

Along with the empirical and spectral model methods of wave prediction, a parallel effort to develop numerical models of the physical processes of wave generation and growth has been underway since the 1950s. These models are all based on a numerical integration of the spectral energy balance equation (Eq. 6.1). Reviews of the development and status of these models are given by Cardone and Ross (1979) and The SWAMP Group (1985).

In physical terms, the left-hand side of Eq. 6.1 is a source term that represents the energy input at each frequency (S_{in}), the energy transfer across the spectrum from the input frequency to other spectral frequencies by nonlinear wave–wave interactions (S_{nl}), and the energy dissipation at each frequency (S_{ds}). For a directional spectrum, each of these components would depend on direction relative to the wind, position along the fetch in the direction of wave propagation, and time. The right-hand side of Eq. 6.1 represents the resulting temporal and convective wave spectral growth. Thus, application of this equation over a spatial grid with respect to time would yield the growing wave directional spectrum as a function of position and time.

The energy input term commonly employs the mechanisms proposed by Phillips (1957, 1958, 1960) and Miles (1957) that were briefly discussed in Section 6.1.1. The resulting wave growth defined by these mechanisms has been calibrated by field and wind wave tank data to yield reasonably reliable input terms for model application. But S_{in} does not account for the total wave growth that occurs. The input mechanism produces a sharp spectral peak which the wave–wave interaction process modifies particularly by a transfer of wave energy to the growing steep forward face of the frequency spectrum (longer wave periods). Energy transfer by wave–wave interaction accounts for the "overshoot" of the eventual equilibrium value of the spectrum (see Fig. 6.4).

Exact computations employing the complex three-dimensional nonlinear equations that define the wave–wave interaction are too difficult for inclusion in numerical models so empirical models of this process are employed. The energy dissipation term includes wave breaking by the higher-frequency components of the spectrum and, for shallow water, would include bottom effects. Other effects must also come into play, but general knowledge of processes involved is somewhat qualitative, so these effects are also empirically included in the numerical models.

Figure 6.18 is a typical plot of the one-dimensional energy balance at a point in time and space as the wave field is growing. The net input at any frequency in the spectrum would be given by $S_{in} + S_{nl} + S_{ds}$. The example shown in the figure would produce a net growth and stronger than average growth in the lower frequency portion of the spectrum (as depicted in Fig. 6.4).

A large number and variety of numerical wave prediction models are available. (See The SWAMP Group, 1985 for a brief discussion of 10 of these models.) Commonly, they solve Eq. 6.1 in finite difference form on a space grid over which the variable wind field and growing wave field move. Discrete spectral direction and frequency interval components are used and the model is integrated in a space and time-stepping fashion. The input information might be the upper elevation isobaric pattern or the resulting geostrophic wind pattern from which the surface wind field is determined. Or the surface wind field or wind friction velocity pattern might be specified. Different spectral forms (e.g., PM or JONSWAP) with different directional spreading functions are used to define the growing wave field. Different relationships might be used for S_{in},

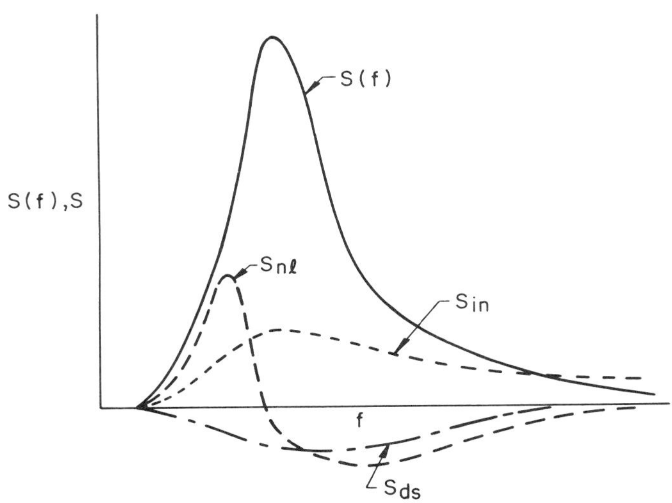

Figure 6.18 One-dimensional energy balance for a growing wave field.

S_{nl}, and S_{ds}. And wave refraction and bottom dissipation effects may be included.

Output from the numerical model could be the resulting directional spectrum at any grid point and time, or more practically a plot of H_{m0}, T_p and the dominant wave direction at selected points at the downwind end of the fetch as a function of time.

Many organizations have these models in operation for important lake and ocean areas. They can then be used to develop long-term wave statistics for a coastal region from historic weather data (e.g., Jensen, 1983) or to predict the wave field at a given coastal site for a particular storm or collection of storms (e.g., Mynett et al., 1983). This is the area where wave prediction improvements are mostly being focused. But continued improvements will heavily depend on the collection of additional improved data sets to produce a better understanding of the various components of Eq. 6.1.

6.6 SWELL DECAY AND TRANSFORMATION

6.6.1 Decay of Swell in Deep Water

As waves propagate away from the region where active wave generation is taking place, significant changes will occur. The area where waves change from sea to swell will naturally be quite complex. The downwind edge of the fetch will only be roughly delineated by a gradual change in wind speed and/or direction. But there will be a point where wave growth will change to wave decay.

As swell propagate out from the generating area, two factors—dispersion and angular spreading—will have the most profound effect on the wave characteristics. Waves are period and, to a much lesser extent, height dispersive in deep water. So the longer period waves in the spectrum (which are generally of lower amplitude) will move ahead of the main body of higher waves having periods near the spectral peak. The shortest period waves will lag behind. The resulting spread of different period components of the spectrum will increase with travel distance. Packets of deep water waves travel at the group celerity ($C_g = C/2 = gT/4\pi$). So, for example, the 5- and 10-sec period components of a wave spectrum would separate by 35.6 h in arrival time after traveling a distance of 1000 km across the ocean. This travel would be along a great circle route. The approach of waves from a distant storm is often indicated by the low longer period "forerunners" of the storm that precede the steeper dominant components of the spectrum.

Also, at the end of a fetch, the wave spectrum has a frequency-dependent spread of wave directions. So, as the swell propagate they spread laterally, causing a continuing reduction of energy density with distance from the storm for each of the spectral components. This is the dominant cause of the decrease in wave height as swell travel across the ocean.

Other factors will also cause a reduction in wave height as swell propagate. These include adverse winds or even wave action against relatively still air (Section 3.3.1), and internal friction and possible internal dissipation by turbulence that exists in the water through which the waves propagate. These effects cause the shorter steeper waves to attenuate the most. Consequently, the shortest waves in a storm spectrum, which arrive latest at some distant point, are usually well attenuated and frequently masked by larger waves from some later or closer source. And, finally, ocean currents may cause wave refraction, which affects the swell height and direction as discussed in Section 5.1.4.

Little quality data exists to develop methods for predicting the changes that occur as swell propagate. The need to develop wave design guidance dictates a stronger interest in the characteristics and behavior of waves being actively generated. Although, swell from some distant source superimposed on a local storm may be a significant wave event at a particular site. Bretschneider (1952b) presented some empirical data on the significant height and period of swell at the end of a decay distance D (realizing that these are hard to define owing to wave dispersion). He plotted the following relationships:

$$\frac{D}{H_D} = f\left(\frac{D}{H_F}, \frac{D}{F}\right) \tag{6.67}$$

$$\frac{D}{gT_D^2} = f\left(\frac{D}{gT_F^2}, \frac{D}{F}\right) \tag{6.68}$$

where the subscripts F and D refer to conditions at the end of the active fetch and the decay distance respectively. The data used to plot these relationships were quite limited. Bretschneider (1958) stated that Eqs. 6.67 and 6.68 should have included the fetch width and the wind speed, but the limited data and its scatter precluded this.

The general trend is for the significant height to decrease and the significant period to increase with decay distance (as expected from the factors discussed above). Bretschneider's data showed, for example, that for a shorter fetch and higher wind speed versus a longer fetch and lower wind speed that will produce the same significant height at the end of the fetch, the former will suffer greater height decay at a given decay distance than the latter. This is because the former fetch/speed combination will generate steeper significant waves (i.e., shorter significant periods) which will, in turn, decay faster.

6.6.2 Swell Propagating Across Intermediate/Shallow Water

The common design practice for analyzing the change in wave characteristics as waves propagate from deep to shallow water is to select a representative wave height, period, and direction (e.g., H_{m0} and T_p having the dominant wave direction) and treat this as a monochromatic wave that shoals, refracts, and diffracts as discussed in Chapters 3–5. But shoaling effects are dependent on the wave period, and refraction and diffraction depend on both the incident period and direction. So, the true changes that occur as a directional wave spectrum propagates from deep to shallow water will be very dependent on the frequency and direction distribution in the spectrum. Each frequency/direction component of the spectrum will shoal, refract, and diffract differently.

For a directional wave spectrum the effective shoaling/refraction coefficient is

$$(K_r)_s = \left[\frac{1}{(m_0)_s} \sum_{0}^{\infty} \sum_{-\pi}^{\pi} S(f, \theta) K_s^2 K_r^2 \Delta f \Delta \theta \right]^{1/2} \tag{6.69}$$

where

$$(m_0)_s = \sum_{0}^{\infty} \sum_{-\pi}^{\pi} S(f, \theta) K_s^2 \Delta f \Delta \theta$$

In Eq. 6.69 K_s is the shoaling coefficient for each frequency component Δf and K_r is the refraction coefficient for each component $\Delta f \Delta \theta$. Since wave shoaling is frequency dependent, it affects the spectral form and needs to be included for a complete representation of refractive effects. The directional spectrum would be broken into frequency and direction components, the shoaling and refraction analysis would be carried out for each component, and the results would be recombined to determine $(K_r)_s$.

Application of Eq. 6.69 requires an elaborate effort. To somewhat simplify the effort a simpler $\cos^2\theta$ distribution (Eq. 6.37) which is not frequency dependent can be employed with only a few frequency segments from the spectrum. This may thus employ say six directional components and four frequency components for a total of 24 refraction shoaling analyses that are recombined to produced the desired analysis.

To demonstrate the effect of a directional spectrum on wave refraction, Goda (1985) gives the results of an analysis for a coastline with straight shore-parallel contours so the refraction coefficient for each period and offshore direction component can easily be determined analytically (Eq. 5.11). He employed a modified Bretschneider spectrum with a directional distribution given by Eqs. 6.38–6.40 and $S_{max} = 10$, 25, and 75. A range of dominant offshore directions was considered. Some results for the common range of offshore approach angles are shown in Table 6.2.

For the conditions shown in the table (but not for all conditions), the monochromatic refraction coefficient is higher than the spectral refraction coefficient. As would be expected, the difference is less as S_{max} increases.

Refraction causes wave crests to reorient so that the angle between the crest and bottom contours decreases as the water depth decreases. The amount of reorientation is greater for larger deep water wave crest approach angles. The

TABLE 6.2 Refraction Coefficients at d/L_0 = 0.05 for Monochromatic and Spectral Waves Having Different Offshore Approach Directions

	Offshore Direction		
S_{max}	0°	20°	40°
10	0.94	0.93	0.87
25	0.97	0.955	0.88
75	0.99	0.97	0.90
Monochromatic	1.00	0.98	0.91

consequence is that the refraction of a directional wave spectrum approaching the shore will result in a narrowing of the directional spread of energy in the spectrum.

In a similar fashion to Eq. 6.69 the effective diffraction coefficient for a directional spectrum would be

$$(K_d)_s = \left[\frac{1}{(m_0)_s} \sum_{0}^{\infty} \sum_{-\pi}^{\pi} S(f, \theta)\, K_d^2\, \Delta f \Delta\theta \right]^{1/2} \tag{6.70}$$

where

$$(m_0)_s = \sum_{0}^{\infty} \sum_{-\pi}^{\pi} S(f, \theta)\, \Delta f \Delta\theta$$

and K_d is the diffraction coefficient for each frequency and direction component.

Goda (1985) also constructed diffraction diagrams for a semiinfinite breakwater and a breakwater gap (like Figs. 5.10 and 5.12 but for directional wave spectra) employing the spectrum and directional spreading conditions that he used for the refraction analysis. The wave spectra approached normal to the breakwater axis and S_{max} values of 10 and 75 were used. The effective diffraction coefficient was determined using Eq. 6.70.

At a given point in the lee of the breakwater there was a shift in the peak spectral frequency as well as different diffraction coefficients than those that developed for monochromatic waves. A shift in peak spectral frequency should be expected because at a given point in the lee of the breakwater, the value of r/L (and thus K_d) would be different for different frequencies in the spectrum. In many instances, the discrepancy between the spectral and monochromatic wave analyses for diffraction coefficient was quite large with the monochromatic analysis often giving a significant underestimation of the resulting wave height. Comparisons with some available field data on refracted wave conditions behind a single breakwater at a coastal port indicated that the spectral approach yielded much better results.

REFERENCES

American National Standards Institute (1972), ''American National Standard Building Code Requirements for Minimum Design Loads in Buildings and Other Structures,'' Publication A58.1, New York.

Battjes, J. A. (1972), ''Set-up due to Irregular Waves,'' *Proceedings, 13th Conference on Coastal Engineering*, American Society of Civil Engineers, Vancouver, pp. 1993–2004.

Bea, R. G. (1974), ''Gulf of Mexico Hurricane Wave Heights,'' *Proceedings, Offshore Technology Conference*, Houston, Paper 2110.

Bouws, E., Gunther, H., Rosenthal, W., and Vincent, C. L. (1985), ''Similarity of the Wind Wave Spectrum in Finite Depth Water, Part I—Spectral Form,'' *J. Geophys. Res.*, **90,** 975-986.

Bretschneider, C. L. (1952a), ''Revised Wave Forecasting Relationships,'' *Proceedings, 2nd Conference on Coastal Engineering*, Council on Wave Research, University of California, Berkeley, pp. 1-5.

Bretschneider, C. L. (1952b), ''The Generation and Decay of Wind Waves in Deep Water,'' *Trans. Am. Geophys. Union,* **33,** 381-389.

Bretschneider, C. L. (1954), ''Generation of Wind Waves Over a Shallow Bottom,'' Technical Memorandum 51, U. S. Army Beach Erosion Board, Washington, DC.

Bretschneider, C. L. (1957), ''Hurricane Design Wave Practices,'' *J. Waterw. Harbors Div., Am. Soc. Civ. Eng.*, May, 1-33.

Bretschneider, C. L. (1958), ''Revisions in Wave Forecasting: Deep and Shallow Water,'' *Proceedings, Sixth Conference on Coastal Engineering*, Council on Wave Research, University of California, Berkeley, pp. 1-18.

Bretschneider, C. L. (1959), ''Wave Variability and Wave Spectra for Wind-Generated Gravity Waves,'' Technical Memorandum 118, U. S. Army Beach Erosion Board, Washington, DC.

Bretschneider, C. L. and Reid, R. O. (1954), ''Modification of Waves Height Due to Bottom Friction, Percolation, and Refraction,'' Technical Memorandum 45, U. S. Army Beach Erosion Board, Washington, DC.

Burcharth, H. F. (1980), ''A Comparison of Nature Waves and Model Waves With Special Reference to Wave Grouping,'' *Proceedings, 17th International Conference on Coastal Engineering*, American Society of Civil Engineers, Sydney, pp. 2992–3009.

Cardone, V. J. and Ross, D. B. (1979), ''State-of-the-Art Wave Prediction Methods and Data Requirements,'' *Ocean Wave Climate*, M.D. Earle and A. Malahoff, Ed., Plenum, New York, pp. 61–91.

Cartwright, D. E. and Longuet-Higgins, M. S. (1956), ''The Statistical Distribution of the Maxima of a Random Function,'' *Proc. R. Soc. London*, *Series A*, 212–232.

Cavanie, A., Arhan, M., and Ezraty, R. (1976), ''A Statistical Relationship Between Individual Heights and Periods of Storm Waves,'' *Proceedings, Conference on Behavior of Offshore Structures*, Trondheim, Norway, pp. 354–360.

Chakrabarti, S. K. and Cooley, R. P. (1971), ''Statistical Distribution of Periods and Heights of Ocean Waves,'' *J. Geophys. Res.* 1363–1368.

Collins, J. I. (1967), "Wave Statistics from Hurricane Dora," *J. Waterw. Harbors Div., Am. Soc. Civ. Eng.*, May, 59–77.

Collins, J. I. (1970), "Probabilities of Breaking Wave Characteristics," *Proceedings, 12th Conference on Coastal Engineering*, American Society of Civil Engineers, Washington, DC, pp. 399–412.

Cote, L. J., Davies, J. O., Marks, W., McGough, R. J., Mehr, E., Pierson, W. J., Ropek, J. F., Stephenson, G. and Vetter, R. C. (1960), "The Directional Spectrum of a Wind Generated Sea as Determined from Data Obtained by the Stereo Wave Observation Program," Meteorological Paper, College of Engineering, New York University, Vol. 2, pp. 1–88.

Darbyshire, J. (1952), "The Generation of Waves By Wind," *Proc. R. Soc. London, Series A*, 299–328.

Donelan, M. A. (1980), "Similarity Theory Applied to the Forecasting of Wave Heights, Periods and Directions," *Proceedings, Canadian Coastal Conference*, National Research Council, Canada, pp. 47–61.

Earle, M. D. (1975), "Extreme Wave Conditions During Hurricane Camille," *J. Geophys. Res.*, 377–379.

Forrestal, G. Z. (1978), "On the Statistical Distribution of Wave Heights in a Storm," *J. Geophys. Res.*, 2353–2358.

Goda, Y. (1970), "Numerical Experiments on Wave Statistics with Spectral Simulation," *Port Harbor Res. Inst. Rep.*, **9,** 3–57.

Goda, Y. (1974), "Estimation of Wave Statistics from Spectral Information," *Proceedings, Ocean Waves Measurement and Analysis Conference*, American Society of Civil Engineers, New Orleans, pp. 320–337.

Goda, Y. (1975), "Irregular Wave Deformation in the Surf Zone," *Coastal Eng. Jpn.*, **18,** 13–26.

Goda, Y. (1976), "On Wave Groups," *Proceedings, Behavior of Offshore Structures* 1976 Conference, Trondheim, Vol. 1, pp. 115–128.

Goda, Y. (1985), *Random Seas and the Design of Maritime Structures*, University of Tokyo Press, Tokyo.

Goda, Y. and Suzuki, Y. (1975), "Computation of Refraction and Diffraction of Sea Waves with Mitsuyasu's Directional Spectrum," Technical Note, Port and Harbor Research Institute.

Goodnight, R. C. and Russel, T. L. (1963), "Investigation of the Statistics of Wave Heights," *J. Waterw. Harbors Div. Am. Soc. Civ. Eng.*, May, 29–54.

Hasselmann, D. E., Dunkel, M., and Ewing, J. A. (1980), "Directional Wave Spectra Observed During JONSWAP 1973," *J. Phys. Oceanogr.*, **10,** 1264–1280.

Hasselmann, K., Barnett, T. P., Bouws, E., Carlson, H., Cartwright, D. E., Enke, K., Ewing, J. A., Gienapp, H., Hasselmann, D. E., Kruseman, P., Meerburg, A., Muller, P., Olbers, D. J., Richter, K., Sell, W. and Walden, H. (1973), Measurement of Wind-Wave Growth and Swell Decay During the Joint North Sea Wave Project (JONSWAP), Report, German Hydrographic Institute, Hamburgh.

Hasselmann, K., Ross, D. B., Muller, P., and Sell, W. (1976), "A Parametric Wave Prediction Model," *J. Phys. Oceanogr.*, **6,** 200–228.

Horikawa, K. (1978), *Coastal Engineering: An Introduction to Ocean Engineering*, Wiley, New York.

Hughes, S. A. (1984), "The TMA Shallow-Water Spectrum Description and Applications," Technical Report CERC 84-7, U. S. Army Waterways Experiment Station, Vicksburg, MS.

Hughes, S. A. and Borgman, L. E. (1987), "Beta-Rayleigh Distribution for Shallow Water Wave Heights," *Proceedings, Coastal Hydrodynamics 87*, American Society of Civil Engineers, Newark, DE, pp. 17–31.

Hurdle, D. P. and Stive, R. J. H. (1989), "Revision of SPM 1984 Wave Hindcast Model to Avoid Inconsistencies in Engineering Applications," *Coastal Eng.*, **12,** 339–351.

Ibrageemov, A. M. (1973), "Investigation of the Distribution Functions of Wave Parameters During Their Transformation," *Oceanology*, 584–589.

Ippen, A. T. (1966), *Estuary and Coastline Hydrodynamics*, McGraw-Hill, New York.

Jensen, R. E. (1983), "Atlantic Coast Hindcast, Shallow-Water, Significant Wave Information," WIS Report 9, U. S. Army Waterways Experiment Station, Vicksburg, MS.

Kinsman, B. (1965), *Wind Waves*, Prentice Hall, Englewood Cliffs, NJ.

Kitaigorodskii, S. A., Krasitskii, V. P., and Zaslavskii, M. M. (1975), "On Phillips Theory of Equilibrium Range in the Spectra of Wind-Generated Gravity Waves," *J. Phys. Oceanogr.*, **5,** 410–420.

Kuo, C. T. and Kuo, S. T. (1974), "Effect of Wave Breaking on Statistical Distribution of Wave Heights," *Proceedings, Civil Engineering in the Oceans III*, American Society of Civil Engineers, San Francisco, pp. 1211–1231.

LeBlond, P. H. and Mysak, L. A. (1978), *Waves in the Ocean*, Elsevier, New York.

Lindgren, G. and Rychlik, I. (1982), "Wave Characteristic Distributions for Gaussian Waves—Wavelength, Amplitude and Steepness," *Ocean Eng.*, **9,** 411–432.

Longuet-Higgins, M. S. (1952), "On the Statistical Distribution of the Heights of Sea Waves," *J. Mar. Res.*, **11,** 246–266.

Longuet-Higgins, M. S. (1975), "On the Joint Distribution of the Periods and Amplitudes of Sea Waves," *J. Geophys. Res.*, **80,** 2688–2694.

Longuet-Higgins, M. S. (1983), "On the Joint Distribution of Wave Periods and Amplitudes in a Random Wave Field," *Proc. R. Soc. London, Series A*, 241–258.

Medina, J. R. and Hudspeth, R. T. (1990), "A Review of the Analysis of Wave Groups," *Coastal Eng.*, **14,** 515–542.

Miles, J. W. (1957), "On the Generation of Surface Waves By Shear Flows," *J. Fluid Mech.*, **3,** 185–204. (This is followed by three more papers on this topic in the journal, 1959–1962.)

Miles, J. W. (1960), "On the Generation of Surface Waves by Turbulent Shear Flows," *J. Fluid Mech.*, **7,** 469–478.

Mitsuyasu, H. (1972), "The One-Dimensional Wave Spectrum at Limited Fetch," *Proceedings, 13th International Conference on Coastal Engineering*, American Society of Civil Engineers, Vancouver, pp. 289–298.

Mitsuyasu, H., Tasai, F., Subara, T., Mizuno, S., Ohkusu, M., Honda, T. and Rikiishi, K. (1975), "Observations of the Directional Spectrum of Ocean Waves Using a Clover-Leaf Buoy," *J. Phys. Oceanogr.*, **5,** 750–760.

Mitsuyasu, H., Tasai, F., Subara, T., Mizuno, S., Ohkusu, M., Honda, T. and Rikiishi, K. (1980), "Observations of the Directional Spectrum of Ocean Waves Using a Clover-Leaf Buoy," *J. Phys. Oceanogr.*, **10,** 286–296.

Molitor, D. A. (1934), "Wave Pressures on Sea-Walls and Breakwaters," *Trans. Am. Soc. Civ. Eng.*, May, 984.

Mynett, A. E., de Voogt, W. J. P., and Schmeltz, E. J. (1983), "West Breakwater—Sines Wave Climatology," *Proceedings, Coastal Structures '83 Conference*, American Society of Civil Engineers, Arlington, VA, pp. 17–30.

Neumann, G. (1953), "On Ocean Wave Spectra and a New Method of Forecasting Wind-Generated Sea," Technical Memorandum 43, U. S. Army Beach Erosion Board, Washington, DC.

Ochi, M. K. (1982), "Stochastic Analysis and Probabilistic Prediction of Random Seas," *Adv. Hydrosci.*, **13,** 218–375.

Ochi, M. K. and Hubble, E. N. (1976), "Six-Parameter Wave Spectrum," *Proceedings, 15th International Conference on Coastal Engineering*, American Society of Civil Engineers, Honolulu, pp. 301–328.

Patterson, M. M. (1972), "Hindcasting Hurricane Waves in the Gulf of Mexico," *J. Pet. Eng.*, August, 321–328.

Phillips, O. M. (1957), "On the Generation of Waves By Turbulent Winds," *J. Fluid Mech.*, **2,** 417–445.

Phillips, O. M. (1958), "The Equilibrium Range in the Spectrum of Wind-Generated Ocean Waves," *J. Fluid Mech.*, **4,** 426–434.

Phillips, O. M. (1960), "On the Dynamics of Unsteady Gravity Waves of Finite Amplitude, 1. The Elementary Interactions," *J. Fluid Mech.*, **9,** 193–217.

Pierson, W. J. (1954), "An Interpretation of the Observable Properties of Sea Waves in Terms of the Energy Spectrum of the Gaussian Record," *Trans., Am. Geophys. Union*, **35,** 747–757.

Pierson, W. J. (1964), "The Interpretation of Wave Spectrums in Terms of the Wind Profile Instead of the Wind Measured at a Constant Height," *J. Geophys. Res.*, **69,** 5191–5203.

Pierson, W. J. and Moskowitz, L. (1964), "A Proposed Spectral Form for Fully Developed Wind Seas Based on the Similarity Theory of S. A. Kitaigorodskii," *J. Geophys. Res.*, **69,** 5181–5190.

Resio, D. T. and Vincent, C. L. (1977), "Estimation of Winds over the Great Lakes," *J. Waterw. Port Coastal Ocean Div., Am. Soc. Civ. Eng.*, May, 265–283.

Rye, H. (1977), "The Stability of Some Currently Used Wave Parameters," *Coastal Eng.*, **1,** March, 17–30.

St. Denis, M. and Pierson, W. J. (1953), "On the Motions of Ships in Confused Seas," *Trans. Soc. Nav. Arch. Mar. Eng.*, **61,** 280–357.

Sarpkaya, T. and Isaacson, M. (1981), *Mechanics of Wave Forces on Offshore Structures*, Van Nostrand Reinhold, New York.

Saville, T. (1954), "The Effect of Fetch Width on Wave Generation," Technical Memorandum 70, U. S. Army Beach Erosion Board, Washington, DC.

Saville, T., McClendon, E. W., and Cochran, A. L. (1962), "Freeboard Allowance for Waves on Inland Reservoirs," *J. Waterw. Harbors Div., Am. Soc. Civ. Eng.*, May, 93–124.

Scott, J. R. (1965), "A Sea Spectrum for Model Tests and Long-Term Ship Protection," *J. Ship Res.*, **9,** 145–152.

Seymour, R. J. (1977), "Estimating Wave Generation on Restricted Fetches," *J. Waterw. Port Coastal Ocean Eng. Div., Am. Soc. Civ. Eng.*, May, 251–264.

Silvester, R. (1974), *Coastal Engineering I*, Elsevier, Amsterdam.

Smith, J. M. (1991), ''Wind-Wave Generation on Restricted Fetches,'' Miscellaneous Paper CERC-91-2, U. S. Army Waterways Experiment Station, Vicksburg, MS.

Stevenson, T. (1886), *Design and Construction of Harbors, A Treatise on Maritime Engineering*, Black, Edinburgh.

Sverdrup, H. U. and Munk, W. H. (1947), ''Wind, Sea and Swell: Theory of Relations for Forecasting,'' Publication 601, U. S. Navy Hydrographic Office, Washington, DC.

The SWAMP Group (1985), *Ocean Wave Modeling*, Plenum, New York.

Thom, H. C. S. (1960), Distributions of Extreme Winds in the United States,'' *J. Struct. Div., Am. Soc. Civ. Eng.*, April, 11-24.

Thompson, E. F. and Vincent, C. L. (1985), ''Significant Wave Height for Shallow Water Design,'' *J. Waterw. Port Coastal Ocean Eng. Div., Am. Soc. Civ. Eng.*, Sept., 828-842.

Toba, Y. (1978), ''Stochastic Form of the Growth of Wind Waves in a Single-Parameter Representation with Physical Implications,'' *J. Phys. Oceanogr.*, **8,** 494-507.

Tucker, M. J. (1991), *Waves in Ocean Engineering—Measurement, Analysis, Interpretation*, Ellis Horwood, New York.

U. S. Army Coastal Engineering Research Center (1977), *Shore Protection Manual*, U. S. Government Printing Office, Washington, DC.

U. S. Army Coastal Engineering Research Center (1984), *Shore Protection Manual*, U. S. Government Printing Office, Washington, DC.

U. S. Army Corps of Engineers (1955), ''Waves and Wind Tides in Shallow Lakes and Reservoirs—Summary Report,'' Jacksonville District, Project CW-167.

U. S. Army Corps of Engineers (1989), ''Water Levels and Wave Heights for Coastal Engineering Design,'' Engineer Manual 1110-2-1414, Washington, DC.

U. S. Department of Commerce (1979), ''Meteorological Criteria for Standard Project Hurricane and Probable Maximum Hurricane Windfields, Gulf and East Coasts of the United States,'' NOAA Technical Report NWS 23, Washington DC.

Vincent, C. L. and Hughes, S. A. (1985), ''Wind Wave Growth in Shallow Water,'' *J. Waterw. Port Coastal Ocean Eng. Div., Am. Soc. Civ. Eng.*, July, 765-770.

Ward, E. G., Evans, D. J., and Pompa, J. A. (1977), ''Extreme Wave Heights Along the Atlantic Coast of the United States,'' *Proceedings, Offshore Technology Conference*, Houston, Paper 2846.

Weesakul, S. and Charaulakana, S. (1990), ''Comparison of Wave Hindcast Methods for Lower Gulf of Thailand,'' *Proceedings, 22nd International Conference on Coastal Engineering*, American Society of Civil Engineers, Delft, pp. 986-992.

Wilson, B. W. (1955), ''Graphical Approach to the Forecasting of Waves in Moving Fetches,'' U. S. Army Beach Erosion Board, Technical Memorandum 73, Washington DC, April.

Wilson, B. W. (1963), ''Deep Water Wave Generation by Moving Wind Systems,'' *Trans. Am. Soc. Civ. Eng.*, **128,** 104-132.

Wilson, B. W., Chakrabarti, S. K., and Snider, R. H. (1974), ''Spectrum Analysis of Ocean Wave Records,'' *Proceedings, Ocean Wave Measurement and Analysis Symposium*, American Society of Civil Engineers, New Orleans, pp. 87-106.

Young, I. R. (1988), ''Parametric Hurricane Wave Prediction Model, *J. Waterw. Port Coastal Ocean Eng. Div., Am. Soc. Civ. Eng.*, September, 637-652.

7

DESIGN WAVE DETERMINATION

A primary reason to investigate and analyze ocean waves is to select representative wave conditions for use in coastal and ocean design or in marine operations planning. Quite different types of wave information are required for such diverse concerns as the design of an offshore fixed platform for undersea oil recovery, the development of an alongshore sediment transport budget for a section of the coast, or the design of a submerged rubble mound breakwater. So first we review the wide range of design wave information commonly required by considering a representative selection of design situations.

Next we discuss the different wave information sources that are available to the designer. These wave data either come from wave hindcast or wave measurement programs. The wave measurements may be made by the less expensive but generally less reliable method of visual observation or by the typically preferred methods employing mechanical devices. These various methods of wave measurement are described in detail and compared.

Given the basic wave data available for a site, the next step is the selection of the design wave conditions. This involves extreme wave analysis of the available wave height and period data to extrapolate to the longer time range commensurate with the life of the project being designed. The duration of high waves during a storm is also of concern. Extreme wave analysis is presented in detail. However, some structures are placed in sufficiently shallow water that wave breaking will limit the effective height that the waves can achieve irrespective of offshore conditions. This limit on design wave conditions and the factors that affect the long-term breaker characteristics are discussed. Although the focus of this chapter is on the determination of design wave conditions, it should be remembered that the design water level at which waves act is also important. Some aspects of water level change are discussed in chapter 9.

7.1 DESIGN WAVE INFORMATION REQUIRED

The types of wave information required for the design of several classes of coastal and offshore structures as well as for some other coastal design concerns are discussed below. An aim is to provide a broad understanding of the variety of wave information required.

7.1.1 Rubble Mound Structures

A wide variety of rubble mound structures is built in the coastal zone. This includes (1) revetments and seawalls built along the shore on the beach face or to protect an eroding bluff; (2) groins and jetties built perpendicular to the shore that may extend seaward beyond the breaker line for most wave conditions; and (3) breakwaters that are offshore and shore parallel for beach stabilization, as well as breakwaters that are a harbor component and may be in relatively deep water. Thus, depending on the specific structure, the design wave height for a rubble mound structure may be depth limited or it may be the height having a selected return period based on an extreme wave analysis.

For rubble mound structures that experience zero to moderate wave overtopping, the required weight of armor stone unit is, according to the classic Hudson equation (Hudson, 1959), a function of the incident wave height cubed. Most of the empirical data used to support this equation is based on wave tank tests with monochromatic waves. Commonly, for design, the significant height is used; but heights as great as H_{10} or H_5 may be employed (U. S. Army Coastal Engineering Research Center, 1984). Note that the stone dimension is essentially proportional to the wave height and from the Rayleigh distribution $H_5 = 1.4H_s$. The choice of which height to use is based on such considerations as probability of occurrence or return period of the design wave used (see Section 7.6), the degree of confidence in the representativeness of the wave information used to establish the design wave, and the importance of the structure.

Other factors such as the wave period (Ahrens and McCartney, 1975) and wave groupiness (Burcharth, 1979) affect the stability of rubble structures that are not significantly overtopped. But the former is only indirectly included by the choice of a design coefficient that depends on whether the design wave is breaking at the structure, and the latter is not normally considered.

Experiments by Behnke and Raichlen (1984) on the stability of a severely overtopped breakwater showed the initial stability and subsequent damage to be related to the cumulative wave energy (H^2L) to which the structure was exposed. For the irregular waves used in these tests, T_p was used to determine the wave length. For the stability of very low-crested and submerged-sill rubble mound breakwaters, stability is defined by a modified form of the Hudson equation that relates the required stone weight to H^2L (van der Mer and Pilarczyk, 1990) rather than H^3. Also, the probability distribution of both the wave runup and the overtopping rate, as well as the height and period distribution of transmitted waves, depend not only on H_s and T_p but also on the joint probability distribution of height and period. But current design practice only relates runup, overtopping, and transmission to H_s and T_p.

The classic rubble mound structure is designed to maintain its initially constructed shape when subjected to wave attack. Recently, a new class of dynamically stable stone mound structures, called "berm breakwaters" (Baird and Hall, 1984) has been constructed. These structures use a smaller and wider graded stone size that is allowed to adjust its profile in response to wave attack. For a given design storm, experiments (van der Meer, 1988) have shown that the resulting profile geometry of the structure face can be related to H_s, T_p, and the number of waves that attack the structure. This would require a quantification of the duration of high waves during a design storm.

The direction of incident wave attack on a rubble mound structure will also affect its stability (Losada and Gimenez-Curto, 1982 and Christensen et al., 1984). Although wave refraction diagrams might be constructed to establish the design wave height distribution along the structure, the wave obliquity is usually not considered in a rigorous way in most rubble mound structure designs.

7.1.2 Framed Structures

This is a class of structures consisting of three-dimensional frames made of cylindrical members and typically constructed seaward of the breaker zone. The individual component cylindrical members are very small in diameter compared to the incident wave length. Consequently, wave loadings are calculated by the well-known Morison equation (Morison et al., 1950). The instantaneous wave force is composed of time-dependent drag and inertia components, which in turn depend on the water particle velocity squared and the particle acceleration respectively (see Sarpkaya and Isaacson, 1981 for a thorough discussion and application of the Morison equation).

The simpler approach to calculating wave loadings is to first determine an appropriate extreme wave height. This typically involves an analysis of available wave information to select a significant wave height with a return period commensurate with the design life of the structure. Then, employing the Rayleigh distribution relationships, a design wave height such as $H_{0.1}$ is determined from this extreme significant wave height. Using this design wave height and either a wave period representative of extreme storms or a representative range of possible periods, the time-dependent particle velocities and accelerations at points along the structure are calculated to determine the time-dependent wave loadings, some appropriate nonlinear wave theory is used.

This approach complements the more detailed stochastic approach in which the extreme design wave spectrum (usually one-dimensional) is used. This approach is important for the dynamic analysis of framed structures and the investigation of structural vibrations. The linear wave theory is used to determine the particle velocity and acceleration spectra from the design wave spectrum. The Morison equation is then used to determine the force spectrum for the structural component of interest (see Borgman, 1972 and Sarpkaya and Isaacson, 1981). Sarpkaya and Isaacson (1981) briefly discuss the effects of directional wave spectra and wave nonlinearity on the resulting forces.

A brief mention of "freak" or extremely large breaking or near breaking waves is appropriate at this point. These waves are believed to account for some of the vessel sinkings and platform damage that have occurred. These extreme waves are believed to be caused by a "particular random phase relationship between waves" and by certain bottom and opposing current effects on waves (Kjeldsen and Myrhaug, 1980). The directionality of wave spectra is important because components having different directions can combine for an instant to produce extreme waves with unusually high particle velocities and accelerations. Special laboratory techniques are being developed to generate and study the effects of these waves (Kjeldsen, 1982). Some progress has also been made on predicting where and when these waves might occur.

7.1.3 Vertical-Faced Structures

Particularly in Japan, but in other countries too, large vertical-faced monolithic concrete breakwaters are commonly constructed. They are often placed in a relatively wide range of water depths so that they are exposed to unbroken, breaking, or broken waves at the design condition. The pressure load on the vertical face must be determined as input to a structure sliding stability analysis. (See Goda, 1985 for a summary of the methods for pressure load calculation.)

For structures in sufficiently deep water that the design wave is unbroken, Goda (1985) recommends that the loading calculations use a wave height equal to $1.8H_s$ for the design storm level. In a design storm, particularly one of long duration, there will be a few waves of greater height than $1.8H_s$, but considering the accuracy of load calculation formulas and the fact that a few waves will not cause significant sliding, this design height is recommended. If judgment dictates, Goda says the design engineer might use a value between 1.6 and $2.0H_s$ for design.

The worst loading is caused by waves that break on the structure. These waves break somewhat seaward of the structure and plunge forward to hit the structure midway through the length of forward movement when breaking. This plunge distance depends on the fronting bottom slope and incident wave conditions (see Section 7.6). Thus a comparison of the calculated breaking height, at the point where the wave that hits the structure breaks, with the design incident wave height will determine whether design calculations should be for a breaking or unbroken wave.

7.1.4 Moored Floating Structures

Moored floating breakwaters are commonly used to provide protection for vessels in small craft harbors and for a variety of other purposes, including temporary construction operations. To be successful they must significantly reduce the transmission of incident waves so that transmitted wave heights are below some limiting value. The key to their deployment is that the coefficient of transmission (transmitted wave height/incident wave height) primarily depends on the incident wave period. For common floating breakwater configurations, satisfactory transmission coefficients are only achieved for wave periods around 2.5–3.5 sec or less. This severely limits the fetch/duration/wind speed combinations for which floating breakwaters are viable (see Sorensen, 1990).

A second floating breakwater design concern is the mooring load that develops because of wave action. This is primarily dependent on the incident wave height (e.g., see Giles and Sorensen, 1979). Thus, for an adequate design, the designer should know the wave spectrum for the design storm. The resulting H_s or some higher wave such as H_5 or H_1 would be used to calculate mooring loads and the total spectrum would allow a determination of the energy transmitted at the range of active wave periods.

Moored floating structures typically have six modes of oscillation (the three linear and three angular directions). If a portion of the incident wave spectrum significantly excites one of these modes of oscillation, the resulting mooring load can be significantly increased. For a large moored vessel, the structure resonant periods are typically higher than the periods in the wind wave portion of the spectrum. However, longer period waves generated by wave grouping and perhaps amplified by harbor resonance could cause serious problems for large vessels (see Bruun, 1989).

7.1.5 Beach Processes

Beach processes are driven primarily by wind-generated waves acting at a mean water level which is controlled by the tide, storm surge, and longer-term sea level variations. The functional design of most shore stabilization structures (e.g., groins, detached breakwaters, and jetties) requires that gross and net alongshore sediment transport rates be predicted on an annual and often shorter-term basis. The commonly used "SPM formula" (U. S. Army Coastal Engineering Research Center, 1984) for predicting surf zone sediment transport rates relates the transport to the alongshore energy flux factor. This, in turn, is a function of the significant wave height and crest orientation at the breaker line (or the deep water significant height and direction as well as the significant or spectral peak period so a refraction analysis can be conducted).

Thus, to make sediment transport rate predictions for a site, the wave climate must be determined from hindcasts or measurements taken for at least a year and preferably for several years. Ideally, the determination of wave height and direction (and period if in deep water) should be made every 12 or 24 h to accumulate an adequate data base for transport rate calculations. Practically, data sets often contain data taken less frequently. For visual wave observations made as part of the U. S. Army Corps of Engineers Littoral Environment Observation program, it is recommended that observations should minimally be made four times a week over a one-year period (U. S. Army Coastal Engineering Research Center, 1986).

Besides longer-term alongshore transport rate determinations, shorter-term beach events are also of interest—events that occur during a design storm. These include, for example, the beach profile recession and related toe scour at the base of a shoreline structure that will occur during a single storm or a series of closely timed storms, and the movement of a slug of alongshore sediment transport that will occur during a major storm. The latter is important for the design of sediment bypassing systems. The time-dependent pattern of significant wave heights and periods as well as the dominant nearshore wave directions during the design storm would need to be known.

A major consideration in the design of stable beaches is their use for recreation. Some attention should be given to the affect of structures and the nearshore beach profile on the resultant day-by-day wave climate during the recreation season (see Pratte, 1987). Dally (1990) has done a rudimentary analysis using a planar beach profile and linear wave theory to develop an algorithm for predicting conditions that are desirable for surfing. From annual wave statistics at a site, the percentage of time during a year at which waves are suitable for surfing can be estimated. Wave statistics for a site may also be useful to establish the value of the site for recreational bathing.

7.1.6 Harbor Design

Several aspects of the design of harbors and marinas are strongly impacted by the incident wave conditions. These include the exterior protective structures, moored vessel motion which depends on the vessel size, mooring arrangement and fender system characteristics, and the ease of vessel navigation into and through the harbor.

Harbor protective structure design depends on the action of wind-generated waves; but the movement and mooring of vessels depends on a longer period portion of the incident wave spectrum as discussed above. Analysis of the effects of longer period wave agitation can be carried out by physical and/or numerical models of the harbor and adjacent sea. Of particular interest are the resonant periods and related patterns of water motion in the harbor that may be excited by exterior wave action.

Whereas wind wave analysis for the design of protective structures requires an extreme wave analysis, navigation and mooring investigations also require an analysis of the amount of time that the entrance cannot be navigated or unloading operations cannot proceed owing to excessive wave activity. And the design of interior harbor structures and the mooring of very small vessels may be affected as much by vessel-generated waves as by the wind waves that penetrate the harbor. They may also be affected by waves generated by wind acting on open water inside of the harbor.

7.2 WAVE INFORMATION SOURCES

Ultimately, we would want to have the directional wave spectrum precisely measured every few hours for several years at the point or points of interest. Depending on how and why an actual wave data set is obtained, it will somehow fall short of this ultimate goal. There may not be a real need for all of this data, the cost for obtaining this much data would likely be prohibitive, and there may be insufficient time to obtain the data before it is needed for a site investigation or project design.

Wave data sets primarily come from three sources: (1) hindcasts from wind records or from weather charts from which the wind fields can be determined, (2) visual observations made from ships that are moored or traveling offshore or by observers along the coast, and (3) wave gaging programs conducted for specific projects or as general monitoring programs established by various national agencies. Each of these sources is further discussed in the following sections and two examples of specific projects are given.

7.2.1 Wave Hindcasts

Procedures for wave prediction were discussed in the previous chapter. With long-term wind records from a local airport or the weather charts for a major storm, wave hindcasts can be made. There are also national programs, for example, the U. S. Army Corps of Engineers' Wave Information Study (WIS), which employed numerical wave hindcasts to develop an extensive set of wave statistics for the coasts of the United States (U. S. Army Coastal Engineering Research Center, 1989).

The WIS program was conducted for the following five U. S. coastal regions: the Atlantic, the Gulf of Mexico, the northern Pacific, the southern Pacific, and the Great Lakes. Hindcasts were made from weather maps that produced wind fields and then the resulting wave fields at 3-hour intervals for the 20-yr period from January 1956 to December 1975 (Jensen, 1983b). These hindcasts typically included three phases. First waves were hindcast every 3 hours at a few selected deep water sites along the coast. Using these results as a boundary condition at the seaward edge of the continental shelf and considering refraction, diffraction, shoaling, and bottom friction, second phase forecasts were made for a larger number of locations at the edge of the nearshore zone. The final phase, employing the same wave transformation effects as in the second phase, carried the waves into the 10-m depth and provided information at approximately 20-km intervals along the coast. The first two phases produced directional wave spectra and the final phase produced significant height and period plus dominant direction information. Reports are available (e.g., Jensen, 1983a) that tabulate wave height, period, and direction data for the 20-km coastal intervals over the 20-yr period.

For the Atlantic and Gulf coasts tropical storms were excluded, so an additional set of hindcasts were made for 25 hurricanes in the Gulf and 43 hurricanes in the Atlantic that occurred from 1956 to 1975.

7.2.2 Visual Wave Observation Programs

Basic wave characteristics such as the average period, wave height (assumed to be the significant height), and dominant wave direction as well as other parameters such as the wind speed and direction and the nearshore breaker type and surf zone width can be measured by visual observation. Measurement programs have been established both for deep ocean conditions and conditions along the coast.

An example of a coastal visual wave measurement program is the Littoral Environment Observation (LEO) program of the U. S. Army Corps of Engineers (Schneider, 1981). Data have been collected by volunteer observers from government and private agencies since 1968 at over 200 coastal stations. The U. S. Army Coastal Engineering Research Center is responsible for tabulating, archiving, and distributing the data to users. A network computer-based data retrieval system (U. S. Army Coastal Engineering Research Center, 1986) has been established to make the data available to government and nongovernment users. In addition to coastal wave parameters such as the wave breaker height, period, direction, and type, a variety of other littoral zone data are collected and reported.

At sea, wave and meteorological data are collected by both weather ships stationed at fixed positions throughout the ocean and by voluntary observers on maritime vessels that travel the various shipping lanes throughout the world's oceans (e.g., see deGraauw, 1986). Weather ships remain on station through all weather conditions, whereas transiting maritime vessels try to avoid rough conditions. The oceans have been divided into zones and a different maritime country has taken responsibility for collecting and processing the ship's wave and meteorological data from each zone. For the United States, for example, data are published by the National Climate Data Center (U. S. Naval Weather Service Command, 1976). These data include tabulated summaries of wind and wave conditions for grid cells on the order of a hundred or more miles square spaced over the ocean.

7.2.3 Wave Measurement Programs

The mechanical measurement of ocean waves at a particular site, depending on the instrumentation used (see Section 7.4), will yield either the one-dimensional or the directional wave spectrum. In either case, the cost in physical effort and money is relatively large. Owing to a variety of causes, the percentage of time for which worthwhile data can be collected is typically much less than 100%.

Sources of measured wave data are quite diverse. Data may be collected by oil companies in preparation for the development of an offshore oil recovery program, by corporations as part of a coastal development program, as part of special studies (e.g., the JONSWAP Program), or as a component of a national program to define the wave climate for a country's coastal waters. Examples of some national wave climate measurement programs include Wilson and Baird (1972)—Canada, Kruseman (1974)—the Netherlands, Takahashi and Soejima (1974)—Japan, and Hemsley (1984)—the United States. A detailed catalog of 550 sources of worldwide instrumental wave measurements is discussed by Draper (1980).

7.2.4 Some Project Examples

To give some sense of how various wave data sources may be integrated for the accomplishment of a given project, two examples—Sines, Portugal and Saint John Deep, Canada—are briefly discussed below.

Sines, Portugal. The dolos-armoured stone mound breakwater at the port of Sines in Portugal is exposed to direct wave attack from the Atlantic Ocean. This breakwater, one of the largest in the world, was very seriously damaged during a storm in February, 1978 (Baird et al., 1980). Additional damage occurred during storms in December 1978 and February 1979. As part of the effort to understand the specific cause of damage and to redesign the structure, extensive studies of the wave climate at the site, including the particular storms that did the damage, were carried out (Mynett et al., 1983).

Using the numerical wave hindcasting models from the U. S. Army Corps of Engineers Wave Information Study, wave conditions were hindcast for 20 major storms during the period 1956–1980, including the 1978 and 1979 storms that damaged the breakwater. Directional wave spectra were calculated at 6-h intervals during the strength of each storm. Both sea and swell components during the storms were observed in the spectra that were hindcast at the site. The narrow-banded swell components yield significant wave grouping which can have a significant impact on structure stability. The hindcast directional spectra and derived significant heights were compared, where possible, with ship observations and wave gage measurements made during the storms. Monochromatic numerical calculations and physical model studies employing spectral waves were conducted to evaluate the effect of refraction and shoaling from deep water to the structure site. Extreme wave predictions were made from the hindcasts of the wave conditions from the 20 major storms.

Saint John Deep, Canada. A wave climate study was carried out in conjunction with the design of a proposed deep sea terminal at Saint John, New Brunswick, Canada (Khanna and Andru, 1974). The site is on the Bay of Fundy, which opens through the Gulf of Maine to the Atlantic Ocean. The basic objective of the study was to determine expected extreme wave heights for the site.

For a 1-yr period wind measurements were made at the site and wave measurements were made from a wave gage installed offshore of the site in water about 40 m deep. Owing to malfunctions, wind records were obtained for 61% of the time and the monthly wave data collection varied from 54.2 to 92.4% of the time. Eighteen years of wind record were available from a nearby airport. A comparison of the wind rose for the 18 yr and the one year when wind data were collected at the site as well as the wind rose for that year at the site indicated that differences were not large, so the 18 yr of wind records could be used for wave hindcasting. Wave hindcasts for the site were prepared using the Sverdrup–Munk–Bretschneider method. A third source of wave data was visual wave observations made at a weather ship in the Gulf of Maine. Refraction diagrams for dominant wave periods were constructed from the Gulf of Maine to the site to adjust the visually observed wave heights to site conditions. The extreme wave height projections were then based on the year of measured wave data with the hindcast and visually observed data sets used to reinforce the measured wave projections.

7.3 VISUAL WAVE MEASUREMENTS

Visual wave measurements made along the coast or at sea include the wave height, period, and direction. The height would be estimated without a vertical scale located at the point of observation. Some objects in view might be used by the observer to help in the wave height estimate. The wave period might typically be determined by timing a number of waves and dividing by this number to report an average period. A compass for offshore measurements and a protractor zeroed on the mean shoreline orientation for coastal measurements might be used to estimate the wave direction.

Investigations have been conducted to compare visually determined wave heights and periods with values measured by a wave gage (De Graauw, 1986; Jardine, 1979; Schneider and Weggel, 1980; Soares, 1986). The general consensus concerning observed wave heights is that individual values may show significant differences from measured values, but long-term statistical compilations of observed wave height data are more representative of measured conditions. Since the deviations between average observed and measured wave heights differed with the magnitude of the wave height, regression equations relating the observed height to the measured significant height have been developed (see Jardine, 1979 and Soares, 1986).

Particularly for swell, the observed average wave period should be quite representative of the true value. For actively generated seas, this would be less so, but the measured value of wave period (and significant height) is itself somewhat dependent on wave record analysis procedures (see Chapter 6). Schneider and Weggel (1980), considering observed wave periods along the coast, found that the visually estimated period, on the average, was higher than the peak period of the measured wave spectrum. They attribute this to the observers' failure to include smaller waves.

7.4 INSTRUMENTAL WAVE MEASUREMENTS

A wide variety of instruments is employed to make wave measurements. Most commonly, the water surface time history (e.g., Fig. 6.1) is measured at a point of interest. This would then be analyzed by the methods described in Chapter 6 to produce wave height/period distributions or the one-dimensional wave spectrum. In recent years, a number of directional wave gages have been developed and employed to measure the directional wave spectrum. Gages for one-dimensional or directional wave measurements may operate from above the water surface, at the water surface, or from a point below the water surface. And some remote systems have been developed to measure wave direction and refraction/diffraction patterns by providing a picture of the wave crest pattern over a broad area of the sea surface. Good general discussions of ocean wave instrumentation can be found in National Research Council (1982), Ribe and Russin (1974), and Tucker (1991). Most of the common types of instrumentation in use are briefly discussed below.

7.4.1 One-Dimensional Wave Gages

The three most common types of wave gages are the staff gage, the pressure gage, and the accelerometer buoy gage. The other gages discussed in this section are variations on these three.

Staff Gages. In principle, a staff gage is a vertical staff that penetrates the water surface and electronically senses and records the time-dependent water surface position. The staff must cover the entire range of possible water surface positions. Staff gages typically require mounting on a rigid structure such as a pier or offshore platform in a position such that the structure does not interfere with the wave motion at the gage. They may also be mounted on a spar buoy if the damping and resonant period of the buoy are appropriately designed.

Electronically, staff gages function by sensing a change in voltage owing to the variable resistance, capacitance, or inductance of a vertical wire probe (see Ribe and Russin, 1974). Or they may be a step-resistance gage, where a series of individual resisters along the gage are set at precise positions and detect the water surface motion as it rises or falls past an exposed point on the individual resister.

Depending on design details and gage location, some of the following difficulties may develop: marine fouling or ice may prevent the probe from functioning, lightning may disturb gage electronics, water splashing and runup may cause erroneous results, and there may be tampering or damage from floating objects. The rigid support on which the staff gage is mounted will also support the necessary equipment for recording the surface profile signal.

Pressure Gages. As previously discussed in Section 2.4.5, a pressure sensor that is placed below the water surface can be used as a wave gage. The gage may be mounted on a stable base placed on the sea floor in sufficiently shallow water or on the leg of a pile structure. For bottom-mounted gages the signal may be recorded internally and periodically retrieved or it may be carried to a shore recorder by a cable. Rather than measure the total static plus wave-induced dynamic pressure, more accuracy can be obtained by employing a differential pressure transducer. The transducer measures the difference between the instantaneous pressure and a damped longer-term average pressure that represents the mean hydrostatic pressure. Since pressure gages avoid the free surface they are less exposed to tampering and damage from floating objects, but they may still suffer fouling and damage from dragged anchors and other bottom disturbances. Also, pressure gages are more practical than surface piercing staff gages where there is a large tide range or a significant storm surge occurs.

Some wave theory must be employed to convert the pressure record to a wave-by-wave water surface record or the surface wave spectrum. Both linear and higher-order theories have been used (Grace, 1978 and Hsiang et al., 1986), but the procedures require significant effort to employ. A good review of several past efforts to evaluate the efficacy of pressure gages and linear wave theory to transfer pressure measurements to surface wave information is given by Bishop and Donelan (1987). From the linear wave theory we can relate the time-dependent pressure variation $p(t)$ to the surface amplitude by

$$p(t) = \rho g K_{\mathrm{p}} \eta$$

where K_{p}, the pressure response function, is given by

$$K_{\mathrm{p}} = \frac{\cosh k(d + z)}{\cosh kd}$$

The wave number k $(=2\pi/L)$ is calculated from the wave period which is determined from the pressure record. The wave spectrum may be derived by computing the spectrum of the pressure record and then converting to the surface spectrum using the pressure response function. Pressure gages have a natural filtering capability in that the shorter waves go undetected (i.e., where the water depth is approximately greater than half of the wave length).

Pressure sensors have been employed to measure waves in deep water by suspending a pressure transducer from a buoy at a sufficient depth so that the wave-induced dynamic pressure are completely attenuated. Then the pressure gage only detects the changing hydrostatic pressure as it moves up and down in response to the buoy following the passing wave surface fluctuation.

Accelerometer Buoys. A surface following buoy, which may be moored in relatively deep water, can be employed to measure water surface fluctuations. An accelerometer in the buoy measures the time-dependent vertical acceleration and this is integrated twice to determine the time-dependent surface displacement. The double integration is done electronically and either the acceleration or surface displacement records may be stored in the buoy for later retrieval or transmitted to shore directly or via a satellite link.

Commercially available accelerometer buoys are about a meter or less in diameter and have been moored in water up to 200 m deep. A common problem is damage from vessels that hit or tie up to the buoy. As with pressure gages, they are not significantly affected by large tide or storm surge ranges.

Shipborn Wave Recorders. Ocean-going weather ships are used for both visual wave observations and wave measurements employing wave gages installed on the ship. A shipborn wave recorder developed by the British Institute of Oceanographic Sciences has been described by Tucker (1956). It consists of a pair of combined pressure and accelerometer units that are installed on each side of the ship, approximately at the pitch axis of the ship. The accelerometers measure the heave of the ship and the pressure sensors respond to the shorter part of the wave spectrum. Employing a detailed analysis procedure, the wave signal can be determined from the ship's vertical movement and the relative surface displacement measured by the pressure gages.

Photo-Pole Gages. Staff gages for nearshore shallow water wave measurement often suffer from nonlinearity and calibration difficulties. They are limited in measuring breaking waves where there is significant air entrainment in the water or waves where there is a heavy suspended sediment concentration such as in estuaries. A simple device that eliminates these problems consists of a vertical pole with a scale and a movie or video camera that produces a picture that can be analyzed frame by frame. No electrical power source is needed and the gage is self-calibrating and linear. Ebersole (1987) has employed six synchronized cameras and 15 poles spaced across the surf zone to study wave decay across the surf zone. The author uses a video camera that can be analyzed at 30 frames a second for both lab and field wave measurements. The major drawback to this approach is the amount of time required to analyze the photo record, so a photo-pole gage is most useful for research efforts rather than synoptic measurement of wave climate.

Sonar/Acoustic Gages. A variety of devices that employ the sonar/radar concept of timing reflected waves have been designed for employment above and below the water surface (see Ribe and Russin, 1974). A key to the design of these devices is the width of the acoustic or electromagnetic beam being used. The wider the beam width is, the larger the lower limit on the size of wave that can be detected.

7.4.2 Directional Spectra Wave Gages

The two primary methods in use for measuring directional wave spectra at a particular location are: (1) array-type measurements, where point measurements are made at at least three points over a grid and (2) buoy measurements, where combined measurements of surface elevation and slope are made at a point by instrumentation on a buoy. Some remote sensing schemes employing light and radio waves have also been employed to measure directional wave spectra. Panicker (1974) gives a review of these methods and related data analysis.

Point-Array Gages. A horizontal array of point wave gages, situated in a line or preferably over a three-dimensional grid and taking simultaneous measurements, can be used to measure the directional wave spectrum. Various arrangements and spacings have been used. The optimal spacing depends on the dominant wave length being measured, so when a wide range of wave frequencies is to be measured, additional gages providing various combinations of spacings are preferred. Gage spacings equal to about a half or less than half of the dominant wave length are optimal. Typically, 3–6 gages have been used in field wave measurement programs. Both pressure and staff gages have been used to form the arrays.

Analysis of the multigage data requires an intercomparison of the measurements made at adjacent gages. Rather than measure independent absolute pressures, Bodge and Dean (1984) developed a pressure gage array where four differential pressure gages, each measuring the pressure difference between the center and a point on the periphery of the array, and a center total pressure gage were employed. Combinations of wave gages and current meters or an array of current meters alone may also be used for measurement of ocean wave spectra (e.g., see van Heteren et al., 1988). Suzuki (1969) used a point wave gage in combination with a bottom-mounted sphere that measured the wave force in two orthogonal directions to measure directional spectra.

Heave–Pitch–Roll Buoys. A moored buoy that simultaneously measures the water surface elevation and the slope in two orthogonal directions can be used to determine the directional wave spectrum (i.e., a heave–pitch–roll buoy). As with the array gage developed by Bodge and Dean, the spectrum is calculated from the surface elevation and two slope components.

A number of directional wave measurement buoys, ranging in weight from less than a 100 kg to a few thousand kilograms, are available commercially. Allender et al. (1989) report on an extensive field evaluation of six of these gages and generally concluded that most systems produced good quality directional wave data.

Remote Methods. Some remote sensing schemes, employing optical or radio wave techniques, have been used to measure directional wave spectra. Optical techniques include the use of dual-photo stereo photography and the analysis of single photographs for the reflection of sunlight. Holthuijsen (1983) describes a stereo photographic procedure that employs a pair of synchronized cameras carried by a pair of aircraft to take simultaneous overlapping sea surface photographs from which instantaneous surface contours may be determined. The directional spectrum then can be determined from a series of these contour sets.

Airborne or land-based radio transmitters can be used to measure wave directional spectra (see Panicker, 1974). The backscatter of the reflected radio waves will be in resonance for ocean waves having a length of half of the radio wave length. A directional scan of the resulting backscatter of radio waves can thus provide information on the directional wave spectrum.

While these remote sensing techniques are useful for some areas of wave research, at their present stage of development they are not commonly used for the accumulation of long-term wave statistics.

7.4.3 Wave Direction Measurements

Standard X-band radar, operated from an elevated shore position, has been used synoptically to depict offshore wave crest patterns (see Heathershaw et al., 1980 and Mattie and Harris, 1979). The existence of multiple wave trains and the nearshore wave refraction and diffraction patterns that develop can be well defined. Small ripples on the water surface significantly improve the radar's capacity to define the wave crests and the lower limit of wave length that can be detected is limited by the resolution power of the radar. Although similar information can be obtained from standard air photographs, radar can operate at night and in storms and does not require the use of an airplane. Hirakuchi and Ikeno (1990) describe a method where the radar signal is digitized to routinely define the mean wave direction and period of incident waves.

7.5 EXTREME WAVE ANALYSIS

Wave data collected at a site, either by wave gage measurements or by hindcasts from historical meteorological data, will only produce a relatively short record of wave conditions at the site. The design of coastal and ocean structures is usually based on an extreme wave condition that is likely to occur during a longer period of time than the duration of the measured or hindcast wave record. So the available wave record must commonly be extrapolated to this longer period to develop the design wave conditions.

The primary concern is usually to determine an extreme wave height, although an appropriate wave period or duration of high waves may also be of concern. (Remember that the maximum wave height depends on the duration of higher waves—see Eq. 6.6.) We commonly define a return period or recurrence interval T_r as the average time interval (in years) during which the specified wave height might be expected to be equaled or exceeded. The return period does not imply an actual sequence of events. A 50-yr return period means that the event has a 2% probability of occurring or being exceeded in any year. Two 50 yr events could occur in the same year or month.

The return period wave height employed to establish a design wave depends on the design life of the structure—the economic life and possibly the longer expected functioning life of the structure. It may also depend on other concerns, including the need for an additional factor of safety. A structure protecting a nuclear installation may have a planned life of less than a hundred years, but a much higher return period wave might be used for its design. Selection of a design wave return period is obviously somewhat subjective. Common design lifes for major coastal structures are 50 or 100 yr.

7.5.1 Extreme Wave Heights

The general approach to determining the height and related characteristics of an extreme wave having a given return period is as follows:

1. Tabulate and order by size a representative value such as the significant height for each of the measured/hindcast events.
2. From this, determine and plot the cumulative probability distribution of heights on graph paper based on a selected probability distribution.
3. Extrapolate this plot to obtain the return period significant wave height desired.
4. From the Rayleigh distribution, determine the H_n value desired.
5. With this wave height and an appropriate wave period, a nonlinear wave theory can be employed to calculate wave characteristics of interest.

Ideally, the cumulative probability distribution graph used to plot the data will fit the data and consequently yield a straight line that can easily be extrapolated to the desired return period. It appears (Ochi, 1982) that there is no probability distribution function which perfectly fits long-term significant wave height data so it is necessary to try a number of commonly used distributions and employ the one that provides the best fit. The probability distributions most commonly employed are the lognormal, the Gumbel or Fisher–Tippet I, the Fretchet or Fisher–Tippett II, and the Weibull (see Isaacson and Mackenzie, 1981). These probability distributions are tabulated in Table 7.1. In Table 7.1 $P(H)$ is the probability that the wave height is equal to or less than H. The parameters α, β, and γ are coefficients that are determined empirically to make the data best fit the distribution. Specifically, α is a shape parameter, β is a parameter that controls the spread along the H axis, and γ is a parameter which locates the position of the probability density function.

Only the Weibull distribution includes all three empirical coefficients. This increases the adaptability of this probability distribution to fitting empirical

TABLE 7.1 Probability Distributions Commonly Used for Extreme Wave Height Analysis

Distribution	Cumulative Probability
Log normal	$P(H) = \frac{1}{\sqrt{2\pi}} \int_0^H \frac{1}{\alpha H} \exp\left[-1/2 \left(\frac{\ln H - \beta}{\alpha}\right)^2\right] dH$
Gumbel	$P(H) = \exp\left\{-\exp\left[-\left(\frac{H - \gamma}{\beta}\right)\right]\right\}$
Frechet	$P(H) = \exp\left[-\left(\frac{H}{\beta}\right)^{-\alpha}\right]$
Weibull	$P(H) = 1 - \exp\left[-\left(\frac{H - \gamma}{\beta}\right)^{\alpha}\right]$

data. But it also increases the effort required to fit the data, as one of the three parameters must be estimated before the data can be plotted. This parameter is then reestimated until the best fit is obtained. Goda (1990), for example, assumed $\alpha = 0.75$, 1.0, 1.4, and 2.0 as four Weibull distributions which he then test fitted to his data.

With appropriate graph paper, data that fit each of these distributions will plot as a straight line. For the lognormal distribution, arithmetic-normal graph paper would suffice. The $P(H)$ term would plot on the normal distribution axis and $\log(H)$ would plot on the arithmetic axis. For the other distributions normal arithmetic scale paper could be used with the values plotted as shown in Table 7.2. Commonly, H is plotted as the abscissa and $P(H)$ as the ordinate.

Khanna and Andru (1974) plotted their data for Saint John Deep (as discussed in Section 7.2.4) on lognormal, Gumbel, and Weibull distribution graphs and found that the Weibull plot best fit a straight line which was then extrapolated to yield a 100-yr return period wave height. Other data sets might fit better on lognormal or Gumbel distribution graphs. For applications employing

TABLE 7.2 Plotting Scales for Various Probability Distributions

Distribution	Wave Height, H	Probability, $P(H)$
Gumbel	H	$-\ln\{-\ln[P(H)]\}$
Frechet	$\ln(H)$	$-\ln\{-\ln[P(H)]\}$
Weibull	$\ln(H - \gamma)$	$\ln\{-\ln[1 - P(H)]\}$

these distributions see Isaacson and MacKenzie (1981), Jasper (1956), Muir and El-Shaarawi (1986), Nolte (1973), and Ochi (1982).

Depending on the wave data source, one may have significant heights based on a short (e.g., 20 min) recording taken every few hours or daily, or the peak significant wave height values for individual storms or for the worst storm during a longer period such as a month or year. (See Goda, 1990, Muir and El-Shaawari, 1986, and Nolte, 1973 for a discussion of these different possibilities.) The cumulative probability can be related to the return period by

$$\frac{T_r}{r} = \frac{1}{1 - P(H)} \tag{7.1}$$

where r is the time interval (in years) between successive data points. For example, for average significant heights determined from daily wave measurements r is $1/365$. Thus, from Eq. 7.1, if a wave has a 2% chance of occurring or being exceeded in a given year, $P(H) = 0.98$ and $T_r = 50$ yr.

The first step in plotting data is to assign a value of T_r or $P(H)$ to each data point. If there are N data points ranked in order of decreasing size m from $m = 1$ to N (i.e., $m = 1$ for largest wave height and N for smallest wave height), the most commonly used formula for assigning a probability to a data point is (see Gumbel, 1958)

$$P(H) = 1 - \frac{m}{N + 1} \tag{7.2}$$

For annual data this yields

$$T_r = \frac{N + 1}{m} \tag{7.3}$$

So, for example from Eq. 7.3, the largest wave height from 25 annual values would plot with a return period of 26 yr.

Equation 7.2 has a slight bias depending on the probability distribution used, so some authors have proposed variations that include a coefficient that depends on the specific distribution (see Gringorten, 1963 and Goda, 1990). But most wave data analysts continue to use Eq. 7.2.

Consider an example where wave data were collected every 6 h for 2 yr, and analyzed to yield 2920 values of the significant wave height. These values were then ordered by size and T_r determined for each value of significant height [$r = \frac{1}{4}(365)$ and $T_r = 2.0007$ yr for the largest measured significant wave height]. Rather than plot all 2920 values, only enough values need to be plotted to define the data trend. This has been done in Figure 7.1 for a Gumbel distribution.

The points on Figure 7.1 do not perfectly fit a straight line. A line has been fit to the data by eye and extended to yield an estimated 50-yr return period

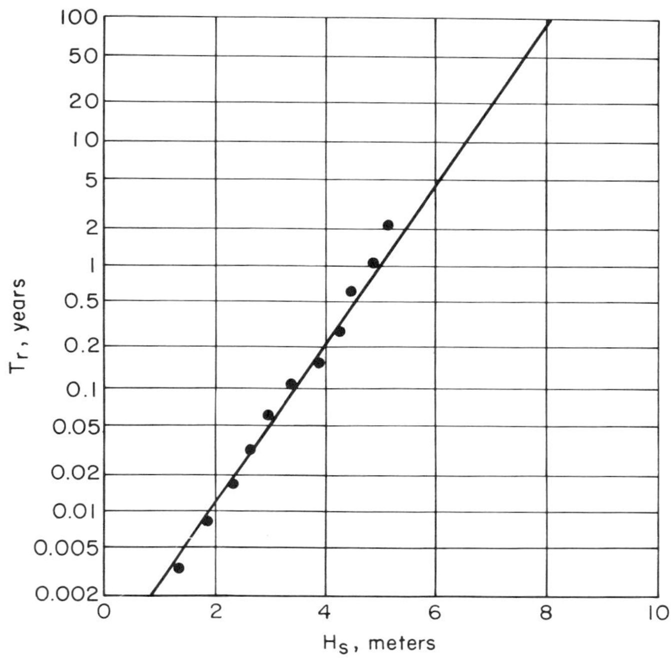

Figure 7.1 Gumbel plot of H_s versus T_r.

significant wave height of 7.7 m [and $H_1 = 1.67\ (7.7) = 12.1$ m]. To determine the 50-yr H_{max} one needs to know the significant height and the number of waves in the design storm (Eq. 6.6). Tucker (1991) recommends assuming the length of the storm is 3 h and an average period associated with the extreme storm (see Section 7.5.2) to estimate the number of waves.

These data would also be plotted on graph paper developed for the other probability distributions and the distribution yielding the best fit would be used. Although most analysts fit a line by eye, three analytical methods (see Isaacson and Mackenzie, 1981)—the method of moments, the method of least squares and, the method of maximum likelihood—have been used. Goda (1990) recommends use of the least-squares method. He then calculates the correlation coefficient for the data, the highest correlation coefficient value suggesting the best fit line for extrapolation.

The data set used in the given example above includes two complete years of wave gage measurements. As discussed earlier, it is unlikely that full years of gage data can be obtained, so the irregular data may not be fully representative of the given year. Also, wave climate can vary significantly from year to year, so a 2-yr data set is shorter than would be desired. Wave records can often be extended, for example, by supplementing with hindcast data.

Another concept that is helpful in the selection of a design wave height is the encounter probability E. This may be defined as the probability that a wave height having a return period T_r will be equaled or exceeded during some other period of time L. This period of time might be the design life of the structure. Borgman (1963) shows that

$$E = 1 - \left(1 - \frac{r}{T_r}\right)^{L/r} \tag{7.4}$$

If $T_r^2/Lr >> 1$ (which is usually the case), Eq. 7.4 can be replaced by Eq. 7.5 which is independent of r and easier to apply.

$$E = 1 - e^{-L/T_r} \tag{7.5}$$

Applying Eq. 7.5, the probability of a 50-yr return period wave occurring any year is 0.02, as discussed above, and the probabilities are 0.181, 0.632, and 0.865 of the wave occurring in any 10-, 50-, or 100-yr period, respectively. If it is desired that a design wave height have a 20% chance of occurring ($E = 0.20$) in a 50-yr period, one must design for a wave having a return period of 224 yr.

7.5.2 Other Extreme Wave Considerations

Although the dominant concern in extreme wave analysis is the wave height, it may also be necessary to determine a related extreme wave period for certain design situations. If sufficient long-term data on wave period are available, one can use the same procedure as outlined above for wave heights. Zero-upcrossing average or spectral peak periods are tabulated for each measured or hindcast event and fit to a probability distribution to determine the period having the same return period as the wave height return period being considered.

A simpler analysis, particularly if sufficient wave period data are not available, is to select a wave period or range of periods that relate to the extreme design wave height. A lower limit of wave period is set by the steepness limits for breaking of the design wave height in deep water. Battjes (1970) recommends that this lower limit wave period be given by

$$T > \left(\frac{32\pi}{g} H_s\right)^{1/2} \tag{7.6}$$

where H_s is the significant height for the return period wave of interest. Then, wave refraction, structure stability analyses, and so on are done employing the wave period given by Eq. 7.6 and an appropriate range of larger periods to determine the effects of possible wave periods. Naturally, a good dose of experience is needed to select this range of periods and to evaluate resulting wave effects.

Ochi (1982) observed that long-term wave period data often follow a log-normal probability distribution. Assuming that the long-term significant height data also follow a log-normal distribution, he was able to present a bivariate log-normal probability distribution for wave height and period. From this different combinations of wave height and period can be determined that have the same probability of occurrence or return period. Some of these combinations, of course, might not occur, owing to wave breaking limits. See Houmb and Overvik (1976) and (Ochi, 1982) for further discussions of this bivariate approach.

Often long-term wind data may be more available than long-term wave data. Rather than do wave hindcasts for each wind data point and then an extreme wave analysis, it is less cumbersome to do a return period analysis of the wind data to develop extreme design wind conditions and then perform the appropriate wave hindcasts. When doing the long-term extreme wind analysis, the data is subdivided by directions if fetches are different in each direction. Also, conditions must generally be fetch limited so wind duration effects may be ignored.

Thom (1971, 1973) developed extreme wave predictions for the world's oceans by working with the more available wind velocity data. First he developed extreme wind velocity distributions employing a modified Frechet distribution that combines the effects of tropical and extratropical storms. Then, employing these relationships, he empirically related the wind probability distribution parameters to wave probability distribution parameters to develop long-term mixed Frechet significant wave height probability distributions for the world's oceans.

For example, for the design of rubble mound structures or for the prediction of beach profile recession during a storm, it may be necessary to know the duration of extreme wave conditions during the design storm. Smith (1988) investigated the duration of major storms by working with the U. S. Army Corps of Engineers 20 yr of 3-hr wave hindcasts (see Section 7.2.1). The duration of high waves was defined as the period of time that the significant wave height exceeded a selected threshold wave height. Smith considered using the wave steepness (H/gT^2) or the "wave severity" (H^2L) to define wave conditions but, after evaluating these parameters for data at some sites, he decided it was best to just employ the significant height. Then employing an approach similar to that used for wave height, duration versus cumulative probability plots were developed using Gumbel and Weibull distributions. Regression analysis was used for curve-fitting the data.

Smith also used the data set to evaluate extreme significant wave heights. As might be expected, there was little correlation between extreme durations and extreme wave heights.

7.6 WAVE BREAKING

If the extreme wave analysis produces a design wave for a nearshore structure that will break before it reaches the structure, the breaking conditions (rather than the previously discussed extreme wave characteristics) will define the structure's design wave. Breaking will influence the height distribution of the wave spectrum attacking the structure as the higher waves will break first. Particularly if a steeper nearshore bottom slope fronts the structure, the controlling depth for wave breaking will not be the water depth immediately in front of the structure but a deeper seaward depth where the wave that impinges on the structure commences to break. Wave reflection from the structure may impact the breaking process, causing lower waves to break further seaward than might otherwise occur. The last factor is difficult to quantify and is usually not considered in most design applications.

Wave breaking in shallow water is very dependent on the water depth, which may change over the life of the project. Bottom scour or deposition owing to local shore-normal and shore-parallel beach processes, and mean water level changes owing to the tide or storm surge as well as sea level rise because of atmospheric warming, may all significantly alter design wave conditions during the life of a structure. Kyper and Sorensen (1985) report on a study concerning the stability of a stone mound seawall that has no fronting beach. During the past decade or more the frequency of wave-induced damage to this seawall has been multiplying owing to increased wave attack because the water depth in front of the structure is increasing. Extrapolating historic nearshore erosion rates and adding expected values of future sea level rise, they found that the average water depth (below mean low water) in front of the seawall is expected to increase from the present average value of 1.1 m to about 2 m during the next century. During this period the design significant wave height would correspondingly increase from a present value of 4.6 m to a value of between 5.4 and 6.8 m depending on the rate of sea level rise assumed.

Section 3.5 mentioned that a number of empirical relationships have been developed to allow one to determine the breaking wave height and water depth as a function of the fronting beach slope and the incident deep water wave height and period. Figures 7.2 and 7.3, modified from the U. S. Army Coastal Engineering Research Center (1984) and based on studies by Goda (1970) and Weggel (1972), are commonly used examples of these relationships. These curves apply to monochromatic waves. For elemental design analysis, given H_s or H_1 and T_p in deep water, one can apply these curves to determine the nearshore breaking wave height and then the water depth at which the wave would break. If the wave breaks far seaward of the structure, one could try successively lower deep water heights to find the height of wave that will break at the structure and use this for the design wave height. Note that Figure 7.2 also gives the breaker type.

The heaviest loading on a structure will be caused by a plunging wave that breaks seaward of the structure and hits the structure midway through the breaking process. So breaking conditions should not be evaluated for the design water depth right at the structure but for a depth some distance seaward of this point that is related to the breaker plunge distance. Based on laboratory data Goda (1985) suggests that this distance will be approximately $5H_s$. The U. S. Army Coastal Engineering Research Center (1984) recommends that the breaker

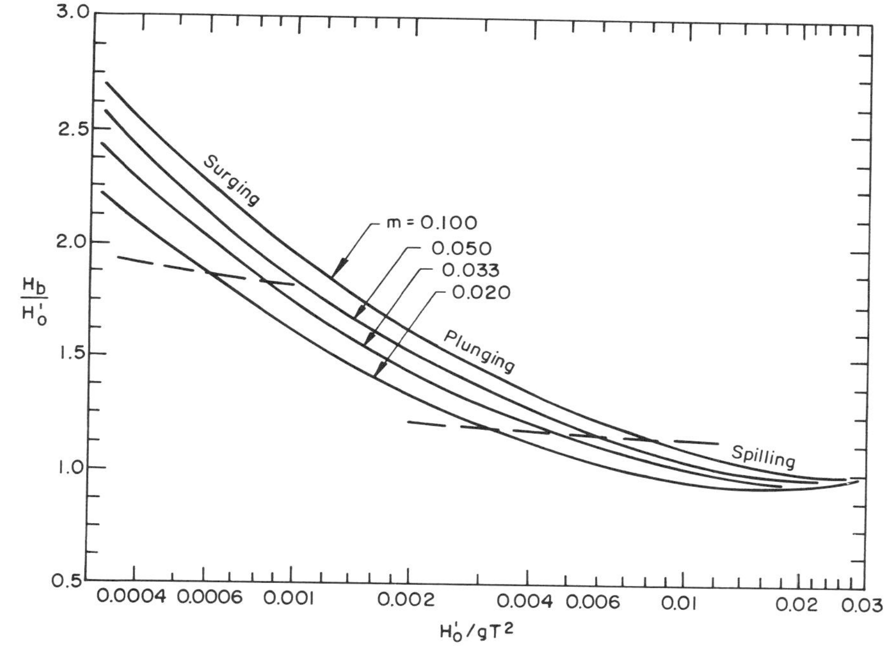

Figure 7.2 Wave breaker height as a function of bottom slope and wave period and deep water height (U. S. Army Coastal Engineering Research Center, 1984).

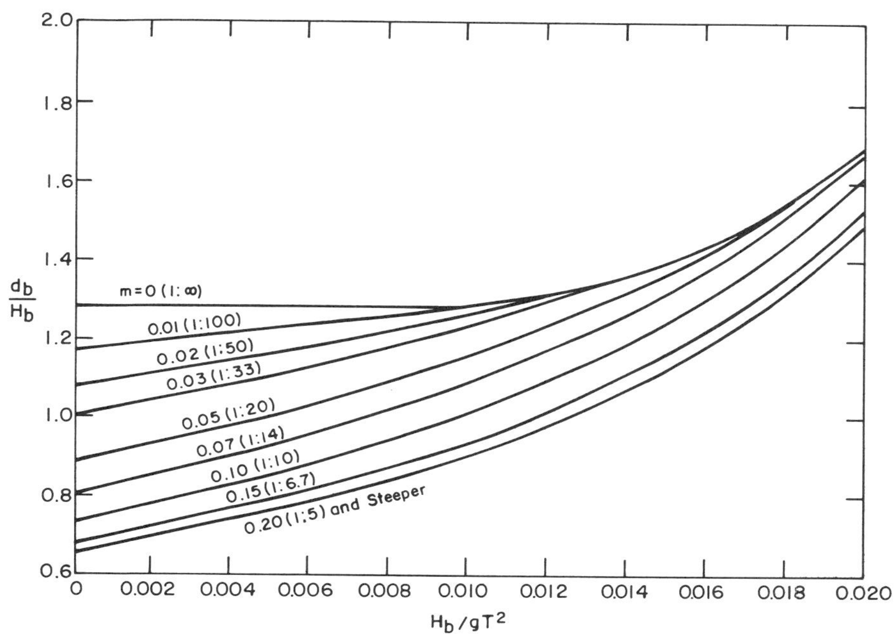

Figure 7.3 Wave breaking depth as a function of bottom slope and wave period and deep water height (U. S. Army Coastal Engineering Research Center, 1984).

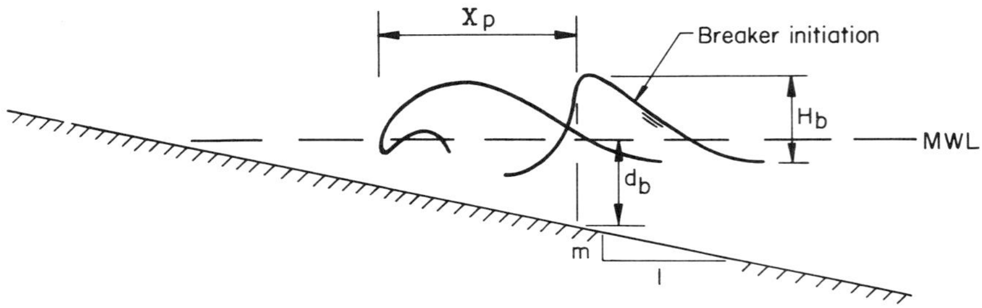

Figure 7.4 Definition of terms for breaking wave plunge distance.

travel distance X_p for a plunging breaker (see Fig. 7.4) be given by

$$X_p = (4.0 - 9.25m)H_b \qquad (7.7)$$

where m is the fronting bottom slope and H_b is the plunging breaker height. Half of X_p would be used as the distance seaward of the structure for the breaking point of the wave. The distance given by Eq. 7.7 will be somewhat less than the distance recommended by Goda. Other authors (see Smith and Kraus, 1991) concur that Eq. 7.7 somewhat underestimates the value of X_p.

Smith and Kraus (1991) studied wave breaking, including the breaker travel distance for plunging waves, on both plane slopes and slopes with a superimposed bar. They found that X_p/H_b related to the offshore Iribarren number or surf similarity parameter (see Section 3.7) defined as

$$I_{r0} = \frac{m}{\sqrt{H_0/L_0}} \qquad (7.8)$$

For plane slopes

$$\frac{X_p}{H_b} = 3.95\,(I_{r0})^{-0.25} \qquad (7.9)$$

and for barred profiles

$$\frac{X_p}{H_b} = 0.63\,(I_{r0})^{-1} + 1.81 \qquad (7.10)$$

Comparison of Eqs. 7.9 and 7.10 will show that for the same slope a bar will produce a shorter plunge distance.

When a spectrum of waves enters the nearshore zone, breaking occurs over a wide area as waves of different steepness break at different points in the surf zone. Goda (1975, see Section 6.2.3) developed a model for irregular wave

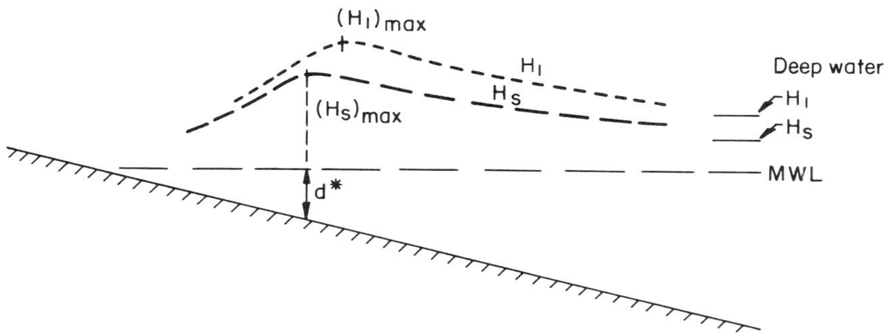

Figure 7.5 Variation of H_1 and H_s toward the shore and through the surf zone.

shoaling that gives the values for H_s and H_1 into and through the surf zone. The model assumes that waves have a Rayleigh distribution in deep water and that broken waves reform at a lower height. A schematic plot of this model yields the results shown in Figure 7.5, where H_s and H_1 increase as the wave propagates toward the shore to a peak where breaking commences, and then decrease. The value of H_1 peaks slightly before H_s because the highest waves generally break first. Thus, the maximum H_s and H_1 values would represent the limiting breaking heights.

Employing Goda's model, Seelig (1980) developed Figure 7.6, where the

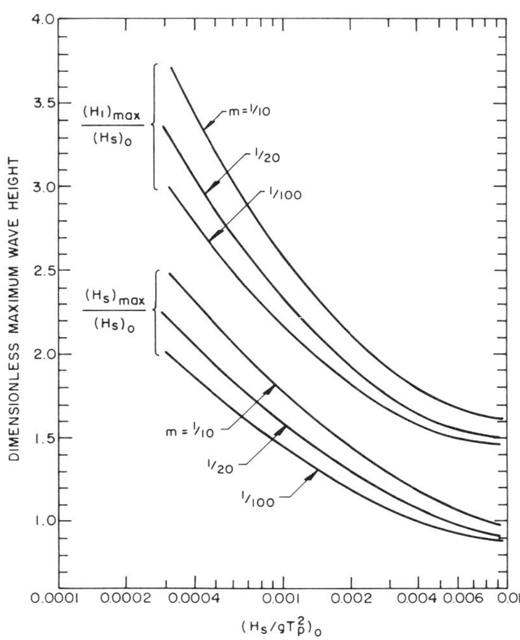

Figure 7.6 Maximum values of H_1 and H_s based on Goda's model (from Seelig, 1980).

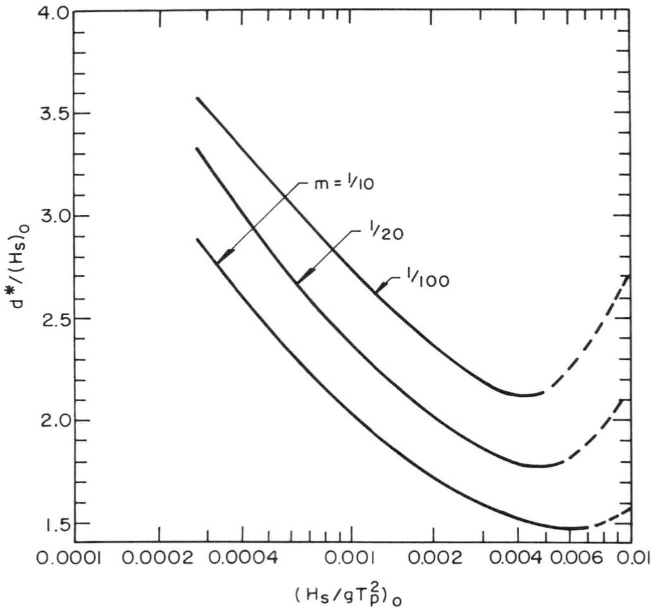

Figure 7.7 Water depth where maximum significant height occurs based on Goda's model (from Seelig, 1980).

maximum values of H_s and H_1 (nondimensionalized by dividing by the deep water significant height) are given as a function of the deep water wave steepness and the nearshore bottom slope. The maximum values of H_s and H_1 both decrease as the wave steepness increases and the beach slope decreases. The maximum value of H_s occurs at a depth d^* (see Fig. 7.5) and the maximum value of H_1 occurs slightly seaward of this point. Figure 7.7, which is also from Seelig (1980), can be used to determine d^*.

Based on comparisons of calculated values with limited laboratory and field data, Seelig (1980) found that Goda's model gives useful estimates of $(H_s)_{max}$ and $(H_1)_{max}$. Thornton et al. (1984), using a more extensive field and laboratory data set for comparison, found that Goda's model reasonably predicts $(H_1)_{max}$ but slightly overpredicts $(H_s)_{max}$. They found that using Figures 7.2 and 7.3, which are based on monochromatic waves, to determine the breaking values for H_s and H_1 yielded overly conservative results, particularly for low steepness waves.

REFERENCES

Ahrens, J. and McCartney, B. (1975), "Wave Period Effect on the Stability of Riprap," *Proceedings, Civil Engineering in the Oceans III Conference*, American Society of Civil Engineers, Baltimore, pp. 1019–1034.

Allender, J. and eight other authors (1989), "The Wadic Project: A Comprehensive Field Evaluation of Directional Wave Instrumentation," *Ocean Eng.*, **16,** 505–536.

Baird, W. F., Caldwell, J. M., Edge, B. L., Magoon, O. T., and Treadwell, D. D. (1980), "Report on the Damages to the Sines Breakwater, Portugal," *Proceedings, 17th International Conference on Coastal Engineering*, American Society of Civil Engineers, Sydney, pp. 3063–3077.

Baird, W. F. and Hall, K. R. (1984), "The Design of Breakwaters Using Quarried Stones," *Proceedings, 19th International Conference on Coastal Engineering*, American Society of Civil Engineers, Houston, pp. 2580–2591.

Battjes, J. A. (1970), "Long-Term Wave Height Distribution at Seven Stations Around the British Isles," National Institute of Oceanography, Report A44, England.

Behnke, D. L. and Raichlen, F. (1984), "Breakwater Armor Displacements Thresholds: A Possible Correlation with Cumulative Wave Energy," *Proceedings, 19th International Conference on Coastal Engineering*, American Society of Civil Engineers, Houston, pp. 2760–2772.

Bishop, C. T. and Donelan, M. A. (1987), "Measuring Waves with Pressure Transducers," *Coastal Eng.*, **11,** 309–328.

Bodge, K. R. and Dean, R. G. (1984), "Wave Measurement with Differential Pressure Gages," *Proceedings, 19th International Conference on Coastal Engineering*, American Society of Civil Engineers, Houston, pp. 755–769.

Borgman, L. E. (1963), "Risk Criteria," *J. Water. Harbors Div., Am. Soc. Civ. Eng.*, August, 1-35.

Borgman, L. E. (1972), "Statistical Models for Ocean Waves and Wave Forces," *Advances in Hydroscience*, Vol. 8, Academic, New York, pp. 139–181.

Bruun, P. (1989), *Port Engineering*, 4th ed., Vol. 1, Gulf Publishing, Houston.

Burcharth, H. (1979), "The Effect of Wave Groupiness on On-Shore Structures," *Coastal Eng.*, **2,** 189–199.

Christensen, F., Broberg, P. C., Sand, S., and Tryde, P. (1984), "Behavior of Rubble-Mound Breakwaters in Directional and Unidirectional Waves," *Coastal Eng.*, **8,** 265–278.

Dally, W. R. (1990), "Stochastic Modeling of Surfing Climate," *Proceedings, 22nd International Conference on Coastal Engineering*, American Society of Civil Engineers, Delft, pp. 516–529.

DeGraauw, A. (1986), "Wave Statistics Based on Ship's Observations," *Coastal Eng.*, **10,** 105–118.

Draper, L. (1980), "Sources of Measured Wave Data," *Proceedings, 17th International Conference on Coastal Engineering*, American Society of Civil Engineers, Sydney, pp. 372–380.

Ebersole, B. A. (1987), "Measurement and Prediction of Wave Height Decay in the Surf Zone," *Proceedings, Coastal Hydrodynamics Conference*, American Society of Civil Engineers, Newark, DE, pp. 1–16.

Giles, M. L. and Sorensen, R. M. (1979), "Determination of Mooring Loads and Wave Transmission for a Floating Tire Breakwater," *Proceedings, Coastal Struc-*

tures '79 Conference, American Society of Civil Engineers, Alexandria, VA, pp. 1069–1085.

Goda, Y. (1970), ''A Synthesis of Breaker Indices,,'' *Trans. Jpn. Soc. Civ. Eng.*, **2,** Tokyo, 227–230.

Goda, Y. (1975), ''Irregular Wave Deformation in the Surf Zone,'' *Coastal Eng. Jpn.*, **18,** 13–26.

Goda, Y. (1985), *Random Seas and the Design of Maritime Structures*, University of Tokyo Press, Tokyo.

Goda, Y. (1990), ''On the Methodology of Selecting Design Wave Height,'' *Proceedings, 22nd International Conference on Coastal Engineering*, American Society of Civil Engineers, Delft, pp. 899–903.

Grace, R. A. (1978), ''Surface Wave Heights from Pressure Records,'' *Coastal Eng.*, **2,** 55–68.

Gringorten, I. I. (1963), ''A Plotting Rule for Extreme Probability Paper,'' *J. Geophys. Res.*, **68,** 813–814.

Gumbel, E. J. (1958), *Statistics of Extremes*, Columbia University Press, New York.

Heathershaw, A. D., Blackley, M. W. L., and Hardcastle, P. J. (1980), ''Wave Direction Estimates in Coastal Waters Using Radar,'' *Coastal Eng.*, **3,** 249–267.

Hemsley, J. M. (1984), ''US Army Corps of Engineers Field Wave Gaging Program,'' *Proceedings, 19th International Conference on Coastal Engineering*, American Society of Civil Engineers, Houston, pp. 304–315.

Hirakuchi, H. and Ikeno, M. (1990), ''Wave Direction Measurement Using Marine X Band Radar,'' *Proceedings, 22nd International Conference on Coastal Engineering*, American Society of Civil Engineers, Delft, pp. 703–715.

Holthuijsen, L. H. (1983), ''Stereophotography of Ocean Waves,'' *Applied Ocean Research*, Vol. 5, October, pp. 204–209.

Houmb, O. G. and Overvik, T. (1976), ''Parameterization of Wave Spectra and Long-Term Joint Distribution of Wave Height and Period,'' *Proceedings, Behavior of Offshore Structures Conference*, Trondheim, pp. 144–169.

Hsiang, W., Dong-Young, L., and Garcia, A. (1986), ''Time Series Surface-Wave Recovery from Pressure Gage,'' *Coastal Eng.*, **10,** 379–393.

Hudson, R. L. (1959), ''Laboratory Investigation of Rubble Mound Breakwaters,'' *J. Waterw. Harbors Div., Am. Soc. Civ. Eng*, September, 93–121.

Isaacson, M. and MacKenzie, N. G. (1981), ''Long-term Distributions of Ocean Waves,'' *J. Waterw. Port Coastal Ocean Eng. Div., Am. Soc. Civ. Eng*, May, 93–109.

Jardine, T. P. (1979), ''The Reliability of Visually Observed Wave Heights,'' *Coastal Eng.*, **3,** 33–38.

Jasper, N. H. (1956), ''Statistical Distribution Patterns of Ocean Waves and of Wave-induced Ship Stresses and Motions, with Engineering Applications,'' *Trans. Soc. Nav. Arch. Mar. Eng.*, **64,** 375–432.

Jensen, R. E. (1983a), ''Atlantic Coast Hindcast, Shallow-Water Significant Wave Information,'' WIS Report 9, U. S. Army Waterways Experiment Station, Vicksburg, MS.

Jensen, R. E. (1983b), "Methodology for the Calculation of a Shallow Water Wave Climate," WIS Report 8, U. S. Army Waterways Experiment Station, Vicksburg, MS.

Khanna, J. and Andru, P. (1974), "Lifetime Wave Height Curve for Saint John Deep Canada," *Proceedings, Conference on Ocean Wave Measurement and Analysis*, Vol. 1, American Society of Civil Engineers, New Orleans, pp. 301–319.

Kjeldsen, S. P. (1982), "2- and 3-Dimensional Deterministic Freak Waves," *Proceedings, 18th International Conference on Coastal Engineering*, American Society of Civil Engineers, Cape Town, pp. 677–694.

Kjeldsen, S. P. and Myrhaug, D. (1980), "Wave-Wave Interactions, Current-Wave Interactions and Resulting Extreme Waves and Breaking Waves," *Proceedings, 17th International Conference on Coastal Engineering*, American Society of Civil Engineers, Sydney, pp. 2277–2303.

Kruseman, P. (1974), "Wave and Wave Climate Study in the North Sea," *Proceedings, Ocean Wave Measurement and Analysis Conference*, Vol. 1, American Society of Civil Engineers, New Orleans, pp. 1–12.

Kyper, T. N. and Sorensen, R. M. (1985), "The Impact of Selected Sea Level Rise Scenarios on the Beach and Coastal Structures at Sea Bright, N.J.," *Proceedings, Coastal Zone '85 Conference*, American Society of Civil Engineers, Baltimore.

Losada, M. and Gimenez-Curto, L. (1982), "Mound Breakwaters Under Oblique Wave Attack; A Working Hypothesis," *Coastal Eng.*, **6,** 83–92.

Mattie, M. G. and Harris, D. L. (1979), "A System for Using Radar to Record Wave Direction," Technical Report 79-1, U. S. Army Coastal Engineering Research Center, Ft. Belvoir, VA.

Morison, J. R., Johnson, J. W., O'Brien, M. P. and Schaaf, S. A. (1950), "The Forces Exerted by Surface Waves on Piles," *Pet. Trans. Am. Inst. Min. Eng.*, **189,** 145–154.

Muir, L. R. and El-Shaarawi, A. H. (1986), "On the Calculation of Extreme Wave Heights: A Review," *Ocean Eng.*, **13,** 93–118.

Mynett, A. E., de Voogt, W. J. P., and Schmeltz, E. J. (1983), "West Breakwater—Sines, Wave Climatology," *Proceedings, Coastal Structures '83 Conference*, American Society of Civil Engineers, Arlington, VA, pp. 17–30.

National Research Council (1982), "Proceedings, Workshop on Wave Measurement Technology," NRC Marine Board, Washington, DC.

Nolte, K. G. (1973), "Statistical Methods for Determining Extreme Sea States," *Proceedings, 2nd International Conference on Port and Ocean Engineering Under Arctic Conditions*, University of Iceland, pp. 705–742.

Ochi, M. K. (1976), "On Long-Term Statistics for Ocean and Coastal Waves," *Proceedings, 16th International Conference on Coastal Engineering*, American Society of Civil Engineers, Hamburg, pp. 59–75.

Ochi, M. K. (1982), "Stochastic Analysis and Probabilistic Prediction of Random Seas," *Adv. Hydrosci.*, **13,** 218–375.

Paniker, N. N. (1974), "Review of Techniques for Directional Wave Spectra," *Proceedings, Conference on Ocean Wave Measurement and Analysis*, Vol. 1, American Society of Civil Engineers, New Orleans, pp. 669–688.

Pratte, T. P. (1987), ''Ocean Wave Recreation,'' *Proceedings, Coastal Zone '87 Conference*, American Society of Civil Engineers, pp. 5386–5398.

Ribe, R. L. and Russin, E. M. (1974), ''Ocean Wave Measuring Instrumentation,'' *Proceedings, Conference on Ocean Wave Measurement and Analysis*, Vol. 1, American Society of Civil Engineers, New Orleans, pp. 396–416.

Sarpkaya, T. and Isaacson, M. (1981), *Mechanics of Wave Forces on Offshore Structures*, Van Nostrand Reinhold, New York.

Schneider, C. (1981), ''The Littoral Environment Observation (LEO) Data Collection Program,'' Coastal Engineering Technical Aid 81-5, U. S. Army Waterways Experiment Station, Vicksburg, MS.

Schneider, C. and Weggel, J. R. (1980), ''Visually Observed Wave Data at Pt. Mugu, Calif.,'' *Proceedings, 17th International Conference on Coastal Engineering*, American Society of Civil Engineers, Sydney, pp. 381–393.

Seelig, W. N. (1980), ''Maximum Wave Heights and Critical Water Depths for Irregular Waves in the Surf Zone,'' Coastal Engineering Technical Aid 80-1, U. S. Army Coastal Engineering Research Center, Ft. Belvoir, VA.

Smith, E. R. and Kraus, N. C. (1991), ''Laboratory Study of Wave Breaking Over Bars and Artificial Reefs,'' *J. Waterw. Port Coastal Ocean Eng. Div., Am. Soc. Civ. Eng.*, July/August, 307–325.

Smith, O. P. (1988), ''Duration of Extreme Wave Conditions,'' *J. Waterw. Port Coastal Ocean Eng. Div., Am. Soc. Civ. Eng.*, January, 1–17.

Soares, C. G. (1986), ''Assessment of the Uncertainty in Visual Observations of Wave Heights,'' *Ocean Eng.*, **13,** 37–56.

Sorensen, R. M. (1990), ''The Deployment of Floating Breakwaters: Design Guidance,'' *Proceedings, 12th International Coastal Society Conference*, San Antonio.

Suzuki, Y. (1969), ''Determination of Approximate Directional Spectra for Coastal Waves,'' Port and Harbor Research Institute Report, Japan, Vol. 8, pp. 43–101.

Takahashi, T. and Soejima, T. (1974), ''Coastal Wave Observation and Information for Engineering Use in Japan,'' *Proceedings, Ocean Wave Measurement and Analysis Conference*, Vol. 2, American Society of Civil Engineers, New Orleans, pp. 62–81.

Thom, H. C. S. (1971), ''Asymptotic Extreme-Value Distribution of Wave Heights in the Open Ocean,'' *J. Mar. Res.*, **29,** 19–27.

Thom, H. C. S. (1973), ''Extreme Wave Height Distributions Over Oceans,'' *J. Waterw. Harbors and Coastal Eng. Div., Am. Soc. Civ. Eng.*, August, 355–374.

Thornton, E. B., Wu, C. S., and Guza, R. T. (1984), ''Breaking Wave Design Criteria,'' *Proceedings, 19th International Conference on Coastal Engineering*, American Society of Civil Engineers, Houston, pp. 31–41.

Tucker, M. J. (1956), ''A Shipborne Wave Recorder,'' *Trans. Inst. Nav. Arch.*, **98,** 236–246.

Tucker, M. J. (1989), ''An Improved 'Battjes' Method for Predicting the Probability of Extreme Waves,'' *Appl. Ocean Res.*, **11,** 212–218.

Tucker, M. J. (1991), *Waves in Ocean Engineering—Measurement, Analysis, Interpretation*, Ellis Horwood, New York.

U. S. Army Coastal Engineering Research Center (1984), *Shore Protection Manual*, U. S. Government Printing Office, Washington, DC.

U. S. Army Coastal Engineering Research Center (1986), *Coastal Engineering Notebook*, CETN 1-8, (LEO) Littoral Environment Observations.

U. S. Army Coastal Engineering Research Center (1989), *Coastal Engineering Notebook*, CETN 1-23 Sea-State Engineering Analysis System (SEAS).

U. S. Naval Weather Service Command (1976), "Summary of Synoptic Meteorological Observations," National Climate Data Center, Ashville, NC.

van der Meer, J. W. (1988), "Rock Slopes and Gravel Beaches Under Wave Attack," Communication No. 396, Delft Hydraulics Laboratory, Delft.

van der Meer, J. W. and Pilarczyk, K. W. (1990), "Stability of Low-Crested and Reef Breakwaters," *Proceedings, 22nd International Conference on Coastal Engineering*, American Society of Civil Engineers, Delft, pp. 1375–1388.

van Heteren, J., Keijser, H., and Schaap, B. (1988), "Comparison Wave Directional Measurement Systems," *Appl. Ocean Res.*, **10,** 129–143.

Weggel, J. R. (1972), "Maximum Breaker Height," *J. Waterw. Harbors Coastal Eng. Div., Am. Soc. Civ. Eng.*, November, 529–548.

Wilson, J. R. and Baird, W. F. (1972), "A Discussion of Some Measured Wave Data," *Proceedings, 13th Conference on Coastal Engineering*, American Society of Civil Engineers, Vancouver, pp. 113–130.

8

WAVE–STRUCTURE INTERACTION

Usually, the most concentrated change that occurs to waves as they propagate forward takes place when they interact with a variety of structures, including beaches. Our focus herein is on the changes in wave characteristics that occur rather than the effect the waves have on the particular structure. Included in our considerations are both waves that have not broken as well as broken waves in the nearshore surf zone.

Some aspects of wave–structure interaction have previously been considered. Wave refraction and diffraction were considered in detail in Chapter 5, and basic aspects of wave reflection and runup were presented in Sections 2.5.3, 3.7, 3.8, and 5.4. Wave reflection and runup as well as wave overtopping on and transmission past structures is covered in greater detail in this chapter.

Owing to their nature, much of what is known about these processes is based on experimental data collected mostly at reduced scale from wave flume experiments. (Some progress is being made in the application of numerical models to the analysis of some structure-induced wave transformations.) A large volume of experimental data, which is quite useful in the design of structures, is available. Only selected results are presented to demonstrate processes and resulting trends. Other valuable data sets are referenced.

A common theme of most of the empirically derived wave–structure interaction data is that earlier basic data sets were collected in laboratory experiments with monochromatic waves. Subsequently, many of the studies were repeated using irregular spectral waves for a more realistic evaluation of the particular phenomenon. Irregular wave studies had to await the development of adequate irregular wave generators, and often more importantly, adequate means to measure the more complex runup and overtopping of irregular waves. In many cases, insufficient irregular wave experimental results are available so predictions of the behavior of irregular waves are based on analytical models constructed from monochromatic wave experiments.

Figure 8.1 is a schematic example of some of the specific wave–structure interactions that are of interest. A train of waves approaching the shore from offshore first encounters a cluster of vertical cylindrical piles. Wave-induced water particle orbital motion past the piles will lead to the dissipation of wave energy at the piles. Also a small amount of wave energy will be reflected seaward from the piles. Consequently, after the waves propagate past the pile cluster they have the same period and continuous phase, but are slightly reduced in amplitude.

At the floating breakwater wave reflection and dissipation also occur and yield a further reduction of the wave amplitude as the waves enter the lee of the breakwater. But, owing to the articulation of the floating breakwater, a small portion of the wave energy reflection and transmission is in the form of waves generated seaward and landward by the motion of the floating breakwater. These regenerated waves may be out of phase with the waves that are directly reflected and transmitted and they may have a different period, depending on the dynamic characteristics of the breakwater.

When the waves next reach the front face of the stone mound breakwater, they will break, run up the breakwater face, and, if the runup is sufficiently high, overtop the breakwater, causing a jet of water to plunge into the water at the lee of the breakwater. This plunging jet will regenerate waves in the lee of the structure that will likely have a portion of their energy at periods that are different from the incident wave period range. If the stone mound structure is sufficiently porous, wave energy may also propagate through the structure to combine with the waves regenerated by wave overtopping.

The resulting wave system in the lee of the stone breakwater will then break and run up on the beach face. The breaking, runup, and water seepage into the beach face will ultimately dissipate the remaining wave energy except for a small portion of wave energy that will reflect seaward owing to the wave downrush on the beach face.

If we consider the floating breakwater in a little more detail, the incident wave power will equal the reflected wave power plus the transmitted wave power plus the power dissipated by the breakwater. For a monochromatic wave, if the water depth is constant and assuming the wave period remains constant,

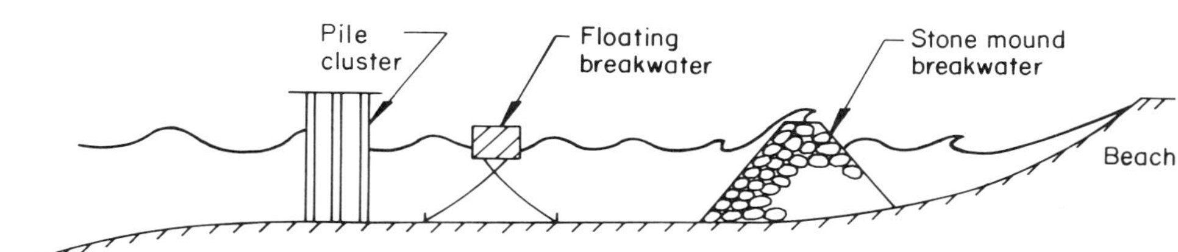

Figure 8.1 Wave–structure interaction examples.

this yields (from Eqs. 2.32 and 2.35), for a unit wave crest width,

$$(H_i)^2 = (H_t)^2 + (H_r)^2 + \frac{P_d n \gamma L}{8T}$$

or

$$1 = \left(\frac{H_t}{H_i}\right)^2 + \left(\frac{H_r}{H_i}\right)^2 + \frac{P_d n \gamma L}{8T(H_i)^2}$$

where H_i, H_r, and H_t are the incident, reflected, and transmitted wave heights and P_d is the dissipated wave power. The equation can be rewritten as

$$1 = C_t^2 + C_r^2 + \frac{P_d n \gamma L}{8T(H_i)^2} \tag{8.1}$$

where C_t and C_r are the wave transmission and wave reflection coefficients. These coefficients are the most commonly used definitions of wave reflection or transmission for a structure. If the wave energy dissipation is small or can be neglected, if the water depth is constant fore and aft of the structure, and if the wave period remains constant, then evaluation of one coefficient gives the value of the other. To evaluate the reflection coefficient from wave flume tests with monochromatic waves, the incident and reflected wave heights must be separated out of the combined surface time history record of the superimposed waves measured in front of the structure. A method for doing this is described in Section 2.5.3.

When irregular waves are used, different definitions of the wave reflection and transmission coefficients are needed. The incident, transmitted, and reflected components are now defined in terms of their total spectral energy (i.e., m_0 or the mean squared displacement of the water surface, see Eq. 6.15). A commonly used laboratory procedure (Goda and Suzuki, 1976) for measuring incident and reflected spectral energy components employs three wave gages in front of the structure and produces three estimates of the incident and reflected wave amplitude spectra (i.e., the incident and reflected wave amplitudes at each spectral frequency point). The three estimates are typically close, so an average value can be used. The square root of the sum or average of the square of these amplitudes might then be used to define the reflected to incident wave amplitude ratio (see Seelig, 1980).

In the remainder of this chapter the runup, overtopping, transmission, and reflection of waves that interact with various basic structure forms are considered. The emphasis is on the basic processes involved and the general nature of the results.

8.1 WAVE RUNUP ON STRUCTURES

The runup R of a wave on a slope, defined as the maximum vertical rise of the water above the still water line (see Fig. 3.5), was introduced in Section 3.8. Figure 3.6 shows a typical plot of the primary variables: the relative runup R/H_0' versus the inverse of the plane slope $1/m$ and the incident wave steepness H_0'/gT^2. It also demonstrates the importance of the breaking condition and, for broken waves, the breaker type. But Figure 3.6 is only for a monochromatic wave on a smooth plane slope with the toe of the slope being at a depth greater than $3H_0'$.

In prototype situations, several complexities develop. The slope may be complex rather than planar with a break at or above the still water line, and the slope may be porous and rough such as the face of a riprap revetment. Also, we are now dealing with a spectrum of waves. Individual waves in the spectrum interact in complex ways that depend on their break point and the type of breaking that occurs as well as the phasing between the uprush of one wave and the downrush of the preceding wave. Long wave effects produced by spectral wave grouping also cause a rise and fall of the nearshore water level which, in turn, affects wave runup.

8.1.1 Monochromatic Wave Runup

One of the early formulas for the runup of breaking monochromatic waves on a plane slope was given by Hunt (1959) and was based on a number of European and American laboratory data sets. Hunt's formula is

$$\frac{R}{H} = \frac{\tan \alpha}{\sqrt{H/L_0}} \tag{8.2}$$

where α is the slope angle (i.e., $\tan \alpha = m$). Note that the right-hand side of Eq. 8.2 is just the Iribarren number or surf similarity parameter (I_r, see Eq. 3.9) which, according to Hunt, is equal to the relative wave runup R/H. Equation 8.2, with H being the wave height approaching the slope, has been used for simple wave runup calculations. It includes the plane slope and the incident wave period, but is only for broken waves and neglects the slope conditions (porosity, roughness).

Several monochromatic wave runup experiments have been run for other than planar smooth slopes (e.g., see Battjes, 1974 and U. S. Army Coastal Engineering Research Center, 1984). But the most appropriate way to predict monochromatic wave runup for most slope conditions encountered in practice is to employ the *Shore Protection Manual* (U. S. Army Coastal Engineering Research Center, 1984) approach. This uses curves similar to Figure 3.6 to predict runup on a smooth plane slope and then corrects for slope condition and compound slopes.

Plots of R/H_0' versus H_0'/gT^2 and α (i.e., like Fig. 3.6) are given in the *Shore Protection Manual* for d_s/H_0' equal to 0, 0.45, 0.80, 1–3, and >3 (where d_s is the water depth at the toe of the slope and the first three conditions have a fronting slope of 0.10). The latter two are the most useful and are repeated here as Figures 8.2 and 8.3. Given the plane slope and the deep water unrefracted wave height and period for a slope having its toe submerged at least one wave height, one can apply Figure 8.2 or 8.3 to predict the monochromatic wave runup. Hunt's formula, which employs the wave height at the slope and applies to broken waves, would relate to the right-hand portions of Figures 8.2 and 8.3 (see Fig. 3.6 and related discussion).

The results shown in Figures 8.2 and 8.3 are based on small-scale wave flume tests. Some limited larger-scale tests indicate that the values predicted by these figures should be somewhat increased (Saville, 1987), particularly for the steeper slopes. For slopes of 1 : 1.5 down to 1 : 5 the increase varies from

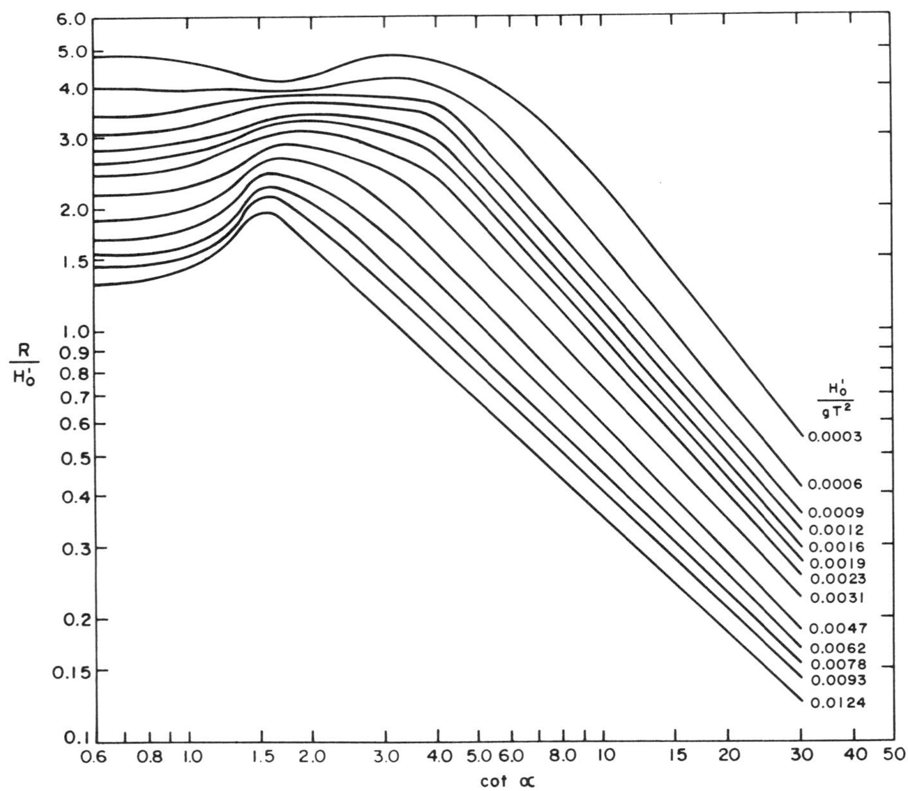

Figure 8.2 Wave runup, smooth impermeable slope, $1 < d_s/H_0' < 3$ (modified from U.S. Army Coastal Engineering Research Center, 1984).

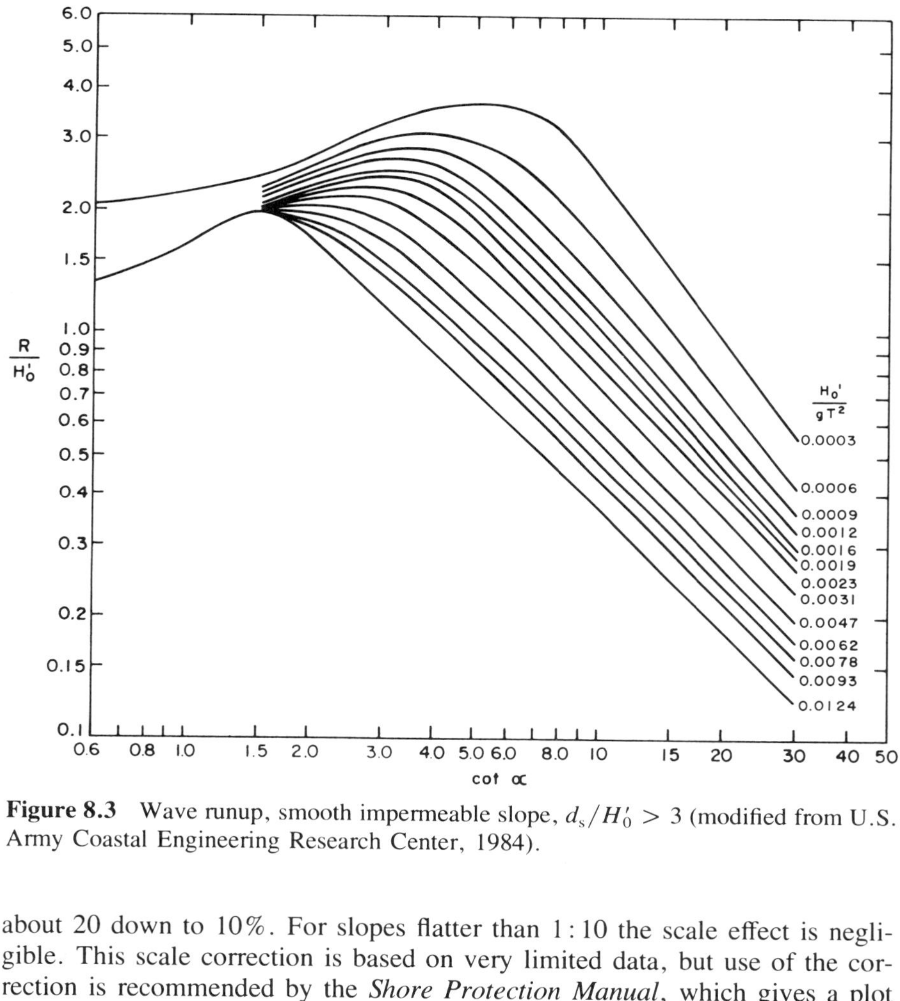

Figure 8.3 Wave runup, smooth impermeable slope, $d_s/H_0' > 3$ (modified from U.S. Army Coastal Engineering Research Center, 1984).

about 20 down to 10%. For slopes flatter than 1:10 the scale effect is negligible. This scale correction is based on very limited data, but use of the correction is recommended by the *Shore Protection Manual*, which gives a plot of recommended scale correction values.

Surface roughness and porosity reduce the runup to less than the value predicted by Figure 8.2 and 8.3. Battjes (1974) has summarized the results of many experiments with monochromatic wave runup on rough and porous slopes by introducing a factor r defined as the ratio of the runup on a rough porous slope to the runup on a smooth impermeable slope.

It has been demonstrated that the value of r for a given surface condition varies with the plane slope and the incident wave steepness. Savage (1959) found that r decreased with decreasing slope and incident wave steepness; Wagner (see Battjes, 1974) also found the same slope effect but an increase in r with increasing wave steepness. The disagreement may result from the different surface conditions tested. However, for design purposes, values of r are typically (Battjes, 1974; U. S. Army Coastal Engineering Research Center,

TABLE 8.1 Runup Reduction Factors *r* for Various Surface Conditions

Condition	*r* Value
Smooth, impermeable	1.0
Fitted stone or concrete blocks	0.85–0.9
Turf	0.85–0.9
One layer, stone rubble	0.8
Rounded stone mound	0.6–0.65
Two or more stone rubble layers	0.5
Tetrapods	0.5

1984) specified for a given surface condition independent of slope and incident wave steepness—see Table 8.1. The values tabulated in Table 8.1 can be used to estimate r for other surface conditions. For slopes consisting of more than one surface condition, the *Shore Protection Manual* recommends that the composite r value be determined by apportioning the individual values by the percent of surface length having that value.

The other condition that may generally be encountered is that of a compound slope—either a change in slope at a point where runup is active or a horizontal berm section near the midpoint of a slope. If the structure is important, model tests of wave runup should be conducted. A method proposed by Saville (1957) for calculating wave runup on composite slopes may be used with caution (see Battjes, 1974), particularly if there is a large horizontal berm in the slope for which the method tends to underestimate the actual runup. When the berm width is greater than about one fourth of the incident wave length and the berm is near the still water level, water sets up on the berm so a reformed wave can travel across the berm to runup on the next slope section.

Saville's method is demonstrated in Figure 8.4 where the actual composite slope is represented by a hypothetical slope that extends from the point of wave breaking to an estimated point of wave runup on the composite slope. The runup on this hypothetical slope is then determined from Figure 8.2 (with appropriate r factor) and compared with the estimated runup on the composite slope to see if the two values agree. If not, the procedure is repeated until agreement is achieved.

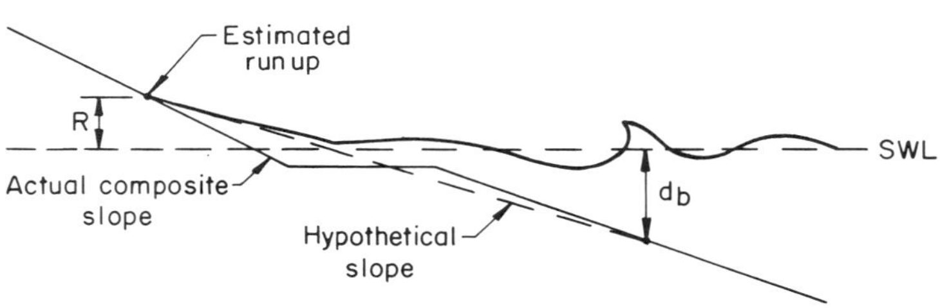

Figure 8.4 Saville's composite slope method.

8.1.2 Irregular Wave Runup

The simplest approach to predicting the runup of irregular waves for prototype design conditions is to select a representative wave height such as H_s, H_1 or H_{max}, and T_p and calculate the runup as if this were an independently acting wave. However, runup of a single wave in an irregular wave train is strongly affected by the effects of the preceding and following waves. For example, a high and long wave proceeded by a relatively lower and shorter wave will run up much higher than if it were preceded by another high and long wave. Small waves following larger waves tend to be obliterated by the downrush of the larger waves, so in a train of irregular waves there are typically fewer runup events than incident waves. These complexities require a more realistic approach to the prediction of irregular wave runup.

Figure 8.5 is a typical plot of the vertical elevation (above the still water level) of the tip of the irregular wave runup tongue on a plane slope, plotted versus time. Most investigators work with the individual peak runup values (R_1) when considering the statistics of wave runup. Some use the peak value between crossings of the mean uprush value (R_2) which is analogous to the zero upcrossing wave height. Unless noted we use the former.

Mase and Iwagaki (1984) investigated the statistics of irregular wave runup on relatively flat slopes (1 : 5–1 : 30). Pierson–Moskowitz spectra with six different peak periods and two levels of wave grouping were used in the wave flume experiments. Their results give some general insight into irregular wave runup. The ratio of the runups defined by R_2 to those defined by R_1 varied between 0.4 and 0.9, the ratio decreasing as the slope became flatter. The ratio of the number of wave runups (R_1) to the number of incident waves varied from just over 20% up to 90% as the Iribarren number (defined in terms of H_0/L_0) increased from 0.15 to 2.0. That is, for flatter slopes and steeper waves there are progressively less runups than incident waves.

The incident wave spectra with the greater amount of wave grouping produced a slightly higher relative maximum runup than the spectra with the lesser grouping. Although the highest wave in a wave record usually appears with a number of other high waves (see Chapter 6), Mase and Iwagaki found that the highest runup in a record often appeared alone. They attribute this to the highest

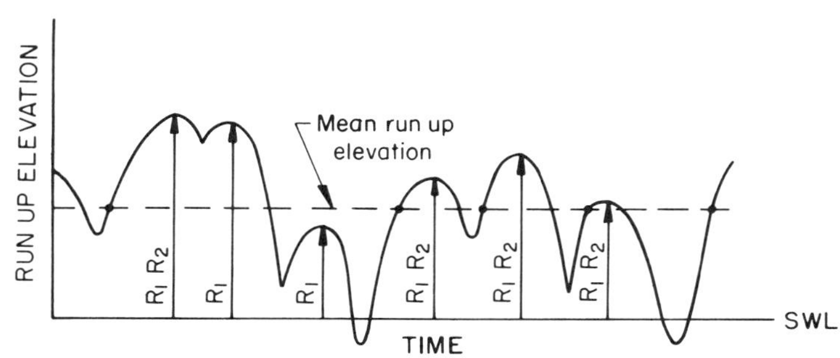

Figure 8.5 Typical wave runup record.

runup occurring when the preceding downrush of the highest runup (i.e., there is more of a wave-to-wave dependence for runup than for propagating waves).

A basic question when considering wave runup statistics is, what frequency distribution might the runup (R_1) values have? With this, as with the distribution of wave heights, we can then predict the runup for any probability value if we know the runup of one probability value. Several experimenters (see Ahrens, 1977a) have suggested that the Rayleigh distribution is appropriate (possibly conservative) for the runup of wind-generated waves. Ahrens (1983) conducted a detailed evaluation of the wave runup distribution on a plane smooth slope. He found that a Weibull distribution (see Table 7.1) could be used to best define the runup distribution where the shape factor α and the scale factor β varied with structure slope and incident wave conditions. But the shape factor varied from about 1 to less than 3 with a mean value around 2, and a shape factor of 2 for a Weibull distribution yields a Rayleigh distribution. Consequently, until more detailed information on the wave runup distributions and their dependencies are available, a Rayleigh distribution is best assumed for design purposes.

The assumption of a Rayleigh distribution for wave runup yields

$$\frac{R_p}{R_s} = \left(\frac{\ln (1/p)}{2}\right)^{1/2} \tag{8.3}$$

where R_p is the runup associated with a particular probability of exceedence p and R_s is the runup of the incident significant wave height as if it were a monochromatic wave. It must be emphasized that R_p is not the average runup of the highest p fraction of the runups, but it is the runup exceeded by the upper p fraction of the runups. The latter is a more useful value for runup analyses (e.g., to determine what fraction of the incident waves will overtop a sloped revetment crest). For example, if the incident significant wave height has a runup elevation of 2.5 m above the still water line on a riprap revetment having a crest elevation of 2 m above the still water line (runup calculated as if the structure crest were high enough to support the runup), then p = 0.28 or 28% of the wave runups would overtop the structure.

Equation 8.3 has been used for design owing to a deficiency of information from model tests and field experiments on the runup of irregular waves on a variety of structure shapes and surface conditions. Where such information is available or can be collected, it is preferable. The offshore profile geometry seaward of the still water line, for example, might cause breaking of some of the highest waves in the spectrum before they reach the structure, break, and runup. This seriously distorts the probability distribution of wave runup.

Recently, a number of laboratory and some field experiments have been conducted on irregular wave runup for a variety of structure shapes and surface conditions and for natural beaches. Wire staff gages placed on the slope or videos of wave motion past markers on the slope are typically used to make the irregular wave runup measurements. Some of the results that are of basic interest are discussed below.

The previously discussed laboratory study of irregular wave runup on smooth plane slopes by Mase and Iwagaki (1984) produced empirical formulas of the form given in Eq. 8.4.

$$\frac{R}{H_{s0}} = a\left(\frac{\tan\alpha}{\sqrt{H_{s0}/L_{s0}}}\right)^{b} \tag{8.4}$$

In Eq. 8.4, H_{s0} and L_{s0} represent the deep water significant wave height and length, and R represents either the maximum, the significant, or the mean runup. For these cases the coefficients a and b have the following values: R_{max}, $a = 2.32$, $b = 0.77$; R_s, $a = 1.38$, $b = 0.70$; R_{mean}, $a = 0.88$, $b = 0.69$. (Note that R_s is the average of the highest third of the wave runups.) Equation 8.4 has the general form of Hunt's formula (and the term in parenthesis is another form of the Iribarren number) but the coefficients a and b are not unity, as they would be for Hunt.

For design purposes concerning wave runup on rough and porous plane slopes one might use Eq. 8.4 with the r values given in Table 8.1. The ratio R_p/R_s might also be obtained from Eq. 8.3.

Ahrens and Heimbaugh (1988) report on wave flume experiments on irregular wave runup on riprap covered plane slopes for two different Army Corps of Engineers field sites. Three different stone sizes and structures slopes ranging from 1:2 to 1:4 were used. Their focus was on the maximum wave runup, a key factor in the selection of the revetment crest elevations. The results were presented by an equation of the form

$$\frac{R_{max}}{H_{m0}} = \frac{aI_r}{1 + bI_r} \tag{8.5}$$

where the Iribarren number is now defined as

$$I_r = \frac{\tan\alpha}{\sqrt{\dfrac{H_{m0}}{L_p}}} \tag{8.6}$$

L_p, as before, is the wave length at the structure toe calculated by linear wave theory using the spectral peak frequency; H_{m0} is also determined at the structure toe. The coefficients $a = 1.154$ and $b = 0.202$ fit quite well all of the data for the range of slopes and stone sizes used in the tests. A runup equation having the form of Eq. 8.5 has also been used to present other experimental results (see Ahrens and McCartney, 1975 and Seelig, 1980).

The establishment of set-back lines and the understanding of beach processes require an ability to predict wave runup on natural beaches. But beach profiles are quite irregular and they change as the incident wave climate changes. Beaches also tend to be much flatter than the slopes of structures for which most wave runup studies have been conducted. Since model studies of mobile beach profiles are not always successful (owing to beach response scale effects) in producing the correct beach profile and consequently the correct wave runup, the best approach to developing a method for predicting runup on beaches would be to conduct prototype field investigations.

Douglass (1990) surveys the scant amount of available guidance for predicting wave runup on beaches. The most useful field data set is from Holman (1986) where maximum runup values are related to the beach slope near the still water line and the incident deep water wave steepness (H_{m0}/L_{p0}). But a reanalysis of Holman's data by Douglass (1992) showed that there was little correlation between the runup and the beach slope, the deep water wave steepness being a better runup predictor. In Holman's experiments the beach slope varied from about 1:5 to 1:16. Douglass (1992) found that the relationship

$$\frac{R_{max}}{H_{m0}} = \frac{0.12}{\sqrt{\dfrac{H_{m0}}{L_{p0}}}} \tag{8.7}$$

best fit the Holman data. He offers no conjecture for the lack of correlation between maximum runup and the beach slope. More data sets taken at other sites are needed to confirm the form of Eq. 8.7 and to shed further light on the coefficient value 0.12.

Finally, brief mention should be made of recent developments in numerical modeling of wave runup on slopes that may be rough and have a nonplanar slope. The models only consider monochromatic waves but, in addition to the runup time history, will also yield the flow velocity field in the runup—a useful result when considering, for example, the stability of a rubble mound structure. Allsop et al. (1988) and Kobayashi et al. (1987) present recent examples of these numerical runup models.

8.2 WAVE OVERTOPPING OF STRUCTURES

If wave runup is sufficiently high, the tongue of water will rise higher than the crest elevation of the structure and produce wave overtopping. For the economic design of a structure it may be cost effective to allow wave overtopping and deal with the consequences. Aesthetic and functional considerations, such as a preference for not blocking an ocean view or not interfering with easy access to a fronting beach, may also make a lower structure more desirable. If the structure is a breakwater or jetty with water in the lee, wave transmission will result. The jet of water passing over the structure may also affect the top and lee side stability of the structure. If the structure has land on the lee side like a revetment, the flooding produced by the overtopping will have to be dealt with by a drainage system that may include a pumping system. The overtopped water may also cause erosion of the land in the lee of the structure.

If the structure fronts a coastal roadway, the allowable overtopping for more frequently occurring waves may have to be limited to allow safe use of the roadway.

Our primary concern is to predict the volumetric flow rate of water overtopping a structure. For irregular waves, both the average and peak flow rates will be of interest. For some purposes, the velocity of flow at the structure crest as the water overtops the structure might also be of interest, but this phenomenon has received little attention in the literature.

If the rate of wave overtopping exceeds the desired limit, a simple expedient is to construct a vertical, or preferably recurved, wall at the crest of the structure to deflect the water. An increase in the structure surface roughness and porosity will reduce the runup and resulting rate of overtopping (and wave reflection too). These considerations, as well as the structure's stability, may dictate the selection of structure surface composition. And any other feature that reduces wave runup, such as a berm section in the structure face, will also be helpful.

The common way to measure wave overtopping the laboratory is to catch the water in a container that is being weighed or has volumetric marks on the container wall. Continuous depth and velocity measurements at the structure crest, which are very difficult to make, have also been used to measure overtopping rates. When accomplished, this technique gives instantaneous overtopping rates rather than longer term averages. It is apparent that field measurements of wave overtopping rates are difficult to achieve, so little good field data on wave overtopping is available.

8.2.1 Monochromatic Wave Overtopping

The most obvious indicator of the amount of wave overtopping is the ratio of the structure freeboard F to the potential wave runup R (see Fig. 8.6), where the freeboard is defined as the vertical elevation of the structure crest above

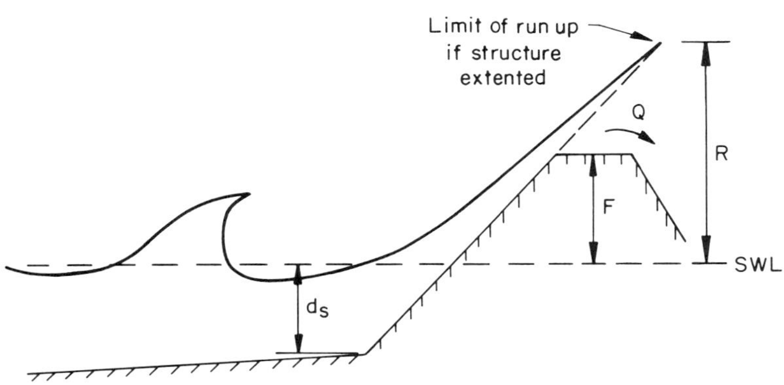

Figure 8.6 Wave overtopping definitions.

the still water level. The overtopping rate Q (total of intermittent volumes per time on a continuous basis) is usually determined per unit length of structure crest. Summaries of most of the laboratory studies on monochromatic wave overtopping are given by Battjes (1974) and Weggel (1976).

The most common relationship used to predict monochromatic wave overtopping of structures was developed by Weggel (1976) using data from a series of Corps of Engineers laboratory studies. Rather than work with the basic factors that control the wave runup, Weggel used the runup as determined in Section 8.1.1 as a basic independent variable. He essentially carried out a curve-fitting exercise employing two variables: the dimensionless crest elevation F/R and the dimensionless overtopping rate $Q^2/gH_0'^3$. This produced a relationship of the form

$$Q = (gQ_0^* H_0'^3)^{1/2} \exp\left[\frac{-0.217}{\alpha} \tanh^{-1}\left(\frac{F}{R}\right)\right] \tag{8.8}$$

where Q_0^* and α are empirical coefficients that depend on the structure geometry, the incident wave steepness H_0'/gT^2 and the relative depth at the structure toe d_s/H_0' (see Section 8.1.1).

Plots giving values for the empirical coefficients Q_0^* and α for a limited number of structure geometries and surface conditions are given in the *Shore Protection Manual* (U. S. Army Coastal Engineering Research Center, 1984). An example of these plots is given in Figure 8.7 which is for a smooth 1 : 1.5 slope and a crest width of 1.5 m. (The value of Q_0^* is the number in parenthesis and other number is the value of α.) The empirical coefficient values at each data point must be interpolated to find the appropriate value for the incident wave steepness and structure toe depth of interest.

The empirical coefficient plots only cover a limited number of structure conditions because of the lack of experiments that have been conducted. Some of the plots are based on a very small data set. However, other than conducting laboratory model tests, use of Eq. 8.8 with a judicious selection of empirical coefficient values is the best available procedure for predicting monochromatic wave overtopping rates. (It should be remembered that the quality of this prediction also depends on the quality of the predicted value for wave runup.) This may be done in a fairly straightforward manner and with some confidence for a plane slope; the runup on a seawall with a concave profile, however, is harder to define and predict for given incident wave conditions.

Another factor that is of major consequence, but not commonly included in wave flume experiments on wave overtopping, is the effect of wind. For most design wave conditions there is a strong onshore wind velocity. If $F > R$ it is unlikely that the wind will cause significant overtopping. When $F < R$ the overtopping action will be enhanced by the wind compared to what it would be with no wind. With a parapet wall that deflects runup upward or gives it a seaward component, the overtopping rate may be small without the wind, but is can be seen that the overtopping rate will be much larger with the onshore wind to carry the water landward.

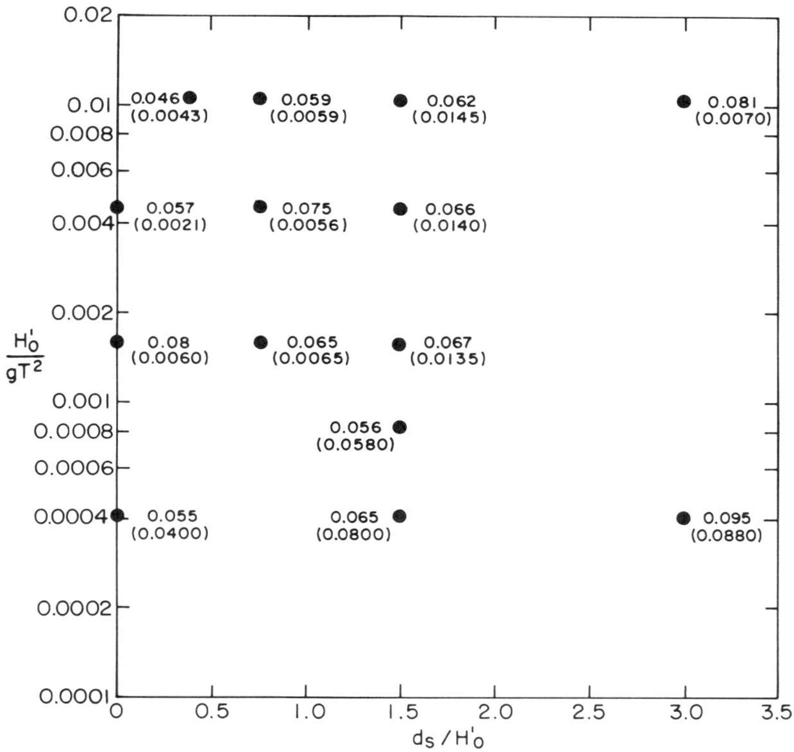

Figure 8.7 Wave overtopping of smooth plane 1 : 15 slope—empirical coefficients (Q_0^*) and α (from U.S. Army Coastal Engineering Research Center, 1984).

Most authors (e.g., Battjes, 1974 and Goda, 1985) believe that an onshore wind will have a significant effect on the wave overtopping rate, but little quantitative information on this affect is available. It is extremely difficult to model the wind effect in the laboratory—there is no reliable modeling law for the wind effect on turbulent surface tension affected flow of water.

The volume of water overtopping a structure during the runup of a wave can be determined by integrating the product of the flow velocity and depth during the overtopping interval. Using the numerical wave runup model of Kobyashi et al. (1987) discussed in Section 8.2, Kobyashi and Wurjanto (1989) were able to develop a numerical wave overtopping model for smooth and rough impermeable slopes. The results achieved with the numerical model for monochromatic wave overtopping of a smooth plane slope compared favorably with experimental data used to develop Eq. 8.8.

8.2.2 Irregular Wave Overtopping

As with wave runup, it is worthwhile to be able to estimate the probability of occurrence of the variable overtopping rates caused by irregular wave runup.

Employing the assumption of a Rayleigh distribution for wave runup, Ahrens (1977b) modified Eq. 8.8 for irregular wave overtopping to yield

$$Q_p = [gQ_0^* (H_0')_s^3]^{1/2} \exp\left[\frac{-0.217}{\alpha} \tanh^{-1}\left(\frac{F}{R}\right)\left(\frac{R_s}{R_p}\right)\right] \tag{8.9}$$

where R_s/R_p is given by Eq. 8.3 and Q_p is the overtopping rate associated with a probability of exceedence p (like R_p). Here $(H_0')_s$ is the incident deep water unrefracted significant wave height. Inherent in the application of Eq. 8.8 is the assumption that Q_0^*, α, and H_0' have constant values for the different components of the runup/overtopping distribution, which is not so but is necessary to make the equation applicable.

Equation 8.9 allows the calculation of the overtopping rate for any probability of exceedence. Ahrens (1977b) calculates overtopping rates for $p = 0.005$ as an indication of an extreme (low probability) rate. For example, for a 1 : 1.5 smooth-sloped structure (Fig. 8.7) with $\alpha = 0.06$ and $F/R_s = 0.5$ the $Q_{0.005}$ is 2.3 times the overtopping rate for the significant wave height as a monochromatic wave.

For the design of a drainage system, the average overtopping rate must be determined. One could use some average wave height and period and calculate the wave overtopping from Eq. 8.8. As a better approach, Ahrens (1977b) integrated Eq. 8.9 numerically using 0.005 intervals to obtain average discharge rates. His results are shown in Figure 8.8, where Q_s is the overtopping rate from Eq. 8.8 for the significant wave height and Q_a is the average irregular wave overtopping rate.

Goda (1985) presents dimensionless plots for wave overtopping rates at vertical-faced structures that were developed using a numerical integration approach somewhat similar to that used by Ahrens, with combined Japanese and Corps of Engineers monochromatic wave overtopping data. The need to first determine the runup was eliminated by Goda as he relates the overtopping rate directly to the incident wave characteristics. Considering all of the caveats that have been mentioned, Eq. 8.9 and Figure 8.8 as well as the work of Goda should be used with extreme caution to determine quantitative overtopping rates. Perhaps they are more useful in obtaining qualitative indications of the effects of design alternatives for coastal structures.

During the past decade some irregular wave laboratory experiments have been conducted to develop irregular wave overtopping prediction procedures. To present the experimental results, the experimenters all relate a dimensionless overtopping rate to a dimensionless freeboard by either a plot of the data or an empirical equation with coefficients that depend on the particular structure geometry and incident wave conditions. Table 8.2 tabulates the parameters used in four of these studies.

Aminti and Franco (1988) conducted a series of wave tank tests on the irregular wave overtopping rate on a typical rubble mound breakwater with a crown wall. They employed each of the overtopping rate-freeboard sets of

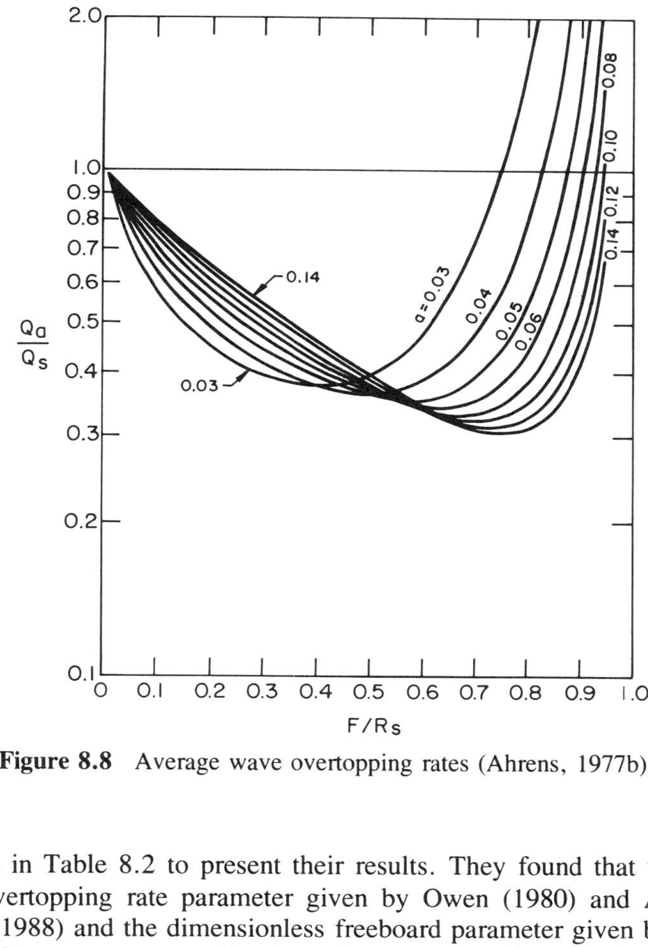

Figure 8.8 Average wave overtopping rates (Ahrens, 1977b).

parameters in Table 8.2 to present their results. They found that the dimensionless overtopping rate parameter given by Owen (1980) and Allsop and Bradbury (1988) and the dimensionless freeboard parameter given by Sawargi et al. (1988) were the most effective and most practical to use for presenting their data. When these parameters were plotted on log-log paper, straight lines resulted, so equations of the form $Q_* = A(F_*)^{-B}$ could be used to present the data. Here Q_* is the dimensionless overtopping rate, F_* is the dimensionless freeboard, and the coefficients A and B depend on the structure geometry. The coefficient A showed relatively wide variation as the structure surface condition and slope varied, whereas the coefficient B varied from just over 2 to less than 4. This power form of equation demonstrates that the average overtopping rate increases rapidly as the freeboard is reduced.

At this time it is not possible to present more general relationships for irregular wave overtopping of structures than those discussed above where the specific data or related empirical equation depends on the details of the structure geometry. The designer must seek out the various model study results to see which most closely fit the structure design under consideration. Or model tests may have to be conducted if the structure is sufficiently important and unique.

TABLE 8.2 Dimensionless Overtopping Parameters Used by Various Authors

Reference	Dimensionless Overtopping Rate	Dimensionless Freeboard
Owen (1980)	$\dfrac{Q_a}{\bar{T}gH_s}$	$\dfrac{F}{\bar{T}\sqrt{gH_s}}$
Ahrens and Heimbaugh (1986)	$\dfrac{Q_a}{\sqrt{gH_{m0}^3}}$	$\dfrac{F}{(H_{m0}^2 L_p)^{1/3}}$
Allsop and Bradbury (1988)	$\dfrac{Q_a}{\bar{T}gH_s}$	$\dfrac{F}{\bar{T}\sqrt{gH_s}}\left(\dfrac{F}{H_s}\right)$
Sawaragi, Deguchi, and Park (1988)	$\dfrac{Q_a}{\sqrt{gL_p H_s}}$	$\dfrac{F}{H_s}$

8.3 WAVE TRANSMISSION PAST STRUCTURES

The design of most coastal structures that have water on the lee side requires that the expected amount of wave transmission be determined. Wave runup and overtopping of stone mound breakwaters will cause wave transmission by the regeneration of waves on the lee side. And energy may be transmitted through the structure if it is sufficiently permeable.

Some stone mound structures are designed with a submerged crest, where wave transmission is accomplished by the direct passage of waves over the structure. There is a decrease in the transmitted energy owing to reflection and dissipation of some of the incident wave energy. For an incident wave spectrum, a portion of the waves may break on the structure to regenerate waves in the lee while another portion of the waves pass directly over the structure. The resulting transmitted wave spectrum is likely more complex than the incident spectrum.

Somewhat like very porous rubble mound structures, structures composed of a cluster of piles allow direct wave transmission. But the water particle motion inside the pile cluster is significantly modified and related energy dissipation results in lower transmitted wave heights. Moored floating structures also allow a significant portion of the transmitted wave energy to directly pass the structure, but a component of the transmitted energy is produced by wave regeneration by the moving structure.

A consequence of the complex nature of most wave transmission processes is that techniques for the prediction of wave transmission rely heavily on empirical studies that are conducted in the laboratory or the field. We consider here some of the structure types for which wave transmission is a common concern. The emphasis is on techniques for predicting wave transmission, usually in the form of a transmission coefficient C_t. Where appropriate we also consider the details of the transmission process and its resulting impact on the nature of the transmitted waves.

8.3.1 Nonsubmerged Stone Mound Structures—Wave Transmission by Overtopping

Besides the complexities that control the amount of wave runup and overtopping of a stone mound structure, the resulting wave transmission by overtopping may, in addition, depend on the lee side slope and surface condition. Several laboratory studies have been conducted on wave transmission by overtopping of stone mound structures. Most have employed monochromatic waves and yielded a plot of the transmission coefficient as a function of incident wave height and period as well as parameters that define the structure geometry. For a discussion of these studies and several of the results see Seelig (1980) and the U. S. Army Coastal Engineering Research Center (1984).

Seelig (1980) conducted a very extensive set of monochromatic and irregular wave transmission experiments for stone mound structures. He reports that a significant portion of the transmitted wave energy had higher frequencies that were harmonics of the incident wave frequency. For irregular waves, the main portion of the transmitted wave spectrum was often more peaked (higher spectral peakedness parameter Q_p), but some spectral energy also appeared at harmonics of the spectral peak frequency. This tendency for a shift of energy to harmonic periods decreased as the amount of wave transmission increased.

Seelig developed a simple formula for predicting the wave transmission coefficient (defined in terms of the incident and transmitted spectral energies) for stone mound breakwaters. This equation, which is sufficiently valid for both monochromatic and irregular waves, is

$$C_t = C\left(1 - \frac{F}{R}\right) \tag{8.10}$$

and it is recommended that the application of Eq. 8.10 be limited to the relative depth (d/gT^2) range of 0.03–0.006. The coefficient C depends on the breakwater crest width B according to

$$C = 0.51 - \frac{0.11B}{d_s + F} \tag{8.11}$$

where $d_s + F$ is the breakwater crest elevation above the sea floor (see Fig. 8.9). Equation 8.11 applies to the range of $B/(d_s + F)$ between 0 and 3.2.

When applying Eq. 8.10 for irregular waves, the wave runup is to be calculated as the monochromatic runup for the mean wave height (i.e., $0.63H_s$). The report by Seelig (1980) contains an extensive set of transmission data based on laboratory tests conducted on 19 different breakwater cross sections. The designer should refer to these data because specific results applicable to a given structure geometry may be available.

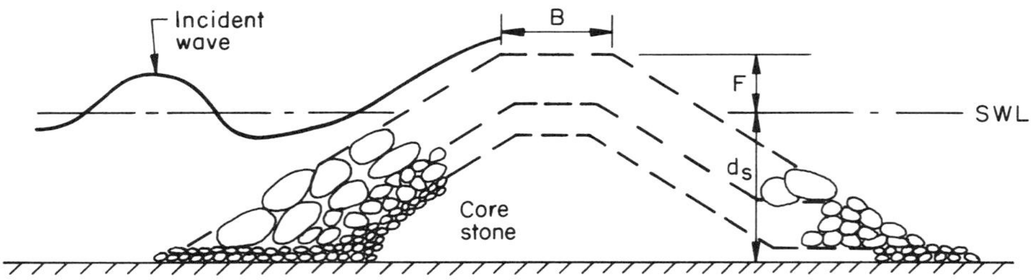

Figure 8.9 Typical stone mound breakwater cross section.

8.3.2 Nonsubmerged Stone Mound Structures—Wave Transmission Through Structure

Stone mound structures are commonly composed of layers that have decreasing stone size toward the center of the structure (Fig. 8.9). The large exterior armor stones need to remain stable under wave attack, whereas the finer core stones must prevent wave transmission through the structure. But some wave transmission through the structure may occur, particularly for long period waves having a low steepness (e.g., the tide would only be slightly reduced by a rubble mound structure, but steep wind waves would have negligible transmission through the structure). Note that although the wave transmission coefficient for energy transmission through a structure decreases as the wave steepness increases, the percentage of wave transmission by overtopping increases with increasing wave steepness.

Madsen and White (1976) developed a numerical procedure for calculating the wave transmission through a layered stone mound structure. Seelig (1980) presented a computer program that simplifies the use of the Madsen and White procedure and he gives comparisons of wave transmission calculated by their procedure with results from his laboratory study. As a consequence, the Madsen and White procedure is recommended for calculation of wave transmission through porous stone mound structures where it is anticipated to be significant. A resulting combined wave transmission coefficient would be given by

$$C_t = \sqrt{C_{tt}^2 + C_{to}^2} \tag{8.12}$$

where C_{tt} and C_{to} are the individually determined coefficients for wave transmission through and over the structure. The development of Eq. 8.12 is analogous to Eq. 8.1 in that wave energy and power are related to the wave height squared.

The Madsen and White (1976) procedure first calculates the wave dissipation due to runup and rundown on the structure seaward face, assuming the wave does not break. Wave reflection is also determined. The remaining wave energy propagates into the structure where turbulent dissipation occurs. Assuming horizontal water particle motion, calculations of energy dissipation are made for a homogeneous rectangular breakwater that is hydraulically equivalent to the actual layered structure; the equivalence being based on the two structures having the same turbulent flow resistance. Application of the procedure requires a knowledge of the incident wave height and period, the water depth, the breakwater layer geometry, and the stone sizes and in-place porosity. Using the mean wave height and spectral peak period for irregular waves, Seelig (1980) found that the Madsen and White procedure could also be used to calculate C_{tt} for irregular waves.

8.3.3 Low-Crested Stone Mound Structures

Although stone mound structures having their crest located at or below the design water level have been built for a long time, they have recently increased in popularity. Their wave transmission coefficients are obviously much higher than nonsubmerged breakwaters, but they are much less expensive to build and they don't obstruct the view. Often they simply consist of a homogeneous wide-graded mass of stone. A recreational advantage is that low-crested breakwaters have a high transmission coefficient for low everyday waves but, as the incident wave amplitude increases, the wave transmission coefficient generally decreases. They have been used in tandem with a conventional nonsubmerged breakwater which is placed in their lee (e.g., Cox and Clark, 1992)—the combined cost of the two structures is less than a single structure having the same operational criteria.

As with the other topics covered in this chapter, a number of wave transmission studies of low-crested breakwaters have been conducted, first with monochromatic waves and then with irregular waves (for summaries see Seelig, 1980 and Van der Meer and Angremond, 1992). Van der Meer and Angremond (1992) collected the available irregular wave laboratory data and, with the addition of their own data, developed a comprehensive procedure for predicting wave transmission coefficients for low-crested structures.

They defined the irregular wave transmission coefficient in terms of the incident and transmitted significant wave heights. Initially, employing the data collected from other authors, they correlated C_t with two parameters that had been used by the other authors: F/H_i and $(F/H_i)(H_s/gT_p^2)^{0.5}$. The second term, which includes the incident wave steepness, did not improve the correlation with the transmission coefficient. For both correlations there was still a significant amount of scatter, although some of the larger scatter was for data collected with low incident wave heights. Figure 8.10 is a plot of the resulting transmission coefficient versus relative freeboard for low-crested breakwaters, based on the data plot from Van der Meer and Agremond (1992). The value of C_t should approach zero at large positive values of dimensionless freeboard and unity at large negative values of freeboard. Figure 8.10 would be appropriate to use for preliminary design of low-crested structures.

To improve the correlation for C_t Van der Meer and Angremond (1992) introduced the median diameter D_{50} of the stone used to construct the structure.

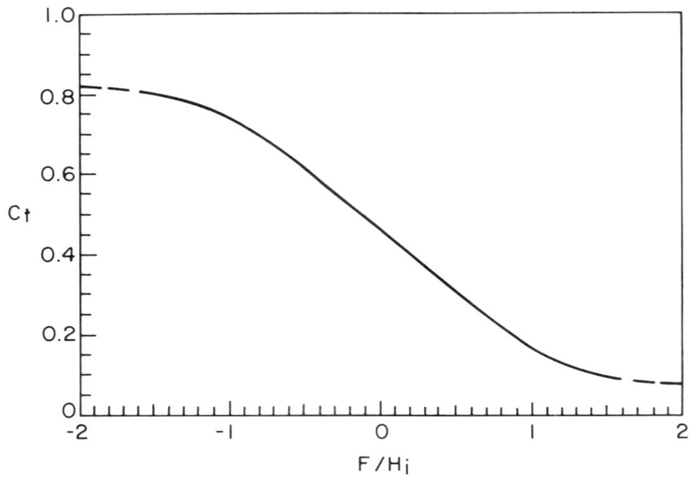

Figure 8.10 Coefficient of transmission for a low-crested stone mound structure (modified from Van der Meer and Angremond, 1992).

(The stone size has a direct relationship with the design wave height for a stable structure.) They then produced a correlation between C_t and F/D_{50} with secondary factors: H_s/gT_p^2, H_i/D_{50}, and B/D_{50}. These correlations are presented in somewhat complex formulas which apply between the limits $1 < H_i/D_{50} < 6$ and $0.01 < 2\pi H_s/gT_p^2 < 0.05$. These limits are based on the range of data tested, but the upper bounds are also physical—the first ($H_i/D_{50} > 6$) is a limit of structure stability and the second ($2\pi H_s/gT_p^2 > 0.05$) is a limit of wave stability owing to breaking. A significant drawback in applying the formulas involving D_{50} is the difficulty of their use with solid rather than stone mound low-crested structures.

One would also expect a shift in the frequencies at which some of the wave energy is concentrated as waves transmit into the lee of a low-crested breakwater. If the waves are sufficiently high and the freeboard positive or slightly negative, the waves will break and regenerate waves which should transfer some energy to higher harmonic frequencies, as discussed in Section 8.3.1. But, even if the waves do not break but are only disturbed as they pass over the structure, a shift will occur. Chandler and Sorensen (1972) investigated the transmission of nonbreaking monochromatic waves that pass over a submerged streamlined offshore bar profile. A portion of the wave energy in the lee of the structure had frequencies that were first, second, and sometimes third harmonics of the fundamental incident frequency.

8.3.4 Floating Breakwaters

Moored floating breakwaters have some distinct advantages over fixed breakwaters: (1) they are more adaptable to water level fluctuations such as those that occur in reservoirs and tidal areas; (2) they are more economical for deep water installations; (3) they are mobile and can be relocated; and (4) they interfere less with water circulation and fish migration. But they also have major limitations: (1) they are articulating structures and thus prone to damage at connecting joints between units and at mooring line connectors; (2) they may break loose and cause damage to other structures in their vicinity; and (3) their performance is extremely dependent on the period of the incident waves. This last factor sets strong limits on where floating breakwaters can be deployed in terms of the wind speed, duration, and fetch to which they may be exposed.

Floating breakwaters function by reflecting wave energy, usually by virtue of a vertical face that is held fairly stable relative to the water particle motion in the incident waves, and by turbulent dissipation of the kinetic energy in the incident waves—either by causing waves to break over the top of the breakwater or by generating turbulence when water particle motion in the waves is disturbed by the structure.

A wide variety of floating breakwater sizes and shapes have been developed (see Hales, 1981). The three most common generic types in use (Fig. 8.11) are the prism, the catamaran, and the flexible assembly (usually interconnected scrap tires). Figure 8.12 shows wave transmission coefficients for a typical

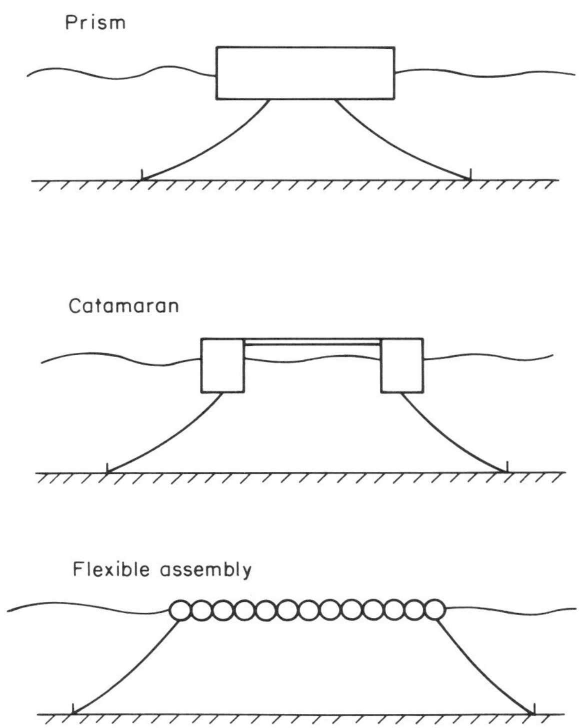

Figure 8.11 Common generic types of floating breakwaters.

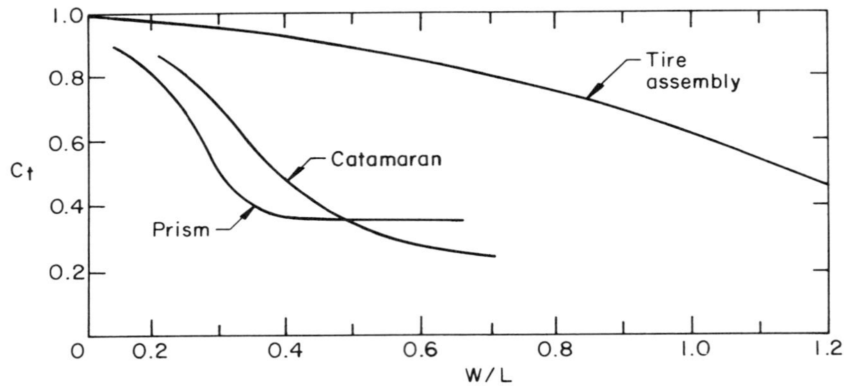

Figure 8.12 Transmission coefficients for three floating breakwaters (from Giles and Sorensen, 1979; Hales, 1981).

representative of each generic type. The transmission coefficient is plotted as a function of the breakwater dimension in the direction of wave propagation W divided by the incident wave length. This is the most common procedure for plotting wave transmission data.

The data yielding Figure 8.12 are all derived from monochromatic wave laboratory experiments. The prism (Hales, 1981) is a concrete box having a width of 4.88 m and a draft of 1.07 m, tested in a water depth of 7.6 m. The catamaran (Hales, 1981) has two pontoons 1.07 m wide with a 1.42-m draft and a total width (W) of 6.4 m. The water depth for the catamaran experiments was also 7.6 m. The scrap tire assembly breakwater (Giles and Sorensen, 1979) is made up of tires set in the "Goodyear module" geometry to yield a width of 12.8 m and a draft of one tire diameter. The water depth for these experiments was 3.96 m. The three breakwaters were all moored with a line fore and aft. Other mooring arrangements or variations in the mooring tautness would somewhat alter the transmission coefficient values.

Note the increase in the wave transmission coefficient as the dimensionless width decreases, that is, for a given breakwater width, as the incident wave period increases. For a given water depth, the longer the wave period, the smaller the relative depth (d/L) and, consequently, the smaller the percentage of wave kinetic energy that is concentrated near the water surface where the floating breakwater is located. (A floating breakwater is more effective for a short wave period that produces a deep water wave than for a long wave period that produces a shallow water wave in the same water depth.)

Sorensen (1990) conducted an analysis to define general wind speed, fetch, and duration guidelines for the deployment of floating breakwaters. Starting with allowable wave heights for vessels moored in a harbor, and employing the wave transmission characteristics of the three breakwaters shown in Figure 8.12, the limiting ranges of wind speed, fetch, and duration were calculated using the wave forecasting relationships in Section 6.5.3. An allowable wave height of 2 ft (0.61 m) based on criteria from the U. S. Army Corps of Engineers (1984) and the Canadian Small Craft Harbors Directorate (1985) was used in the analysis. For each incident wave period (assumed to be T_s), this yields an allowable incident wave height (assumed to be H_s) for each breakwater from Figure 8.12. This, in turn, yields wind speech, fetch, and duration combinations that will produce the allowable significant wave height and period.

The results are depicted in Figure 8.13, which gives the maximum allowable wind speed as a function of fetch and the related required duration, to generate the limiting wave height of 0.61 m behind the particular floating breakwater. The dominant point demonstrated by Figure 8.13 is the limited fetch to which a floating breakwater can be exposed. For example, for a representative design wind speed of 60 mph (26.8 m/sec), the fetch should not exceed 2–3 mi (3.2–4.8 km). Any water body having a much larger fetch would be unsuitable for one of these floating breakwaters owing to excessive wave action in the lee of the structure

Another wave-related concern in the design of floating breakwaters is the wave-induced mooring load on the anchor system, particularly the peak mooring load, which might cause the anchor system to fail. Mooring load data for various types of floating breakwaters are presented in Hales (1981) and Harms et al. (1982). Although wave transmission is primarily dependent on the incident wave period (for a given structure and water depth), the peak mooring load depends primarily on the incident wave height with wave period being of secondary importance (Harms et al., 1982).

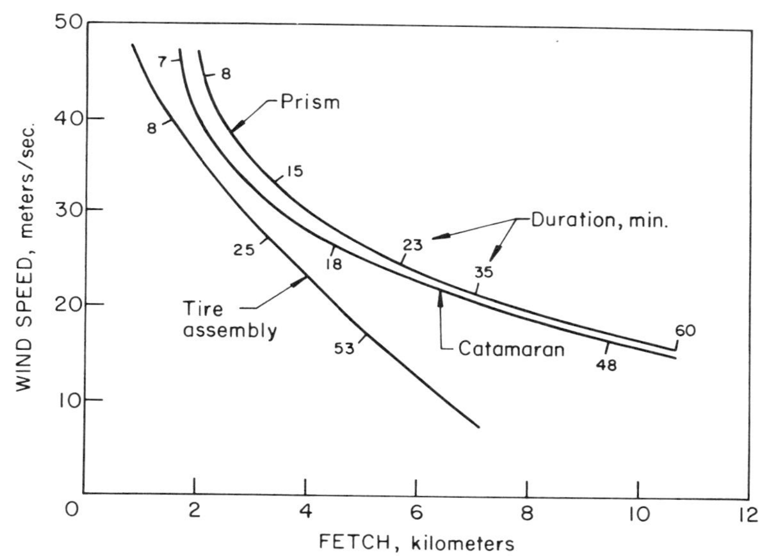

Figure 8.13 Allowable wind conditions for three floating breakwaters (Sorensen, 1990).

8.3.5 Vertical Thin Rigid Barriers

Vertical thin rigid barriers are occasionally used as coastal structures, particularly where the wave loading is relatively small and where wave reflection is not a major concern. Recently, for example, they have been used as breakwaters in small craft harbors where the barrier extends above the water surface but not completely to the bottom (see Gilman and Nottingham, 1992 and Lott and Hurtienne, 1992). The opening on the bottom decreases structure costs and allows for water circulation and fish migration.

Goda (1969) carried out monochromatic wave transmission tests for a thin vertical barrier that extended from the bottom to above or below the water surface. His results are shown in Figure 8.14, where the freeboard may have positive or negative values. He also conducted some experiments with irregular waves and, using a wave transmission coefficient defined in terms of the significant height of the incident and transmitted waves, he found that Figure 8.14 is also applicable for irregular waves (Goda, 1985). As with other overtopped structures, he found that a portion of the transmitted wave energy has first and second harmonic frequencies so that the significant wave period is reduced. For incident irregular waves having a Rayleigh height distribution, the transmitted wave heights had a wider range than the Rayleigh distribution. This is expected when the transmission coefficient is dependent on the incident wave height.

Vertical barriers that do not extend to the bottom allow wave transmission

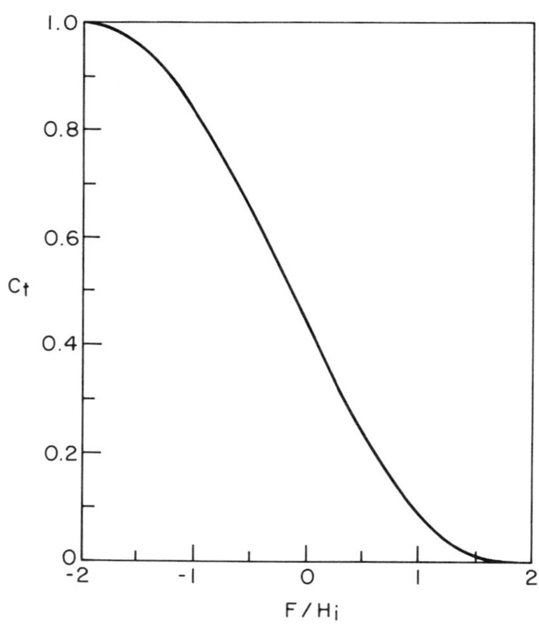

Figure 8.14 Wave transmission for a thin vertical barrier (from Goda, 1969).

under the structure. Typically they are designed so that there is no wave overtopping and employed where the incident wave period is relatively short (d/L is relatively large) so wave transmission is not large. Wiegel (1960) developed a simple theory for wave transmission by assuming that the portion of wave power in the water column below the barrier is the power that transmits past the barrier. For monochromatic waves using the small amplitude wave theory this yields a transmission coefficient given by

$$C_t = \left[\frac{\dfrac{2k(d-h)}{\sinh 2kd} + \dfrac{\sinh 2k(d-h)}{\sinh 2kd}}{1 + \dfrac{2kd}{\sinh 2kd}} \right]^{1/2} \tag{8.13}$$

where $k = 2\pi/L$ and h is the vertical extent of the barrier below the still water surface. As the wave steepness, and thus the water particle velocity increases, the energy dissipation owing to flow separation at the barrier tip should increase. This dissipation is not included in Eq. 8.13. However, limited monochromatic wave experiments by Wiegel (1960) and irregular wave tests by Gilman and Nottingham (1992) indicate that Eq. 8.13 may be used for preliminary design calculations.

8.4 WAVE REFLECTION FROM STRUCTURES

The design of coastal projects often requires that the anticipated reflection of wave energy from any boundary be quantified. Wave reflection in harbors may lead to increased wave agitation and resulting damage to moored vessels, mooring lines, and docking facilities. Vessel navigation may be made more difficult by increased wave steepness in regions of high wave reflection. And wave reflection from structures typically produces greater bottom agitation and scour at the base of the structure. The selection of structure type for a project may partially depend on the structure's wave reflection characteristics.

8.4.1 Shore Structures and Beaches

The wave reflection coefficient for a smooth plane impervious slope depends on the incident wave steepness and the structure slope. A major effect of these two parameters is to determine whether waves break on the slope and dissipate energy that might otherwise reflect. These factors are combined in the Iribarren number given by Eq. 3.9 for monochromatic waves and Eq. 8.6 for irregular waves. If the slope is rough and porous, there should be a reduction of wave reflection compared to the smooth impervious slope owing to increased wave energy dissipation in the runup and rundown on the slope. But note from Eq. 8.1 that if there is no wave transmission past the structure and half of the incident wave energy is dissipated, the value of C_r is 0.71, so the reflected wave height is only reduced 29%.

The most efficient wave absorber is a porous flat slope. But space limitations and widely varying water depths, as may be the case in harbors or reservoirs, often require the use of vertical faced porous structures to reduce wave reflection. Massive concrete blocks with intricate openings to generate turbulence or rows of wave screens consisting of piles or perforated sheets have been used. A judicious choice of screen porosity and spacing (so that the screens are at the point of maximum horizontal water particle motion for the resulting standing wave in the screen system) improves the wave energy dissipation (Allsop and Hettiarachchi, 1988 and Jamieson and Mansard, 1987). For a given wave absorber geometry and water depth, the reflection coefficient would depend on the incident wave steepness.

For sloped coastal structures and beaches a number of empirical equations relating the reflection coefficient and the Iribarren number have been developed from laboratory experiments. These equations most commonly take the form

$$C_r = \frac{aI_r^2}{b + I_r^2} = \frac{a}{1 + \dfrac{b}{I_r^2}} \tag{8.14}$$

where the values of a and b depend primarily on the structure surface condition and to a small extent on the slope and whether monochromatic or irregular waves are used (Seelig and Ahrens, 1981). The general form of this equation is demonstrated by Figure 8.15 for a smooth slope exposed to irregular waves where $a = 1.1$ and $b = 5.7$. The reflection coefficient regularly increases as

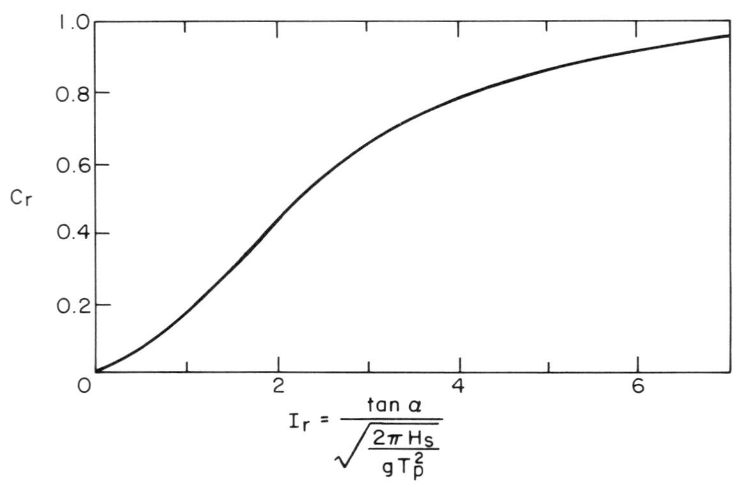

Figure 8.15 Irregular wave reflection, smooth plane slope (Allsop and Hettiarachchi, 1988).

the Iribarren number increases, approaching unity as the Iribarren number approaches a value of 10.

Summaries of wave reflection coefficients for beaches and common types of sloped coastal structures are given in terms of Eq. 8.14 by Allsop and Hettiarachchi (1988) and Seelig and Ahrens (1981). Some of the results are as follows:

Beaches. Laboratory measurements of wave reflection from beaches may suffer from scale effects in the model replication of the prototype beach profile, surface roughness, and porosity. Also, the Iribarren number for a given beach profile depends on the value chosen for the slope of the profile. Consequently, plots of data in the form of Eq. 8.14 show significant scatter. For average reflection coefficients at each Iribarren number Seelig and Ahrens (1981) suggest that $a = 0.5$ and $b = 5.5$ be used.

Stone Mounds. Breakwaters, revetments, and other stone mound slopes have a greatly reduced reflection owing to energy dissipation by the various layers of stone. Seelig and Ahrens (1981) recommended a range of values for a and b depending on the number of stone layers, the relative water depth (d/L), and the ratio of the incident to breaking wave height. For conservative preliminary calculations one can use $a = 0.6$ and $b = 6.6$.

Concrete Armor Units. Allsop and Hettiarachchi (1988) present reflection coefficients for stone mound structures armored with a variety of concrete armor units. Some of their results are:

Structure	a	b
Dolos (monochromatic waves)	0.56	10.0
Cobs (monochromatic waves)	0.50	6.54
Tetrapods (irregular waves)	0.48	9.62
Sheds (irregular waves)	0.49	7.94

Note, that the second form of Eq. 8.14 indicates that C_r approaches the value of a at higher values of I_r. This indicates that peak reflection coefficient values for stone and concrete unit armored mound structures are around 0.5.

8.4.2 Bragg Reflections

Any deviation of the bottom profile geometry from a horizontal plane will cause the reflecting of some wave energy if the waves are transitional or shallow water waves. A sloping bottom need not extend above the mean water line to be a wave reflector. Usually the reflection coefficient for these bottom profile variations is small. However, if the bottom has a sinusoidal profile (or other uniformly spaced undulations), significant wave reflections can occur for selected wave frequencies. These are known as Bragg reflections after a similar phenomenon in optics.

Bragg reflections occur when a resonance develops between the surface waves and bottom undulations. These resonant oscillations cause a portion of the incident wave energy to reflect. The reflections are maximum when the incident wave length is twice the wave length of the bottom undulations. The band width for significant reflection is quite narrow, but can be widened by staggering the spacings of the undulations, which in turn reduces the reflection coefficient for a particular incident wave frequency. Generally, the reflection coefficient increases linearly as the number of undulations increases. Reflection also increases with an increase in the undulation amplitude and a decrease in the water depth. See Davies and Heathershaw (1984) for a theoretical analysis and laboratory experiments on this phenomenon and Mei (1985) for further analytical work. A numerical computation procedure developed by Kirby (1987) allows one to calculate the reflection coefficient for nonsinusoidal undulation geometries. For appropriate undulation geometries and water depths, reflection coefficients can significantly exceed a value of 0.5.

Natural bottom undulations occur in the form of sand waves or nearshore bars generated by wave breaking. Mei (1985) postulates that a nearshore bar system can cause Bragg reflection to set up a standing wave pattern seaward of the bar system that will, in turn, cause the bar system to grow in the seaward direction. Bailard et al. (1990, 1992) investigated the possibility of constructing a series of low height, shore parallel submerged bars as a shore protection device. Their theoretical, laboratory, and field studies indicate that such structures may have merit in reducing shore erosion during storms. The economics and construction feasibilities of such a shore protection device require further investigation.

REFERENCES

Ahrens, J. P. (1977a), "Prediction of Irregular Wave Runup," Coastal Engineering Technical Aid 77-2, U. S. Army Coastal Engineering Research Center, Ft. Belvoir, VA.

Ahrens, J. P. (1977b), "Prediction of Irregular Wave Overtopping," Coastal Engineering Technical Aid 77-7, U. S. Army Coastal Engineering Research Center, Ft. Belvoir, VA.

Ahrens, J. P. (1983), "Wave Runup on Idealized Structures," *Proceedings, Coastal Structures '83 Conference*, American Society of Civil Engineers, Arlington, VA, pp. 925–938.

Ahrens, J. P. and McCartney, B. L. (1975), "Wave Period Effect on the Stability of Riprap," *Proceedings, Civil Engineering in The Oceans III*, American Society of Civil Engineers, Newark, DE, pp. 1019–1034.

Ahrens, J. P. and Heimbaugh, M. S. (1986), "Irregular Wave Overtopping of Seawalls," *Proceedings, Oceans '86 Conference*, Institute of Electronic and Electrical Engineers, Washington, DC, pp. 96–103.

Ahrens, J. P. and Heimbaugh, M. S. (1988), "Approximate Upper Limit of Irregular Wave Runup on Riprip," Technical Report CERC-88-5, U. S. Army Waterways Experiment Station, Vicksburg, MS.

Allsop, N. W. H. and Bradbury, A. P. (1988), "Hydraulic Effects of Breakwater Crown Walls," *Proceedings, Breakwaters '88 Conference*, Institution of Civil Engineers, Eastbourne, UK, pp. 181–183.

Allsop, N. W. H. and Hettiarachchi, S. S. L. (1988), "Reflections from Coastal Structures," *Proceedings, 21st International Conference on Coastal Engineering*, American Society of Civil Engineers, Malaga, Spain, pp. 782–794.

Allsop, N. W. H., Smallman, J. V., and Stephens, R. V. (1988), "Development and Application of a Mathematical Model of Wave Action on Steep Slopes," *Proceedings, 21st International Conference on Coastal Engineering*, American Society of Civil Engineers, Malaga, Spain, pp. 281–291.

Aminti, P. and Franco, L. (1988), "Wave Overtopping on Rubble Mound Breakwaters," *Proceedings, 21st International Conference on Coastal Engineering*, American Society of Civil Engineers, Malaga, Spain, pp. 770–781.

Bailard, J. A., DeVries, J., Kirby, J. T., and Guza, R. T. (1990), "Bragg Reflection Breakwater: A New Shore Protection Method?" *Proceedings, 22nd International Conference on Coastal Engineering, American Society of Civil Engineers*, Delft, pp. 1702–1715.

Bailard, J. A., DeVries, J. W. and Kirby, J. T. (1992), "Considerations in Using Bragg Reflection for Storm Erosion Protection," *J. Waterw. Port Coastal Ocean Eng. Div., Am. Soc. Civ. Eng.*, January/February, pp. 62–74.

Battjes, J. A. (1974), "Wave Run-up and Overtopping," Report of Technical Advisory Committee on Protection Against Inundation, Rijkswaterstaat, The Hague, The Netherlands.

Canadian Small Craft Harbors Directorate (1985), "Guidelines of Harbor Accommodation," Government of Canada, Ottawa.

Chandler, P. L. and Sorensen, R. M. (1972), "Transformation of Waves Passing a Submerged Bar," *Proceedings, 13th International Conference on Coastal Engineering*, American Society of Civil Engineers, Vancouver, pp. 385–404.

Cox, J. C. and Clark, G. R. (1992), "Design Development of a Tandem Breakwater System for Hammond Indiana," *Coastal Structures and Breakwaters*, Institution of Civil Engineers, Thomas Telford, London, pp. 111–121.

Davies, A. G. and Heathershaw, A. D. (1984), "Surface-Wave Propagation over Sinusoidally Varying Topography," *J. Fluid Mech.* **144,** 419–443.

Douglass, S. L. (1990), "Estimating Runup on Beaches: A Review of the State of the Art," Contract Report CERC-90-3, U. S. Army Waterways Experiment Station, Vicksburg, MS.

Douglass, S. L. (1992), "Estimating Extreme Values of Run-up on Beaches," Technical Note, *J. Waterw. Port Coastal Ocean Eng. Div., Am. Soc. Civ. Eng.*, March/April, 220–224.

Giles, M. L. and Sorensen, R. M. (1979), "Determination of Mooring Loads and Wave Transmission for a Floating Tire Breakwater," *Proceedings, Coastal Structures '79 Conference*, American Society of Civil Engineers, Alexandria, VA, pp. 1069–1085.

Gilman, J. F. and Nottingham, D. (1992), "Wave Barriers: An Environmentally Benign Alternative," *Proceedings, Coastal Engineering Practice '92 Conference*, American Society of Civil Engineers, Long Beach, CA, pp. 479–486.

Goda, Y. (1969), "Reanalysis of Laboratory Data on Wave Transmission over Breakwaters," *Port Harbor Res. Inst. Rep.*, **8**(3), 3–18.

Goda, Y. (1985), *Random Seas and the Design of Maritime Structures*, University of Tokyo Press, Tokyo.

Goda, Y. and Suzuki, Y. (1976), ''Estimation of Incident and Reflected Waves in Random Wave Experiments,'' *Proceedings, 15th International Conference on Coastal Engineering*, American Society of Civil Engineers, Honolulu, pp. 828–845.

Hales, L. Z. (1981), ''Floating Breakwaters: State-of-the-Art Literature Review,'' Technical Report 81-1, U. S. Army Coastal Engineering Research Center, Ft. Belvoir, VA.

Harms, V. W., Westerink, J. J., Sorensen, R. M., and McTamany, J. E. (1982), ''Wave Transmission and Mooring Force Characteristics of Pipe-Tire Floating Breakwaters,'' Technical Paper 82-4, U. S. Army Coastal Engineering Research Center, Ft. Belvoir, VA.

Holman, R. A. (1986), ''Extreme Value Statistics for Wave Runup on a Natural Beach,'' *Coastal Eng.*, **9,** 527–544.

Hunt, I. A. (1959), ''Design of Seawalls and Breakwaters,'' *J. Waterw. Harbors Div., Am. Soc. Civ. Eng.*, September, 123–152.

Jamieson, W. W. and Mansard, E. T. P. (1987), ''An Efficient Upright Wave Absorber,'' *Proceedings, Coastal Hydrodynamics Conference*, American Society of Civil Engineers, Newark, DE, pp. 124–139.

Kirby, J. T. (1987), ''A Program for Calculating the Reflectivity of Beach Profiles,'' Report UFL/COEL-87/004, University of Florida, Gainesville, FL.

Kobayashi, N., Otta, A. K., and Roy, I. (1987), ''Wave Reflection and Run-up on Rough Slopes,'' *J. Waterw. Port Coastal Ocean Eng. Div., Am. Soc. Civ. Eng.*, May, 282–298.

Kobayashi, N. and Wurjanto, A. (1989), ''Wave Overtopping on Coastal Structures,'' *J. Waterw. Port Coastal Ocean Eng. Div., Am. Soc. Civ. Eng.*, March, 235–251.

Lott, J. W. and Hurtienne, A. M. (1992), ''Design, Construction and Performance of a Baffled Breakwater,'' *Proceedings, Coastal Engineering Practice '92 Conference*, American Society of Civil Engineers, Long Beach, CA, pp. 487–502.

Madsen, O. S. and White, S. M. (1976), ''Reflection and Transmission Characteristics of Porous Rubble-Mound Breakwaters,'' Miscellaneous Report 76-5, U. S. Army Coastal Engineering Research Center, Ft. Belvoir, VA.

Mase, H. and Iwagaki, Y. (1984), ''Run-up of Random Waves on Gentle Slopes,'' *Proceedings, 19th International Conference on Coastal Engineering*, American Society of Civil Engineers, Houston, pp. 593–609.

Mei, C. C. (1985), ''Resonant Reflection of Surface Water Waves by Periodic Sand Bars,'' *J. Fluid Mech.*, **152,** 315–335.

Owen, M. W. (1980), ''Design of Seawalls Allowing for Wave Overtopping,'' Report EX 924, Hydraulics Research Station Wallingford, UK.

Savage, R. P. (1959), ''Wave Run-up on Roughened and Permeable Slopes,'' *Transactions*, American Society of Civil Engineers, Vol. 124, pp. 852–870.

Saville, T. S. (1957), ''Wave Run-up on Composite Slopes,'' *Proceedings, 6th Conference on Coastal Engineering*, Council on Wave Research, University of California, Berkeley, pp. 691–699.

Saville, T. S. (1987), ''Early Large-Scale Experiments on Wave Run-up,'' *J. Am. Shore Beach Preserv. Assoc.*, Berkeley, 101–108.

Sawaragi, T., Deguchi, I., and Park, S.-K. (1988), "Reduction of Wave Overtopping Rate by the Use of Artificial Reefs," *Proceedings, 21st International Conference on Coastal Engineering*, American Society of Civil Engineers, Malaga, Spain, pp. 335–349.

Seelig, W. N. (1980), "Two-Dimensional Tests of Wave Transmission and Reflection Characteristics of Laboratory Breakwaters," Technical Report 80-1, U. S. Army Coastal Engineering Research Center, Ft. Belvoir, VA.

Seelig, W. N. and Ahrens, J. P. (1981), "Estimation of Wave Reflection and Energy Dissipation Coefficients for Beaches, Revetments and Breakwaters," Technical Paper 81-1, U. S. Army Coastal Engineering Research Center, Ft. Belvoir, VA.

Sorensen, R. M. (1990), "The Deployment of Floating Breakwaters: Design Guidance," *Proceedings, 12th Coastal Society Conference*, San Antonio, TX.

U. S. Army Coastal Engineering Research Center (1984), *Shore Protection Manual*, U. S. Government Printing Office, Washington, DC.

U. S. Army Corps of Engineers (1984), "Hydraulic Design of Small Boat Harbors," Engineer Manual 1110-2-1615, Washington, DC.

Van der Meer, J. W. and Angremond, K. (1992), "Wave Transmission at Low-Crested Structures," *Coastal Structures and Breakwaters*, Institution of Civil Engineers, Thomas Telford, London, pp. 25–41.

Weggel, J. R. (1976), "Wave Overtopping Equation," *Proceedings, 15th International Conference on Coastal Engineering*, American Society of Civil Engineers, Honolulu, pp. 2737–2755.

Wiegel, R. L. (1960), "Transmission of Waves Past a Rigid Vertical Thin Barrier," *J. Waterw. Harbors Div., Am. Soc. Civ. Eng.*, March, 1–12.

9

LONG WAVES

As waves propagate through water of relatively shallow depth (i.e, where the water depth to wave length ratio is generally less than 1:20 or 0.05), their kinematic and dynamic properties asymptotically approach the form we have defined as shallow water waves. These waves may also be termed long waves owing to their great length relative to the water depth. The shallow water wave equations developed in Chapter 2 describe these waves. These equations were developed by employing the asymptotic forms of the appropriate hyperbolic functions that develop as the relative depth d/L decreases toward zero. Alternately, we can start from the known physical characteristics of shallow water waves and directly develop the long wave equations. Owing to the typical applications of the long wave equations, they will be developed in three-dimensional form.

The long wave equations are useful for the study of several types of wave phenomena. A prime example is the propagation of the tide in estuaries and coastal embayments. A tide wave propagating along the coast has its form significantly affected by Coriolis forces. This wave motion, known as a Kelvin wave, can be accounted for by the long wave equations. Another example is the study of tsunami waves as they propagate across the ocean and into coastal waters. Long waves are highly reflected, even by flat and relatively rough boundaries. So harbors and other basins can develop resonance conditions as long waves propagate to and fro in the basin.

The linear wave theory equation for the horizontal component at water particle velocity in shallow water (Eq. 2.28) is

$$u = \frac{\pi H}{Tkd} \cos(kx - \sigma t)$$

which may be transformed to

$$u = \frac{\eta C}{d} \tag{9.1}$$

by employing the linear theory equation for wave celerity, $C = (gd)^{1/2}$. These equations demonstrate that, according to the linear wave theory, the horizontal component of water particle velocity in shallow water is independent of vertical position in the water column, that is, it is essentially constant over the water depth.

As a consequence of the nearly horizontal motion of the water particle velocities in shallow water, the vertical components of particle velocity and acceleration are small relative to their horizontal counterparts. Thus the pressure distribution under a long wave may be assumed to be hydrostatic. That is, the linear theory pressure distribution equation (Eq. 2.30), which may be written

$$p = -\rho g z + \rho g \eta \frac{\cosh k(d+z)}{\cosh kd}$$

takes the form

$$p = -\rho g z + \rho g \eta = \rho g(\eta - z) \tag{9.2}$$

for long waves.

First we assume that the horizontal component of water particle velocity is constant with depth and the vertical pressure distribution is hydrostatic to develop the common form of the three-dimensional long wave equations. Then we apply these equations to a variety of situations to demonstrate their use as well as some results.

9.1 THE LONG WAVE EQUATIONS

The long wave equations include the equation of continuity for three-dimensional flow and the two horizontally directed three-dimensional equations of motion which incorporate the dominant active forces including gravity, a horizontal pressure gradient, Coriolis force per unit mass, and the surface and bottom boundary stresses. In these equations the horizontal force components and resulting flows dominate owing to the long wave assumptions. A Cartesian coordinate system is used with the x and y axes located in the horizontal plane and the z axis vertical and positive upward. The related water particle velocity components are u, v, and w respectively. For a further discussion of the development and application of the long wave equations see Dean and Dalrymple (1984), Dronkers (1964), Stoker (1957), and Wilson (1972).

9.1.1 Equation of Continuity

The desired form of the continuity equation can be developed from the basic three-dimensional conservation of mass equation for incomprehensible flow, which is

$$\frac{\partial u}{\partial x} + \frac{\partial v}{\partial y} + \frac{\partial w}{\partial z} = 0$$

The approach is to integrate over depth, assuming that the horizontal flow velocity components do not vary with depth. The kinematic free surface boundary condition and the bottom boundary condition (see Chapter 2) are then applied to develop the desired result (see Dean and Dalrymple, 1984).

A more physical approach to deriving the desired form of the continuity equation can be carried out by considering Figure 9.1. This depicts a column of water with a rigid impermeable horizontal bottom and a free surface that will rise or fall in response to the flow in and out of the sides. The volumetric flow rates per unit width of vertical section are q_x and q_y, the depth of the water column below the mean water level is d, and the fluctuating free surface elevation for the water column is η. That is,

$$q_x = \int_{-d}^{\eta} u dz$$

and

$$q_y = \int_{-d}^{\eta} v dz$$

From Figure 9.1, the net flow rates into the column in the x and y directions respectively are

$$\frac{-\partial q_x}{\partial x} dx\, dy$$

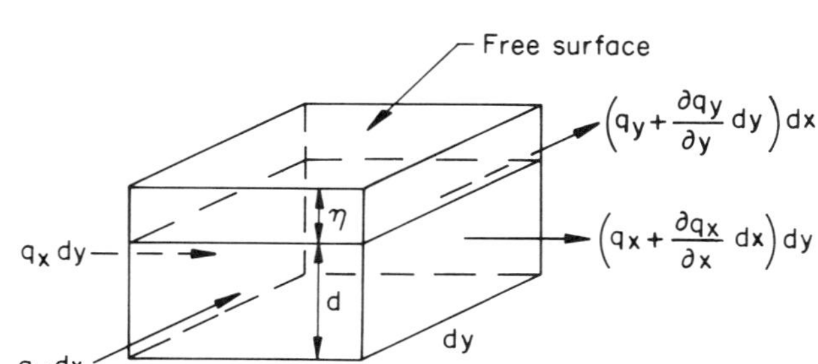

Figure 9.1 Diagram for continuity equation derivation.

and

$$\frac{-\partial q_y}{\partial y} dy\, dx$$

These net horizontal flow rates into the column must equal the net change in volume of the column per unit time owing to the rising or falling free surface given by

$$\frac{\partial \eta}{\partial t} dx\, dy$$

Equating and dividing through by $dx\, dy$ yields

$$\frac{\partial q_x}{\partial x} + \frac{\partial q_y}{\partial y} + \frac{\partial \eta}{\partial t} = 0 \tag{9.3}$$

which is the form of the continuity equation we are seeking.

9.1.2 Equations of Motion

The equations of motion for the two horizontal directions in a viscous fluid may be written

$$-\frac{1}{\rho}\frac{\partial p}{\partial x} + fv + \frac{1}{\rho}\left(\frac{\partial \tau_{yx}}{\partial y} + \frac{\partial \tau_{zx}}{\partial z}\right) = u\frac{\partial u}{\partial x} + v\frac{\partial u}{\partial y} + w\frac{\partial u}{\partial z} + \frac{\partial u}{\partial t} \tag{9.4a}$$

$$-\frac{1}{\rho}\frac{\partial p}{\partial x} + fu + \frac{1}{\rho}\left(\frac{\partial \tau_{xy}}{\partial x} + \frac{\partial \tau_{zy}}{\partial z}\right) = u\frac{\partial v}{\partial x} + v\frac{\partial v}{\partial y} + w\frac{\partial v}{\partial z} + \frac{\partial v}{\partial t} \tag{9.4b}$$

where the terms on the left are the pressure, Coriolis, and viscous shear forces per unit mass acting on the fluid and the terms on the right are the resulting accelerations. Each of the terms in Eq. 9.4 is discussed below.

The horizontal pressure gradient develops from the sloped free surface of the water. Employing Eq. 9.2 we have

$$-\frac{1}{\rho}\frac{\partial p}{\partial x} = -g\frac{\partial \eta}{\partial x} \tag{9.5a}$$

$$-\frac{1}{\rho}\frac{\partial p}{\partial y} = -g\frac{\partial \eta}{\partial y} \tag{9.5b}$$

The resulting terms are independent of z and thus constant over depth. If there is a sufficiently variable atmospheric pressure p_a across the surface (e.g., as in a hurricane), the absolute pressure could be used. Then Eq. 9.2 would become

$$p = \rho g(\eta - z) + p_{\mathrm{a}}$$

and additional pressure gradient terms

$$-\frac{1}{\rho}\frac{\partial p_{\mathrm{a}}}{\partial x}$$

and

$$-\frac{1}{\rho}\frac{\partial p_{\mathrm{a}}}{\partial y}$$

would develop and be included in Eq. 9.4. Often this surface pressure term is not significant or its effects are evaluated separately.

The second term on the left incorporates the effects of the earth's rotation in terms of the Coriolis parameter f, where

$$f = 2\omega \sin \phi \tag{9.6}$$

In Eq. 9.6 ω is the earth's rotational speed (7.28×10^{-5} rad/sec) and ϕ is the latitude of the point where the equations are being applied (positive to the north, negative to the south). Note that Coriolis acceleration is a function of the flow velocity component that acts normal to the direction of interest. It varies in magnitude from zero at the equator to a maximum at the poles. The Coriolis force per unit mass is relatively small, but it can be significant for the large masses of water which are involved in some long wave problems.

The third term on the left represents the resisting and driving forces caused by viscous shear stresses which act along a plane surface. For large-scale flows (large Reynolds number), these shear forces are caused essentially by the convection of turbulence. The two subscripts for each of the shear stresses denote first the axis that is perpendicular to the face on which the stress acts and second the direction of the stress (e.g., τ_{yx} is the stress acting in the x direction on an xz plane—see Fig. 9.2). If the horizontal convection of turbulence is small compared to the vertical convection of turbulence, the terms τ_{yx} and τ_{xy} can be neglected. This is typically the case for the long wave flows we are considering, where velocity gradients are generated primarily by surface and bottom boundary effects.

The right-hand side of Eq. 9.4 is the total fluid acceleration caused by the acting forces. This consists of a convective component (the first three terms) that results from velocity changes along the path of flow and a local component which is the velocity change at a point with time.

To derive the desired long wave form of the equations of motion, Eq. 9.4

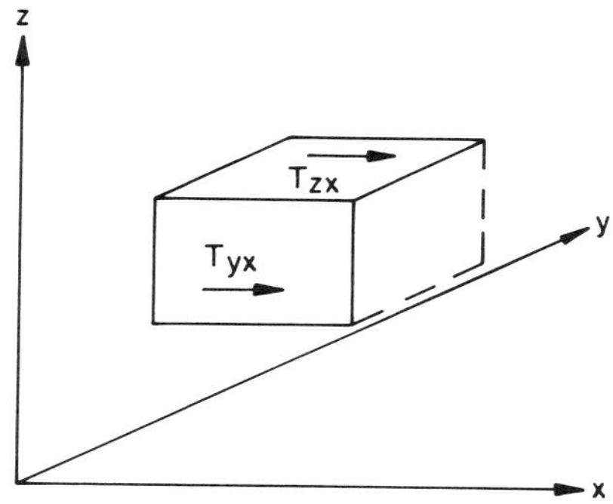

Figure 9.2 The x-directed shear stresses.

is integrated over the depth. Employing Eq. 9.5 and the continuity equation, vertical integration from the bottom to the water surface yields

$$-g(d+\eta)\frac{\partial \eta}{\partial x} + fq_y + \frac{1}{\rho}(\tau_{sx} - \tau_{bx}) = \frac{\partial}{\partial x}\left(\frac{q_x^2}{d+\eta}\right) + \frac{\partial}{\partial y}\left(\frac{q_x q_y}{d+\eta}\right) + \frac{\partial q_x}{\partial t} \tag{9.7a}$$

$$-g(d+\eta)\frac{\partial \eta}{\partial y} + fq_x + \frac{1}{\rho}(\tau_{sy} - \tau_{by}) = \frac{\partial}{\partial x}\left(\frac{q_x q_y}{d+\eta}\right) + \frac{\partial}{\partial y}\left(\frac{q_y^2}{d+\eta}\right) + \frac{\partial q_y}{\partial t} \tag{9.7b}$$

The terms τ_{sx} and τ_{sy} represents the horizontal stress applied at the water surface in the x and y directions. This would commonly be due to the wind acting on the water surface. The terms τ_{bx} and τ_{by} represent the horizontal bottom-induced resistance to the water motion.

Equations 9.3 and 9.7 are common forms of the long wave equations. These equations are nonlinear. Usually, to obtain analytical solutions, they are linearized by neglecting the convective accelerations and neglecting the bottom and surface resistance terms or using some linear form of these resistance terms. Also, if fluctuations of the surface elevation are small compared to the water depth below the still water level, the total depth can be reduced to d.

9.2 TWO-DIMENSIONAL SHALLOW WATER WAVE MOTION

It is of interest to solve the long wave equations developed above for the simple case of a two-dimensional frictionless shallow water wave. Assume that the wave is traveling in the positive x direction. The continuity equation (Eq. 9.3) becomes

$$\frac{\partial q_x}{\partial x} + \frac{\partial \eta}{\partial t} = 0$$

which may be written

$$(d + \eta)\frac{\partial U}{\partial x} + \frac{\partial \eta}{\partial t} = 0$$

where U is the average velocity over the depth given by

$$U = \frac{q_x}{(d + \eta)}$$

If the free surface fluctuation is small relative to the water depth, that is, if we consider a small amplitude wave, the continuity equation becomes

$$d\frac{\partial U}{\partial x} + \frac{\partial \eta}{\partial t} = 0 \tag{9.8}$$

The relevant equation of motion is Eq. 9.7a for the x direction. If we linearize this equation, the convective acceleration terms will drop out. For two-dimensional flow in the x direction $q_y = 0$, so the Coriolis term drops out. Finally, if we neglect surface and bottom shear stresses and make the small amplitude assumption, Eq. 9.7a reduces to

$$-gd\frac{\partial \eta}{\partial x} = \frac{\partial q_x}{\partial t}$$

or

$$-g\frac{\partial \eta}{\partial x} = \frac{\partial U}{\partial t} \tag{9.9}$$

If we differentiate Eq. 9.8 with respect to t and Eq. 9.9 with respect to x and equate equal terms we can eliminate U, yielding

$$gd\frac{\partial^2 \eta}{\partial x^2} = \frac{\partial^2 \eta}{\partial t^2} \tag{9.10}$$

Equation 9.10 is the wave equation where for a long wave $gd = C^2$. This has a solution,

$$\eta = \frac{H}{2}\cos(kx - \sigma t) \tag{9.11}$$

where $H/2$ is the wave amplitude. Equation 9.11 can be substituted into Eq.

9.9, yielding

$$\frac{\partial U}{\partial t} = \frac{gH}{2} k \sin(kx - \sigma t)$$

which can be integrated to yield

$$U = \frac{gH}{2}\left(\frac{k}{\sigma}\right) \cos(kx - \sigma t)$$

or

$$U = \frac{H}{2}\frac{C}{d} \cos(kx - \sigma t)$$

which becomes

$$U = \frac{\eta C}{d} \tag{9.12}$$

Equation 9.12 is identical to Eq. 9.1, the horizontal particle velocity equation for long waves developed from the small amplitude or linear wave theory by taking the shallow water limit.

9.3 KELVIN WAVES

Consider a long wave that propagates in the x direction but is three dimensional. This occurs, for example, when the tide is propagating along a relatively straight coast line (taken as the x direction) and is constrained to move along the coast ($q_y = 0$). Coriolis effects caused by the water particle motion in the direction of wave propagation affect the wave amplitude along its crest (i.e., normal to the direction of wave propagation). This type of wave is known as a Kelvin wave after Lord Kelvin who first derived the related equations.

For a Kelvin wave the continuity equation and x component of the equation of motion are unchanged (Eqs. 9.8 and 9.9). Since q_y is zero and viscous stresses are still being ignored, the equation of motion in the y direction (Eq. 9.7b) reduces to

$$g\frac{\partial \eta}{\partial y} = -fU \tag{9.13}$$

Equation 9.13 simply states that the Coriolis force balances the surface gradient normal to the direction of wave motion. Including Eq. 9.13 in the solution yields

$$\eta = \frac{H}{2} e^{fy/C} \cos(kx - \sigma t) \tag{9.14}$$

and

$$U = \frac{H}{2} \frac{C}{d} e^{-fy/C} \cos(kx - \sigma t) \tag{9.15}$$

where the celerity of a Kelvin wave is still

$$C = \sqrt{gd}$$

Compare with Eqs. 9.14 and 9.15 with Eqs. 9.11 and 9.12 respectively.

Equation 9.14 and 9.15 show that the wave amplitude and water particle velocity decrease exponentially in the positive y direction in the northern hemisphere (where f is positive). The Coriolis force wants to divert the particle motion to the right under the wave crest, but this motion is resisted by the crest elevation gradient. The opposite occurs under the wave trough. The crest and trough amplitudes are depicted in Figure 9.3 where the wave is propagating into the page. Thus, for example, as the tide propagates along a channel in the northern hemisphere, the tide range and reversing tidal current velocities would be greater on the right-hand side of the channel when looking in the direction of wave propagation. In the southern hemisphere, where f is negative, the tide range and reversing currents would be greater on the left when looking in the direction of wave propagation.

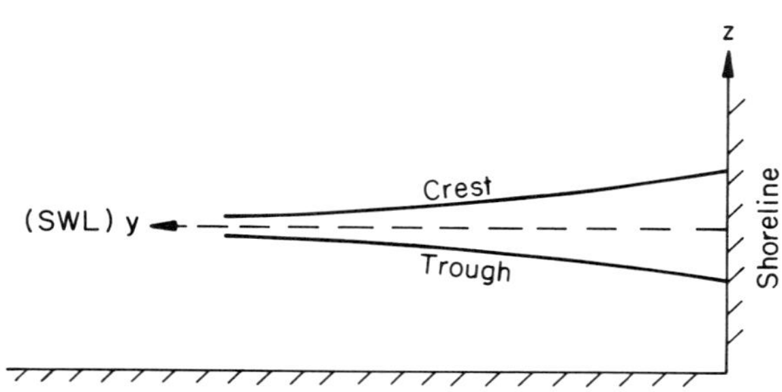

Figure 9.3 Wave crest and trough profiles for a Kelvin wave propagating into the page.

9.4 BASIN OSCILLATIONS

A standing wave (see Section 2.5.3) can form in a closed basin or in a basin open to a larger water body. This resonant condition develops when the standing wave has a period equal to a natural or subharmonic period of the basin. Basically, the standing wave has a length that is equal to the length of the basin or some multiple (0.5, 1, 1.5, etc. for closed basins and 0.25, 0.75, 1.25, etc. for open basins) of the basin length. These first three modes of oscillation are depicted in Figure 9.4. This phenomenon is particularly relevant to long wave action because most basins where resonance is a concern are sufficiently large for the standing waves to be long waves and long waves are highly reflective so a standing wave can develop.

Typical excitation sources for these waves include the tide, seismic activity, atmospheric pressure fluctuations, and shifting wind stresses. Some energy sources may just "pluck" the system, causing an oscillation that subsequently decays exponentially. Energy sources that provide continuous energy input to a system cause an essentially steady oscillation when the rate of energy input and the rate of viscous dissipation balance. If the basin is excited by a wide spectrum of energy it will selectively amplify the resonant modes of the basin.

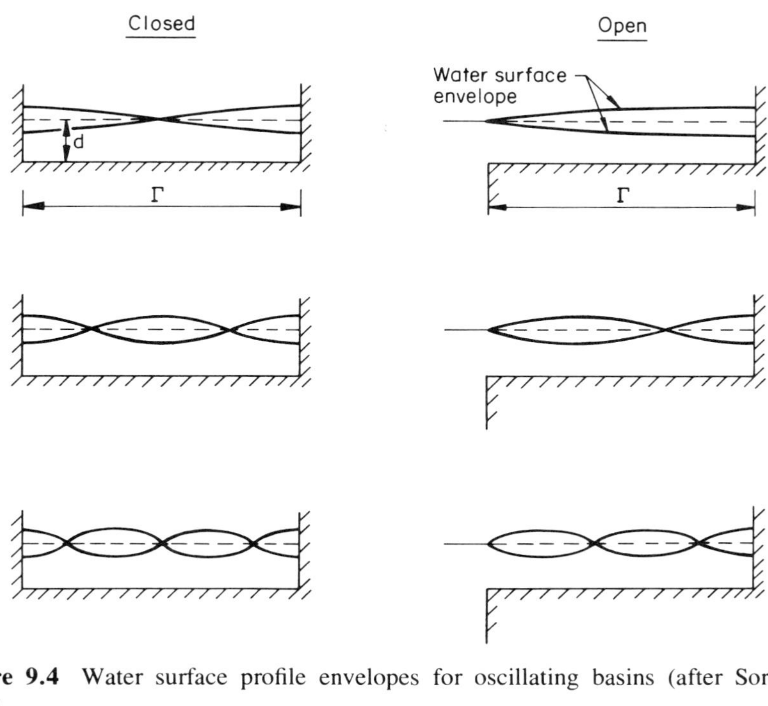

Figure 9.4 Water surface profile envelopes for oscillating basins (after Sorensen, 1978).

Forced basin oscillations at periods other than the resonant periods undergo less (or negative) amplification than occurs at the resonant modes.

9.4.1 Two-Dimensional Basins with Regular Geometry

Consider a two-dimensional basin (x, z) having a constant depth. The standing waves for this basin must be a solution to Eq. 9.10. This yields

$$\eta = H \cos kx \cos \sigma t \tag{9.16}$$

where H is the height of the incident and reflected waves that combine to form the standing wave and is the amplitude of the resulting standing wave. With Eq. 9.16, the horizontal water particle velocities can be derived from Eq. 9.9 (as previously done for progressive two-dimensional waves) to yield

$$U = \frac{HC}{d} \sin kx \sin \sigma t \tag{9.17}$$

In Eqs. 9.16 and 9.17 the values of k and σ depend on the geometry of the basin including its length (Γ) and end boundary conditions. The closed end of a basin is an antinode and the open end is essentially a node if the basin is connected to a sufficiently larger water body. Figure 9.4 shows the water surface envelopes for the three longest period standing waves in a closed and an open basin.

For the closed basins shown in Figure 9.4 the length of the standing wave is equal to Γ/N where $N = 0.5$, 1.0, and 1.5 for the three conditions shown. The resonant period of the closed basins $T_N = L/C$, which for a long wave yields

$$T_N = \frac{\Gamma}{N\sqrt{gd}} \tag{9.18}$$

Thus, in Eq. 9.16, $k = 2\pi N/\Gamma$ and $\sigma = 2\pi/T_N$.

A somewhat similar relationship to Eq. 9.18 can be worked out for two-dimensional open basins. Also, see Wilson (1972) for the equations that give T_N for two-dimensional basins of regular form but nonrectangular shape. Depending in the nature of the excitation, any or several of the resonant modes may be excited. The longer period modes are usually dominant because the shorter period modes are less reflective and thus dissipate more rapidly.

9.4.2 Three-Dimensional Basins with Regular Geometry

The basic wave equation for three-dimensional long wave motion can be developed from the continuity equation and the two equations of motion, in a manner similar to the development of Eq. 9.10 for two-dimensional motion, to yield

$$gd\left(\frac{\partial^2\eta}{\partial x^2} + \frac{\partial^2\eta}{\partial y^2}\right) = \frac{\partial^2\eta}{\partial t^2} \tag{9.19}$$

The standing wave solution to Eq. 9.19 is

$$\eta = H\cos(k_x x)\cos(k_y y)\cos\sigma t \tag{9.20}$$

where the wave numbers k_x and k_y give the standing wave lengths in the x and y directions. As with the two-dimensional basin, the values of k_x, k_y, and σ depend on the basin geometry.

Consider a rectangular closed basin with side dimensions Γ_x and Γ_y as shown in Figure 9.5. Incorporating the appropriate wave numbers and wave angular frequency in terms of the basin dimensions and resonant period yields

$$\eta = H\cos\left(\frac{2\pi Nx}{\Gamma_x}\right)\cos\left(\frac{2\pi My}{\Gamma_y}\right)\cos\left(\frac{2\pi t}{T_{NM}}\right) \tag{9.21}$$

from Eq. 9.20. In Eq. 9.21 the various resonant modes are defined by combinations of N and M which can have the values 0.5, 1.0, 1.5, and so on and T_{NM} is the resonant period of a particular mode. The resonant period is given by

$$T_{NM} = \frac{1}{\sqrt{gd}}\left[\left(\frac{N}{\Gamma_x}\right)^2 + \left(\frac{M}{\Gamma_y}\right)^2\right]^{-1/2} \tag{9.22}$$

Note that when N or M has the value zero, Eqs. 9.21 and 9.22 reduce to the two-dimensional case given by Eqs. 9.16 and 9.18.

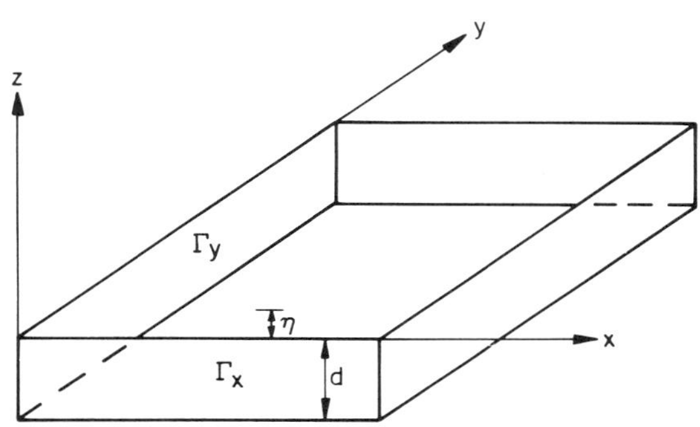

Figure 9.5 Definition sketch for a closed basin.

As an example, consider a basin with a square planform having side lengths Γ (i.e., $\Gamma_x = \Gamma_y = \Gamma$) and the resonant mode $N = M = 0.5$. Then

$$\eta = H \cos \frac{2\pi x}{\Gamma} \cos \frac{2\pi y}{\Gamma} \cos \frac{2\pi t}{T_{11}} \tag{9.23}$$

and

$$T_{11} = \frac{\Gamma}{\sqrt{2}\sqrt{gd}}$$

Nodal lines are located where the surface elevation is zero for all values of t or from Eq. 9.23 at x, $y = \Gamma/2$. The resulting water surface pattern at $t = 0$, T_{11}, $2T_{11}$, and so on is shown in Figure 9.6, where the water surface contours above the still water level are shown by solid lines and those below the still water level are shown by dashed lines. At $t = T_{11}/4$ the water surface is flat; at $t = T_{11}/2$ the solid and dashed lines are reversed; at $t = 3T_{11}/4$ the water surface is flat again, and at $t = T_{11}$ the surface form is the same as it was at $t = 0$. The maximum water surface envelope ranges are $2H$ at the four corners which are antinodes. The nodal lines, where the largest flow velocities occur, are as shown. The flow at these points is normal to the nodal lines.

9.4.3 Coriolis Effects on Basin Oscillations

The developments in the previous section assume that the water particle motions are sufficiently small so that Coriolis effects may be neglected. For longer waves, the effects of Coriolis become significant, particularly when the channel in which the wave travels is sufficiently wide compared to the wave length.

For a standing wave the maximum horizontal velocities occur under a nodal line; at an antinodal line flow velocities are vertical and significantly smaller

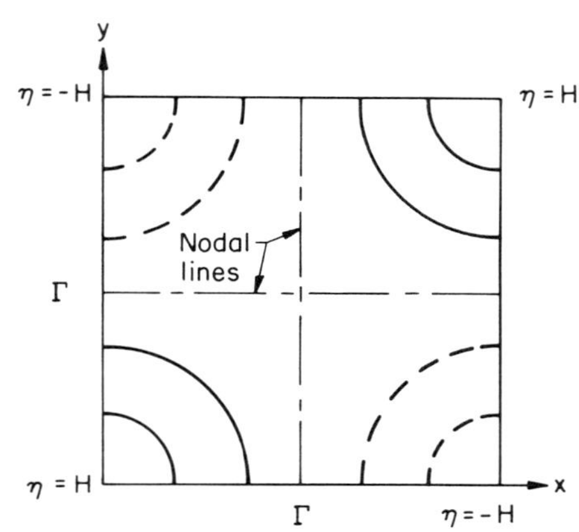

Figure 9.6 Water surface contours, rectangular basin, $t = 0$, $N = M = 0.5$ (after Sorensen, 1978).

(see Fig. 2.6). Thus, as oscillating horizontal flow occurs under a nodal line, the water surface along the line alternately tilts in one direction and then in the other, rather than remain horizontal, as would be the case for a standing wave without Coriolis effects. The nodal line becomes a point, known in tidal wave terminology as an amphidromic point. A transverse oscillation develops which is a quarter period out of phase with the longitudinal oscillation.

If we superimpose two Kelvin waves having opposite directions of travel along the x axis to form a standing wave, the water surface elevation is given by

$$\eta = \frac{H}{2} e^{-fy/C} \cos(kx - \sigma t) - \frac{H}{2} e^{fy/C} \cos(kx + \sigma t) \tag{9.24}$$

At $x, y = 0$ the water surface elevation is zero at all times. This is the amphidromic point. Cotidal lines, which are lines connecting points that have a maximum surface elevation at the same instant, can be determined from Eq. 9.24. Maximum surface elevation occurs when

$$\frac{\partial \eta}{\partial t} = 0$$

or

$$e^{-fy/C} \sin(kx - \sigma t) + e^{fy/C} \sin(kx + \sigma t) = 0$$

which, after some manipulation yields

$$\tanh \frac{fy}{C} = -\frac{\tan kx}{\tan \sigma t} \tag{9.25}$$

For a given standing wave period and water depth, the x, y coordinate positions of cotidal lines can be plotted as a function of time from Eq. 9.25.

Consider a rectangular three-dimensional closed basin. Assume the basin dimensions cause a standing wave pattern like the one shown on the upper left-hand diagram in Figure 9.4 (i.e., $N = 0.5$, $M = 0$). Without Coriolis, there would be a nodal line across the basin width at the half-way point of the basin length (i.e., at $x = 0$). With Coriolis effects the cotidal line pattern will be as shown in Figure 9.7, which is a plan view of the basin. Cotidal lines are curved and rotate counter clockwise around the amphidromic point, completing one cycle in the standing wave period. The largest wave surface envelope ranges are around the periphery of the basin, with the range decreasing inward toward the amphidromic point.

Amphidromic points and cotidal lines occur at several places as the tide propagates across the oceans and enclosed seas (e.g., the north Atlantic Ocean, the North Sea, and the Caribbean; see Sverdrup, Johnson, and Fleming, 1942).

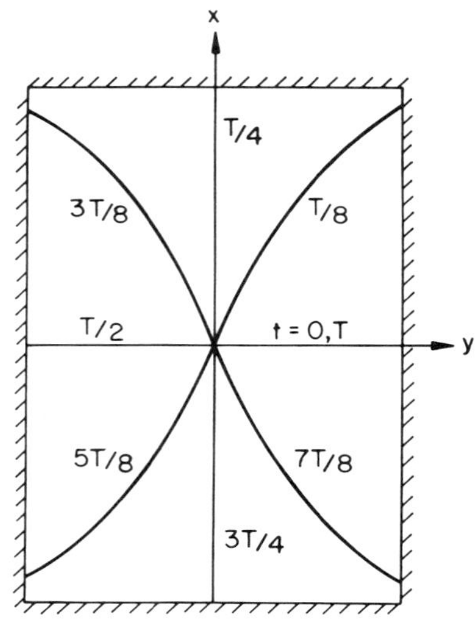

Figure 9.7 Cotidal lines for a rectangular basin.

9.5 EFFECT OF BOTTOM FRICTION

Up to this point we have neglected the effects of bottom friction on wave motion and the resulting wave decay. The wave-induced reversing fluid motion at the bottom generates a bottom shear stress and causes energy dissipation, the rate of energy dissipation depending on the fluid velocity and bottom roughness (see Section 3.3.2). For the long wave field conditions of greatest interest rough turbulent flow can be assumed.

Thus, for the x direction, for example, a bottom stress relationship of the form

$$\tau_b = K_b \rho U^2 \tag{9.26}$$

may be used where the resulting bottom shear stress acts on the fluid in the direction opposite to the flow velocity. The bottom stress coefficient K_b may be written (see French, 1985) in terms of a friction factor f ($K_b = f/8$), a Chezy coefficient C ($K_b = g/C^2$), or a Manning's n ($K_b = gn^2/\phi^2 d^{1/3}$ where ϕ has the value of unity for SI units and 1.49 for English units).

Unfortunately, Eq. 9.26 is nonlinear and the stress direction must be tied to the flow direction as the particle velocity reverses. A common way to overcome these problems is to assume that the flow velocity is given by

$$U = U_m \cos \sigma t$$

where U_m is the peak value of the cyclic flow velocity. Then the bottom stress becomes

$$\tau_b = K_b \rho U_m^2 \cos^2 \sigma t$$

If the $\cos^2 \sigma t$ term is expanded as a cosine Fourier series, the first term in the series will be $(8/3\pi)\cos \sigma t$. Using only this first term as an approximation yields

$$\tau_b = \frac{8}{3\pi} K_b \rho U_m^2 \cos \sigma t$$

or

$$\tau_b = \frac{8}{3\pi} K_b \rho U_m U \tag{9.27}$$

Equation 9.27 is now linear and the bottom stress has the same sign as the flow velocity. The neglected higher-order terms in the Fourier series are non-linear harmonic terms somewhat similar, for example, to the higher-order terms in the Stokes wave theory.

Including bottom stress as given by Eq. 9.27, the two-dimensional x-directed equation of motion becomes (compare to Eq. 9.9)

$$-g \frac{\partial \eta}{\partial x} - \frac{8}{3\pi d} K_b U_m U = \frac{\partial U}{\partial t} \tag{9.28}$$

The continuity equation is unaffected by the inclusion of bottom stress. As before, the equation of motion and the continuity equation yield a wave equation (compare to Eq. 9.10),

$$gd \frac{\partial^2 \eta}{\partial x^2} - \frac{8}{3\pi d} K_b U_m \frac{\partial \eta}{\partial t} = \frac{\partial^2 \eta}{\partial t^2} \tag{9.29}$$

that now includes bottom friction effects.

A solution to Eq. 9.29 for wave motion (see Dean and Dalrymple, 1984) is

$$\eta = \frac{H_I}{2} e^{-k'x} \cos(k''x - \sigma t) \tag{9.30}$$

where at $x = 0$ the wave has an initial wave height H_I and an initial wave length that yields an initial wave number k_I. With $A = 8K_b U_m / 3\pi d$,

$$k' = \frac{k_I A}{2\sigma}$$

and

$$k'' = k_I \left[1 + \frac{1}{8} \left(\frac{A}{\sigma} \right)^2 \right]$$

For the frictionless case ($K_b = 0$) Eq. 9.30 reduces to Eq. 9.11, where the undamped wave height and wave number do not change as the wave propagates. With bottom friction, the wave height decays exponentially in the direction of wave travel. An exponential decay is to be expected, because as the wave height decreases the horizontal velocity decreases, resulting in a decreased bottom stress and rate of energy dissipation that are nonlinear. From the relationship $k' = k_I A/2\sigma$ it can be seen that the rate of exponential decay is proportional to A/σ (as expected, longer waves in shallower water with higher values of K_b decay at a faster rate).

From the equation of motion (Eq. 9.28) we have

$$U = \frac{H_I \sigma}{2d\sqrt{(k')^2 + (k'')^2}} e^{-k'x} \cos(k''x - \sigma t - \epsilon) \tag{9.31}$$

where

$$\epsilon = \tan^{-1} \frac{k'}{k''} \tag{9.32}$$

Equation 9.31 indicates that the peak current velocity is no longer in phase with the wave crest, as would be the case for a frictionless wave. There is a phase lag ϵ (peak current velocity precedes the wave crest) give by Eq. 9.32. Field observations of tidal currents and related water surface elevations have demonstrated that this phase lag does occur.

9.6 SURFACE EFFECTS

The development of the equation of motion points to two surface effects that should be considered. One is a moving, horizontally variable atmospheric pressure on the surface and the other is a horizontal stress caused by the wind. Each is discussed below.

9.6.1 Moving Pressure Disturbance

To this point we have considered free waves; that is, waves that are somehow generated and then propagate free of any additional generating force. Lamb (1945) develops the equations for long waves that are forced by a moving, horizontally variable pressure applied to the water surface (also see Dean and Dalrymple, 1984). Inclusion of a surface pressure of this type in the equations of motion was discussed in Section 9.1.

With the surface pressure defined as p_a, the absolute vertical pressure distribution (from Section 9.1.2) is

$$p = p_a + \rho g(\eta - z)$$

Inserting this in Eq. 9.4a and neglecting the Coriolis, viscous stress, and nonlinear convective acceleration terms yields the x-directed equation of motion

$$-\frac{1}{\rho}\frac{\partial p_a}{\partial x} - g\frac{\partial \eta}{\partial x} = \frac{\partial u}{\partial t} \tag{9.33}$$

Assume that the pressure disturbance is traveling in the x direction with a speed U_p and has the general form

$$\frac{p_a}{\rho} = fct\ (U_p t - x) \tag{9.34}$$

We use a modified form of the continuity equation by employing a coordinate system that travels with the wave. This produces a steady flow situation. Since this is a forced wave, the horizontal particle velocity under the wave crest becomes $u - U_p$. Where the wave surface is at the still water level the unsteady velocity component is vertical, so the horizontal particle velocity in our transformed system becomes $-U_p$. From continuity

$$(u - U_p)(d + \eta) = -U_p d$$

Simplifying,

$$u = U_p \frac{\eta}{d + \eta}$$

or

$$u \approx U_p \frac{\eta}{d} \tag{9.35}$$

With $q_x = ud$, Eq. 9.3 becomes

$$d\frac{\partial u}{\partial x} + \frac{\partial \eta}{\partial t} = 0 \tag{9.36}$$

for x-directed flow. Inserting Eq. 9.35 into Eq. 9.36 yields

$$U_p \frac{\partial \eta}{\partial x} + \frac{\partial \eta}{\partial t} = 0 \tag{9.37}$$

Combining Eqs. 9.33, 9.35, and 9.37 leads to

$$\frac{d}{\rho}\frac{\partial p_a}{\partial x} + (U_p^2 - gd)\frac{\partial \eta}{\partial x} = 0 \tag{9.38}$$

Equation 9.38 can be solved to yield

$$\frac{\eta}{d} = \frac{p_a}{\rho(U_p^2 - gd)} \tag{9.39}$$

which gives the water surface response to a surface pressure disturbance moving at a speed U_p. When $U_p = 0$, we have the hydrostatic condition. When U_p is equal to the celerity of a long free wave, the surface displacement amplitude grows to infinity. This infinite amplification occurs because no viscous damping force was included in the solution. In a real situation the displacement grows until damping is sufficient to yield an equilibrium surface displacement. When U_p is close, but not equal, to the free wave celerity, the surface displacement is less (see Fig. 9.8).

Ewing, Press, and Donn (1954) report on a pressure disturbance that moved across the lower end of Lake Michigan on June 26, 1954. The forward speed of the pressure disturbance matched the speed of a free long wave for the water depths that occur over a sizable section of the lake where the pressure disturbance was traveling. Apparently, the pressure disturbance generated a forced wave which became a free wave when the pressure disturbance moved on. The free wave was observed to reach the shore near the Indiana–Michigan border and then reflect back across the lake to the shoreline near Chicago. The pressure differential for the atmospheric disturbance was about a tenth of a foot of water and the resulting wave height at the shore was about 3–6 ft. This amplification was caused by the combined effects of forced wave growth during generation and shoaling and runup of the wave as the wave approached the shore. Hur-

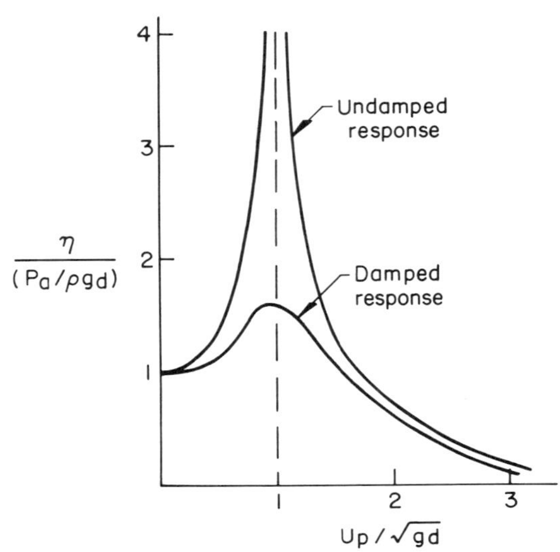

Figure 9.8 Forced-wave amplitude for a moving pressure disturbance.

ricanes moving along a section of the coast having the appropriate depth for their forward speed (so $U_p^2 = gd$) also have an amplified pressure response component.

9.6.2 Surface Wind Stress

A wind-induced stress acting on the water surface contributes to the generation of waves (Chapter 6) and causes a forward mass transport that results in the setup of water against a downwind barrier. We are interested in this setup, commonly called storm surge, which can be evaluated by the long wave equations (Eqs. 9.3 and 9.7).

Wind velocities of interest are fully turbulent, so the resulting surface wind stress would be given by

$$\tau_s = C\rho_a W^2 \tag{9.40}$$

where τ_s is the wind stress, ρ_a is the air density, W is the wind velocity at some reference elevation, and C is a drag coefficient that depends on the surface roughness and related air boundary layer characteristics. Wilson (1960) has evaluated an extensive literature on values presented for the wind stress drag coefficient. He concluded that the most reasonable value for C varied from 1.5×10^{-3} for light winds asymptotically to 2.4×10^{-3} for strong winds.

For storm surge analysis it is more convenient to write Eq. 9.40 in terms of the water density ρ as

$$\tau_s = K_s \rho W^2 \tag{9.41}$$

where the wind stress coefficient $K_s = (1.19 \times 10^{-3})C$ (taking $\rho/\rho_a = 845$). A frequently used relationship for K_s is from a study by Van Dorn (1953):

$$K_s = 1.21 \times 10^{-6} + 2.25 \times 10^{-6}\left(1 - \frac{5.6}{W}\right)^2 \tag{9.42}$$

where the wind speed is in meters per second. This is based on a field study conducted at a small coastal yacht harbor and gives results that are slightly higher than Wilson's recommended values.

Equation 9.7, with the addition of a surface pressure disturbance term (as in Section 9.6.1), is the full equation of motion defining the water surface elevation response to a storm. These equations are normally solved by numerical procedures, given the input meteorological factors and the site boundary conditions. More simply, individual components are calculated assuming a static balance and neglecting continuity (e.g., see Sorensen, 1978). That is, the acceleration terms are neglected and a static balance is written between the surface slope term and the Coriolis, surface pressure, and wind/bottom stress terms separately. The resulting setups caused by each component are then added.

To consider just the wind and bottom stress terms we refer to the very simplified two-dimensional situation depicted in Figure 9.9. The wind is blowing normal to the shoreline, producing a landward current near the surface and a seaward return current near the bottom. The surface and bottom stresses caused by the wind and bottom current respectively both contribute to cause the water surface setup that is shown. There is no flow parallel to the coast, so Coriolis effects may be neglected. If an equilibrium is reached, the local acceleration term may be neglected. Then the x-directed linearized equation of motion becomes

$$-g(d + \eta)\frac{\partial \eta}{\partial x} + \frac{1}{\rho}(\tau_s - \tau_b) = 0 \qquad (9.43)$$

which is just a static balance between the surface slope and the surface/bottom stresses.

The bottom stress effect adds to the surface stress effect to increase setup, but the bottom stress is typically an order of magnitude smaller than the surface stress. Field studies (Saville, 1952; Van Dorn, 1953) indicate that typically $\tau_s > 10\tau_b$. Since the surface and bottom decreases are additive, we can write Eq. 9.43 as

$$-g(d + \eta)\frac{\partial \eta}{\partial x} + \frac{1}{\rho}(\tau_{sb}) = 0 \qquad (9.44)$$

where τ_{sb}, the combined surface and bottom stress, is

$$\tau_{sb} = K_{sb}\rho W^2$$

and

$$K_{sb} = nK_s$$

with n having a value of about 1.1

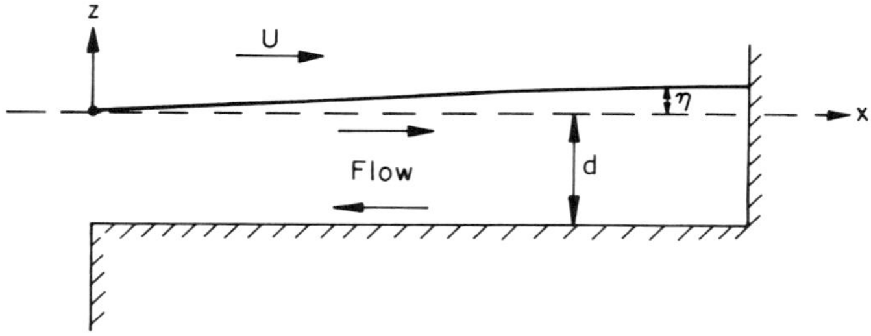

Figure 9.9 Idealized offshore profile and wind setup.

Thus, Eq. 9.43 reduces to

$$\frac{\partial \eta}{\partial x} = \frac{nK_s W^2}{g(d + \eta)} \tag{9.45}$$

which defines the water surface slope at a point in terms of the wind velocity and the local water depth. The surface slope is larger for higher wind speeds acting over shallower water.

Equation 9.45 can be written in finite difference form and solved in steps of length Δx using the average wind speed and water depth over that step. If the depth is constant, as for the simple case shown in Figure 9.9, we can integrate Eq. 9.45 and rearrange to yield

$$\eta = \sqrt{\frac{2nK_s W^2 x}{g} + C} - d \tag{9.46}$$

where the constant of integration C is determined from the boundary conditions. At some seaward point the water is sufficiently deep and/or the wind speed is sufficiently small for the setup to be essentially zero. In Figure 9.9 this point is the seaward edge of the shelf. With this point as the origin to the x axis, $C = d^2$, giving

$$\eta = \sqrt{d^2 + \frac{2nK_s W^2 x}{g}} - d$$

which can be used to calculate the water surface profile.

9.7 LONG WAVES WITH IRREGULAR BOUNDARY CONDITIONS

The preceding considerations of Kelvin waves, basin oscillations, and the effects of surface pressure and surface and bottom stress were all developed for simplified boundary conditions (e.g., constant water depth, uniform wind stress, basins having rectangular plan forms, and excitation by monochromatic waves). For more realistic conditions where the boundary conditions are not uniform, the long wave equations typically must be solved numerically by computer. There is extensive information available on applications for each of these types of problems. Herein our purpose is to present a brief discussion with some examples.

The four problems most commonly solved by numerical solution of the long wave equations are the propagation of tide and tsunami waves, the resonant oscillation of bays and harbors, and coastal storm surge. Tide and tsunami wave analyses are mostly concerned with the propagation and resulting changes that occur in shallow water along the coast and through estuaries and bays.

9.7.1 Storm Surge

The long wave equations in the form given by Eqs. 9.3 and 9.7 are the basis for numerical storm surge models. If the net rate of precipitation minus evaporation defined by P is added, the continuity equation becomes

$$\frac{\partial q_x}{\partial x} + \frac{\partial q_y}{\partial y} + \frac{\partial \eta}{\partial t} = P \tag{9.47}$$

The equations of motion must have wind and bottom stresses terms written in terms of the wind speed and water flow rate respectively. And a surface pressure term must be added. The x component of the wind stress term is commonly written

$$\tau_{sx} = \rho K_s W^2 \cos\theta$$

where K_s is a drag coefficient as given by Eq. 9.42 or some other form, W is the wind speed, and θ is the angle between the wind direction and the x axis. The bottom stress is commonly written in terms of one of the resistance coefficients (f, n, or C). Employing Manning's n this becomes

$$\tau_{bx} = \rho q_x q \left(\frac{n^2 g}{\phi^2 d^{7/3}}\right)$$

where

$$q = \sqrt{q_x^2 + q_y^2}$$

Thus the equations of motion become

$$-gd\frac{\partial \eta}{\partial x} + fq_y - \frac{d\partial p_a}{\rho \partial x} + K_s W^2 \cos\theta - \left(\frac{n^2 g}{\phi^2 d^{7/3}}\right) q_x q = \frac{\partial}{\partial x}\left(\frac{q_x^2}{d}\right) + \frac{\partial}{\partial y}\left(\frac{q_x q_y}{d}\right) + \frac{\partial q_x}{\partial t} \tag{9.48a}$$

$$-gd\frac{\partial \eta}{\partial y} + fq_x - \frac{d\partial p_a}{\rho \partial y} + K_s W^2 \sin\theta - \left(\frac{n^2 g}{\phi^2 d^{7/3}}\right) q_y q = \frac{\partial}{\partial x}\left(\frac{q_x q_y}{d}\right) + \frac{\partial}{\partial y}\left(\frac{q_y^2}{d}\right) + \frac{\partial q_y}{\partial t} \tag{9.48b}$$

Equations 9.47 and 9.48 must be solved for spatially and temporally variable input values of W and p_a, over a basin of variable depth and with specified lateral boundary conditions. Commonly, the Manning coefficient is also made to vary with the water depth. The results produced are the spatially and temporally variable water surface elevation and x- and y-directed flow rates.

Numerous storm surge numerical models having varying degrees of complexity and employing various solution techniques are available in the literature (e.g., see Reid and Bodine, 1968; Sobey, Harper, and Mitchell, 1980; and Coeffe et al., 1984). Depending on the physical situation and the desired complexity of the model these models may employ all or only a portion of the terms in Eq. 9.48. The convective acceleration terms and the Coriolis term may be neglected as they are often relatively small. Boundary conditions are applied in a wide variety of ways, and the internal solution techniques for the numerical equations vary (e.g., see Sobey, 1970).

As an example, consider the hypothetical coastline and interior bay shown in plan view in Figure 9.10. The ocean and bay are separated by a low barrier island but connected by a narrow inlet lined with jetties. A grid has been constructed over the bay and adjacent land sections that might be flooded during a storm. The average water depth is determined for each square in the grid. The continuity equation and the two equations of motion are written in finite difference form for a square in the grid and applied sequentially in space and time to each of the squares in the grid covering the bay. Several different

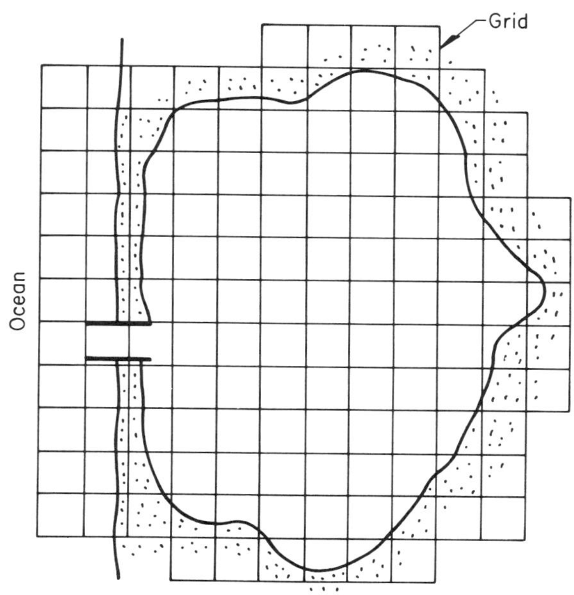

Figure 9.10 Ocean–bay system with superimposed grid.

schemes for doing this have been employed (e.g., see Sobey, 1970). With a finer grid and smaller time step, better resolution is achieved at the cost of larger computer requirements. For example, for hurricane-generated storm surge in Galveston Bay which had a grid covering 40 × 50 mi, Reid and Bodine (1968) used a 2-mi^2 grid size and a 3-min time step.

For an interior square in the grid the continuity equation balances the net flows in and out of the sides with the resulting rise or fall of the water surface elevation in that square. The equations of motion balance the forces acting on the square with the resulting flow acceleration as represented by the temporal and spatial changes in flow rate at the square. At the boundaries, however, the form of the continuity and motion equations must be modified. The sides of grid squares situated on a land/water boundary allow no flow across that side. This would be modified if, by rising to a certain level, the water can flood into an adjacent square that was previously dry. Should the ocean surge level exceed the barrier island level, flow can occur over the barrier island and into the bay. So the grid squares having a side along the barrier island must have a weir formulation to relate the flow rate across the island to the water levels on each side of the boundary. The side of the grid at the inlet entrance must have an orifice-type formulation to account for flow through the entrance, owing to the coastal surge.

Successful application of this storm surge numerical model requires that bottom resistance and weir/orifice coefficients be determined. Preliminary estimates of these coefficients can be made and then subsequently improved by calibrating the model. This involves running the model for known wind/surface pressure conditions and resulting measured surface elevations, and adjusting the coefficients as necessary to improve model predictions. When the model is successfully calibrated, it can be run for any storm conditions to predict the resulting water levels and flow rates at any grid point as a function of time during the storm.

9.7.2 The Tide, Tsunamis, and Basin Oscillations

Tide wave and tsunami propagation as well as basin oscillations can also be modeled by the long wave equations in the form given by Eqs. 9.3 and 9.48. In the latter, the surface pressure and wind stress forcing functions would be omitted. For discussions of long wave numerical models of these phenomena see Camfield (1980), Masch et al. (1977), Parker (1991), and Wilson (1972).

Reid and Bodine (1968) did an initial bottom resistance coefficient calibration of their numerical storm surge model of Galveston Bay by running the model for a spring astronomical tide. Tide records for the adjacent Gulf of Mexico and a number of points in the bay for a period of negligible wind speeds were used. The Gulf tide variation was used as the forcing function and the resulting tide level variations in the bay were calculated for locations where field records were available. Bottom friction factors were adjusted until good agreement was achieved.

Tide and tsunami long wave numerical models are most often employed along shallow coastal and estuarine areas. Here tide records from coastal stations are available to define seaward boundary conditions. The propagation of tsunamis from their source to a local coastal position may be investigated by a numerical refraction model (see Keulegan and Harrison, 1970 and Camfield, 1980). This provides input conditions for the long wave model of nearshore tsunami wave response.

REFERENCES

Camfield, F. E. (1980), "Tsunami Engineering," Special Report 6, U. S. Army Coastal Engineering Research Center, Ft. Belvoir, VA.

Coeffe, Y., Dal Secco, S., Esposito, P., and Latteux, B. (1984), "A Finite Element Method for Storm Surge and Tidal Computation," *Proceedings, 19th International Conference on Coastal Engineering*, American Society of Civil Engineers, Houston, pp. 1209–1224.

Dean, R. G. and Dalrymple, R. A. (1984), *Water Wave Mechanics for Engineers and Scientists*, Prentice-Hall, Englewood Cliffs, NJ.

Dronkers, J. J. (1964), *Tidal Computations in Rivers and Coastal Waters*, North Holland Publishing Co., Amsterdam.

Ewing, M., Press, F., and Donn, W. L. (1954), "An Explanation of the Lake Michigan Wave of 26 June 1954," *Science*, **120,** 1–2.

French, R. H. (1985), *Open-Channel Hydraulics*, McGraw-Hill, New York.

Keulegan, G. H. and Harrison, J. (1970), "Tsunami Refraction Diagrams by Digital Computer," *J. Waterw. Harbors Div., Am. Soc. Civ. Eng.*, May, 219–233.

Lamb, H. (1945), *Hydrodynamics*, 6th ed., Dover, New York.

Masch, F. D., Brandes, R. J., and Reagan, J. D. (1977), "Comparison of Numerical and Physical Hydraulic Models, Masonboro Inlet, North Carolina—Appendix 2, Volume 1 Numerical Simulation of Hydrodynamics (WRE)," General Investigation of Tidal Inlets Report 6, U. S. Army Coastal Engineering Research Center, Ft. Belvoir, VA.

Parker, B. B. (1991), *Tidal Hydrodynamics*, Wiley, New York.

Reid, R. O. and Bodine, B. R. (1968), "Numerical Model for Storm Surges in Galveston Bay," *J. Waterw. Harbors Div., Am. Soc. Civ. Eng.*, February, 33–57.

Saville, T. (1952), "Wind Set-Up and Waves in Shallow Water," Technical Memorandum 27, U. S. Army Beach Erosion Board, Washington, DC.

Sobey, R. J. (1970), "Finite Difference Schemes Compared for Wave-Deformation Characteristics in Modeling of Two-Dimensional Long-Wave Propagation," Technical Memorandum 32, U. S. Army Coastal Engineering Research Center, Washington, DC.

Sobey, R. J., Harper, B. A., and Mitchell, G. M. (1980), "Numerical Modeling of Tropical Cyclone Storm Surge," *Proceedings, 17th International Conference on Coastal Engineering*, American Society of Civil Engineers, Sydney, pp. 725–745.

Sorensen, R. M. (1978), *Basic Coastal Engineering*, Wiley, New York.

Stoker, J. J. (1957), *Water Waves*, Interscience Publishers, New York.

Sverdrup, H. U., Johnson, M. W., and Fleming, R. H. (1942), *The Oceans—Their Physics, Chemistry and General Biology*, Prentice-Hall, Englewood Cliffs, NJ.

Van Dorn, W. C. (1953), "Wind Stress on an Artificial Pond," *J. Mar. Res.*, **12,** 249–276.

Wilson, B. W. (1960), "Note on Surface Wind Stress over Water at Low and High Wind Speeds," *J. Geophys. Res.*, **65,** 3377–3382.

Wilson, B. W. (1972), "Seiches," *Advances in Hydroscience*, Vol. 8, Academic Press, New York, pp. 1–94.

10

LABORATORY INVESTIGATION OF SURFACE WAVES

Much of what is known about the characteristics of surface water waves has been learned from reduced-scale experiments conducted in two-dimensional flumes and three-dimensional basins. (The other two primary sources of our understanding of waves, of course, are theoretical developments and prototype measurements made at sea and nearshore.) Laboratory investigations of waves have been employed to evaluate the results of various wave theory predictions for wave surface profiles, water particle velocities, and so on, as well as to investigate those wave characteristics that are less amenable to theoretical investigation such as wave breaking. Flume and basin experiments have also provided a great deal of information on the impact of waves on beaches, floating and fixed structures, and harbors.

Svendsen (1985) emphasizes that if there is a discrepancy between a theory and the experimental results collected to verify the theory, "it is likely to be due to inaccuracies in the experiment." He makes this statement because he feels that it is "more difficult to make good experiments than it is to make good theories." Whether this is generally true is debatable, but the point is that laboratory wave experiments are a complex undertaking, and the results must be judiciously applied. The generation of realistic waves is a complex task as is their measurement. Unwanted laboratory and scale effects may further diminish the value of experimental results.

Owing to space requirements and a common desire to limit or simplify the types of wave investigations that are conducted, most wave research facilities are two-dimensional. The wave flumes in which they are conducted have a wave generator at one end and a wave absorber at the other end. The waves propagate the length of the tank without refracting or diffracting. Simpler and older wave generators only produce monochromatic waves. But most laboratories now have two-dimensional wave generators that produce wave spectra. Usually only the desired wave spectrum is specified, but some generators attempt to reproduce a given wave surface elevation time history.

For studies of wave action in harbors or along complex shorelines three-dimensional wave basins are required. These often are only fitted with monochromatic wave generators that produce one-directional waves. Many wave basins also have spectral wave generators that generate one-directional waves. During the past two decades, wave generators have been developed that produce multidirectional spectral waves. Waves of this type are required, for example, for realistic studies of the stability of offshore towers exposed to complex seas and for studies of the nearshore refraction and diffraction of storm waves attacking coastal harbors.

The first wave tanks were constructed in the 1800s. Bascom (1980) mentions wave tank studies published in 1825 on the basic behavior of waves including reflection from a vertical wall, surface profile geometries, and particle motions. The studies were done by Ernst and Wilhelm Webber in Germany using a 1-in. wide tank in which waves were generated by drawing fluid up into a tube and releasing it. In the mid-1800s Scott Russell in England studied solitary waves in a flume that were generated by raising a sluice gate to release water and observing the waves as they propagated up and down the tank.

The Beach Erosion Board, the predecessor to the U. S. Army Coastal Engineering Research Center, constructed its first wave tank in 1932 (Quinn, 1977). The tank was 24 ft long, 12 ft wide and had an effective depth of 1.5 ft with a monochromatic wave generator. Within a few years this tank was found to be inadequate and a much longer and deeper tank having essentially the same width was constructed.

In the 1960s wave tanks with irregular wave generators were developed and by the 1970s they were fairly common in larger laboratories. Some of the early spectral wave generators were placed in ship towing tanks to generate deep water wave spectra for seakeeping studies. Now "off the shelf" irregular wave generators are available from several manufacturers.

It is not the intent of this chapter to present a detailed discussion of the many facets of laboratory experiments with waves. The focus is on the scaling requirements for wave experiments, and on the various ways of generating laboratory waves as well as the related laboratory effects that develop when the waves are generated. A brief discussion of waves absorbers concludes the chapter.

10.1 SCALING OF LABORATORY TESTS

Basic wave experiments in flumes and physical hydraulic model studies that employ waves are commonly done at reduced scale so attention must be paid to the appropriate scaling laws. The prototype wave conditions and laboratory wave generating capabilities largely establish a prototype to model length ratio

L_R which must apply to all important lengths (waves, water depths, and structure geometries). But there must also be kinematic and dynamic similarity between model and prototype. These similarity requirements follow from a consideration of the important forces that affect the flow.

Standard fluid mechanics texts (e.g., Vennard and Street, 1982) show that the five most important fluid forces that need to be considered in hydraulic studies are gravity, viscosity, pressure, elasticity, and surface tension. Taking the ratio of each of these to the resultant or inertia force ($\Sigma F = ma$) establishes the respective dimensionless Froude, Reynolds, Euler, Cauchy, and Weber numbers. For similitude of flow these numbers must be equal in model and prototype or their respective forces in the prototype must be sufficiently small to be neglected. One of these dimensionless numbers will automatically be satisfied if the remaining numbers are satisfied or neglectable. This is the Euler number (pressure forces). Thus we have to consider the other four dimensionless numbers.

Water particle motion in a wave represents a balance between gravity, pressure, and inertia (see Chapter 2), so the Froude number

$$F = \frac{v}{\sqrt{gl}} \tag{10.1}$$

is dominant. In Eq. 10.1 v and l represent a significant flow velocity and length scale. From the Froude and Euler numbers certain model scaling ratios can be developed. For example, assuming the same fluid in model and prototype, we have:

Time ratio $= L_R^{0.5}$
Pressure ratio $= L_R$
Force ratio $= L_R^3$
Discharge ratio $= L_R^{2.5}$

The first yields the model-to-prototype wave period ratio and the others, for example, give the ratios for measured pressures, forces, and overtopping rates on a coastal structure.

Using the same fluid (water) in model and prototype and a length ratio other than unity makes it impossible to satisfy both the Froude and Reynolds number criteria simultaneously (see Vennard and Street, 1982). So surface wave experiments need to be run with Froude similarity and sufficiently large model Reynolds numbers to eliminate viscous scale effects (since prototype Reynolds numbers are large). This is usually not a problem when studying wave propagation over typical wave flume distances and with adequate water depths in the flume. Problems do arise, however, when certain wave–structure interaction studies are conducted. For example, when tests of the stability of armor stone for rubble mound structures were conducted Dai and Kamel (1969) found that the model Reynolds number should exceed 4×10^4 where the Reynolds number is defined by

$$R = \frac{\sqrt{gH}\, l_c}{\nu} \qquad (10.2)$$

and l_c is a characteristic linear dimension of the structure armor units. Wave flume measurements of wave loading on vertical pile structures also require that the Reynolds number for flow past the pile be sufficiently large that form drag on the pile is adequately simulated in the model.

Section 2.5.5 discusses the effects of surface tension on surface wave characteristics. If the model wave length is sufficiently long (greater than about 2 cm), surface tension scale effects will be neglectable, as they are for prototype waves. Goda (1985) indicates that the thin film of dust that forms on the water surface in many laboratories will increase the effects of surface tension and recommends a minimum wave period of 0.5 sec to eliminate this concern. For tests with wave spectra, the spectral peak period would need to be sufficiently higher so that the high-frequency portion of the spectrum is free of surface tension effects. Surface tension can also affect wave breaking, so tests where breaking is important should be carefully conducted to avoid this scale effect. Long-wave models may have to be distorted (i.e., have a vertical length ratio that is larger than the horizontal length ratio) so that the water depth is not so small as to cause surface tension scale effects. The effects of scale distortion are further discussed below.

Only rarely will fluid elasticity lead to scale effects. One example of where this might occur is in studies of breaking wave forces on structures. As the front face of a plunging wave hits a rigid structure, water hammer can develop in the water and the resulting force characteristics will then depend on the elasticity of the water and structure (Hudson et al., 1979).

A unique scaling problem is encountered when attempts are made to study sediment movement caused by waves. Using water in the model and model waves of reduced scale requires that the sediment size also be reduced. Sand in the prototype then becomes a subsand size in the model and a different sediment transport regime develops (Kamphius, 1985). Consequently, the conduct of wave–sediment transport experiments becomes more of an art than a science. Several involved testing procedures and related scaling laws have been developed for such experiments (e.g., see Kamphius and Readshaw, 1978; Kamphius, 1985; Kreibel et al., 1986; and Noda, 1972).

Some model studies, such as the investigation of wave action in coastal harbors, require that a large lateral area be modeled. And these models, owing to cost and space limitations, may have to be constructed with limited horizontal dimensions. If an undistorted model is used, water depths in large portions of the model may consequently be very shallow. This can lead to serious viscous and surface tension scale effects. Also, the vertical scale is such that the scaled wave heights in the model are quite small and difficult to measure. This is particularly true for studies of long wave resonance in harbors. A way to deal with these problems is to use a distorted scale model where the vertical scale ratio is different from the horizontal scale ratio.

If a model has a distorted scale, questions arise about the impact on wave reflection, refraction, and diffraction in the model. Sloped boundaries are steeper in the model than in the prototype if the model is distorted. This causes a greater relative wave reflection in the model than in the prototype which must be overcome, for example, by increasing the model slope's roughness and/or porosity.

The impact of model scale distortion on wave refraction and diffraction is more complex. Consider refraction first and define models scales: L_{RV} for the vertical scale ratio and L_{RH} for the horizontal scale ratio. For long waves (shallow water conditions), since $C = (gd)^{0.5}$ the celerity ratio C_R is just governed by the vertical scale ratio and refraction patterns are unaffected by the distorted model scale. Since

$$C_R = \sqrt{L_{RV}}$$

and, by definition

$$C_R = \frac{L_{RH}}{T_R}$$

we have

$$T_R = \frac{L_{RH}}{\sqrt{L_{RV}}} \tag{10.3}$$

which defines the ratio of prototype to model wave periods in a distorted scale refraction model employing long waves.

For the refraction of intermediate or transitional depth waves the full dispersion equation (Eq. 2.10) applies. Thus,

$$C_R = \sqrt{\frac{L_{RH} \tanh\left(\dfrac{2\pi d}{L}\right)_p}{\tanh\left(\dfrac{2\pi d}{L}\right)_m}} \tag{10.4}$$

and refraction can only be correctly modeled if the depth ratio and the wave length ratio are equal [i.e., $(d/L)_R = 1$, see Dalrymple, 1985]. The wave period ratio is given by

$$T_R = \frac{L_{RH}}{C_R} = \sqrt{\frac{L_{RH} \tanh\left(\frac{2\pi d}{L}\right)_p}{\tanh\left(\frac{2\pi d}{L}\right)_m}}$$

which can be shown (Hudson et al., 1979) to yield

$$T_R = \sqrt{L_{RV}}$$

Thus a distorted scale transitional-depth refraction model can be designed. But if diffraction also occurs in the model a conflict develops, making it impossible to scale both refraction and diffraction in the same distorted scale model.

Scaling of wave diffraction is based on the dimensionless ratio of horizontal distance to wave length (see Chapter 5 and Figures 5.9, 5.10, and 5.12). Thus for a distorted scale model, the horizontal scale must be the same as the wave length scale (which differs from the vertical scale in a distorted scale model). This is in contradiction to Eq. 10.4 for intermediate depth waves. For long waves this contradiction does not occur, so distorted scale models studies can be conducted that correctly scale both wave refraction and diffraction.

10.2 LABORATORY WAVE GENERATION

Wave generators vary in complexity from the simple two-dimensional monochromatic wave generator that is basically a piston connected to a motor that can be run at a range of speeds, to the very complex wave generators that are capable of producing directional wave spectra in a basin. We give an overview of the basic characteristics of these different types of generators and discuss related difficulties that arise from the efforts to generate the exact wave systems desired.

10.2.1. Monochromatic Waves

Any device that produces a constant frequency disturbance of the water surface will generate monochromatic waves. Preferably, the device will move the entire water column and immediately produce the desired water particle motion for the wave height/period combination being generated. Otherwise, undesired higher harmonic free waves will be generated that interfere with the desired pure monochromatic wave.

Figure 10.1 shows schematic diagrams of four basic types of monochromatic wave generators—the piston, flap, plunger, and pneumatic generators. The first three consist of a rigid blade driven by a motor. The motor may have a flywheel

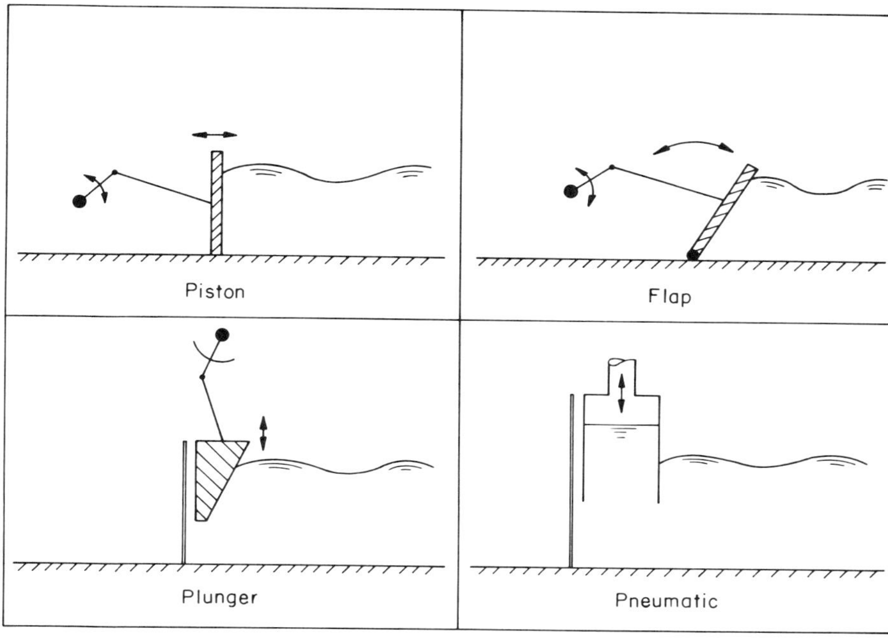

Figure 10.1 Monochromatic wave generators.

with sufficient angular momentum to maintain uniform rotational speed. The wave amplitude and period can be varied by adjusting the stroke and cyclic frequency of the blade. In the pneumatic generator, the water surface displacement is accomplished by cyclically varying the pressure above the water in the pneumatic chamber. Svendsen (1985) briefly discusses 20 types of wave-generating devices, many of which are simple variations of the four types shown in Figure 10.1.

Of the four generator types shown in the figure, by far the most common are the piston and flap. Considering the desired water particle motion, the former would be most effective for generating shallow water waves and the latter for generating deep water waves. Some laboratories have constructed wave generators that can be adjusted to give a blade motion that can be varied between flap and piston motion as the wave relative depth is varied from the deep to shallow water condition. This adjustment may simply be in the support mechanism for the blade, which would be modified as the relative depth of the wave to be generated is changed. Or the blade may have a pair of drive arms, one at the top of the blade and the other at the bottom, whose relative motions can be adjusted for the deep to shallow water range of wave generation. Figure 10.2 shows a typical arrangement.

The design of a wave generator requires that the piston or flap motion be related to the desired wave characteristics produced by the generator. First, the drive motor angular speed range can be selected to match the range of wave

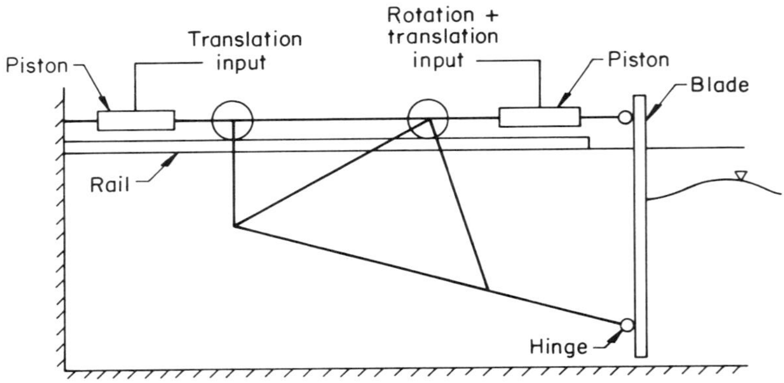

Figure 10.2 Adjustable blade motion wave generator (after D'Angremond and Van Oorschot, 1969).

periods generated. Then a theory is needed to relate the piston or flap stroke to the resulting maximum wave heights generated. Finally, the drive motor power requirements must exceed the power of the largest wave generated, making allowances for blade friction and gap leakage losses and providing excess power so blade motion is uniform.

A simple procedure (see Galvin, 1964) for estimating the height of shallow water waves generated by a given piston or flap blade stroke can be obtained by equating the volume of water displaced by the blade in one half period to the volume of water in one wave length above the still water line. Assuming a sinusoidal water surface profile, this yields the wave height to blade stroke ratio

$$\frac{H}{S} = \frac{2\pi d}{L} \tag{10.5}$$

where S is the blade stroke defined as the total extent of forward to rear motion of a piston blade or half of the forward to rear motion of a flap blade at the still water elevation on the blade. Galvin (1964) found that Eq. 10.5 gave results that were very good for a piston generator, but less so for a flap generator. Some of the discrepancy for the flap generator may have been due to leakage of water between the flap blade and the tank wall, which diminishes the resulting wave height that is generated.

A more effective theory for mechanical wave generation, which covers the range from deep to shallow water, may be derived using procedures similar to those used to derive the small amplitude wave theory (Chapter 2). The same surface and bottom boundary conditions are used and lateral boundary conditions are added. At the far end waves are assumed to propagate to infinity (i.e., no wave reflection). At the wave generator ($x = 0$) a kinematic boundary condition must be applied, namely, that the desired horizontal component of water particle motion along the blade is the same as the blade motion. For a discussion of the resulting wave generator theory and experimental evaluations of this theory see Dean and Dalrymple (1984), Gilbert et al. (1971), and Ursell et al. (1960).

The linearized theory yields the following wave height to blade stroke ratio for a piston generator:

$$\frac{H}{S} = \frac{2\,(\cosh kd - 1)}{\sinh 2kd + kd} \tag{10.6}$$

and for a flap generator,

$$\frac{H}{S} = 8\left(\frac{\sinh kd}{kd}\right)\frac{kd \sinh kd - \cosh kd + 1}{\sinh 2kd + 2kd} \tag{10.7}$$

where S is defined as above for Eq. 10.5.

The values for H/S given by Eqs. 10.6 and 10.7 approach the value given by Eq. 10.5 when $kd = 1$ or less (approaching shallow water). At the deep water limit H/S levels off at a value around 2 for a piston generator and a value around 3.5 for a flap generator.

For a given blade frequency, the equations given above show that the generated wave height increases as the blade stroke increases. This is, of course, limited by wave breaking (see Sections 2.5.2 and 3.5). If the blade stroke is set to generate a wave that would be higher than the breaking limit will allow, the wave will break right off the blade and a wave having a height that is less than the breaking limit will be generated.

Equation 2.35 gives the power per unit width of wave crest in a wave, based on the small amplitude wave theory. This is the average power over the wave that the generator would have to supply to the wave. But, owing to gap leakage and other losses that occur and to the fact that the peak power input during the blade stroke cycle must exceed the average wave power, it is recommended (Snyder et al., 1958) that the generator motor power be at least three times the wave power calculated by Eq. 2.35. This wave power calculation would be based on the desired wave height–period combination that yields the largest value of wave power.

When a wave generator blade is driven by a rotary motor, the blade translates with approximately sinusoidal motion. Whether the blade undergoes pure piston or flap motion (or some in-between form of motion), there will not be a complete match between the blade horizontal velocity and the desired horizontal particle velocity field of the waves being generated. For frequency dispersive waves (deep and intermediate depth waves), besides the desired wave a free second harmonic wave of much lower amplitude and period $T/2$ will be generated (Buhr Hansen and Svendsen, 1974 and Flick and Guza, 1980). This wave will travel down the wave flume superimposed on the primary wave and, since the second harmonic wave is a free wave and travels with a smaller celerity than the primary wave, the combined wave will not have a constant form. Figure 10.3, which shows the water surface time history at successive positions along a wave flume, shows the typical interaction that will result.

Buhr Hansen and Svendsen (1974) conducted experiments to find a way to reduce the amplitude of the free second harmonic wave because this wave can have very undesirable effects in certain types of laboratory experiments (e.g., experiments on beach profile response to wave attack). They found that they were quite successful when they employed a blade motion that is composed of two superimposed sinusoidal motions where the second sine wave was a harmonic of the first and had a phase and amplitude selected to minimize the free harmonic wave. A submerged sill or bar will also disturb incident waves to generate free second harmonic waves. Hulsbergen (1974) found that he could suppress the free second harmonic waves from a wave generator by installing a sill that also generated free second harmonic waves that were of the same amplitude but 180° out of phase of the free harmonic waves from the wave generator.

A wave generator with perfect blade motion to suit the required water particle motion can generate waves that are initially of constant form. However, particularly for steep (Stokes) waves in deep water, any small disturbances will cause instabilities to develop as the waves propagate away from the generator. Benjamin and Feir (1967) showed that when $kd > 1.363$ and there is a disturbance having a frequency slightly higher or lower than the generated frequency, energy is transferred to these disturbance frequencies. The result is similar to the group beating effect developed (see Section 2.4.7) from two waves of close but not equal frequency, and a slow modulation of the wave height along the wave flume is produced. This instability slowly grows as the waves propagate from the wave generator, the instability growth rate being greater for steeper waves. But it requires about 30–50 wave lengths for the instability to be significant, so this problem is not commonly critical except for very short period waves.

Often wave experiments involve structures that have a significant reflection

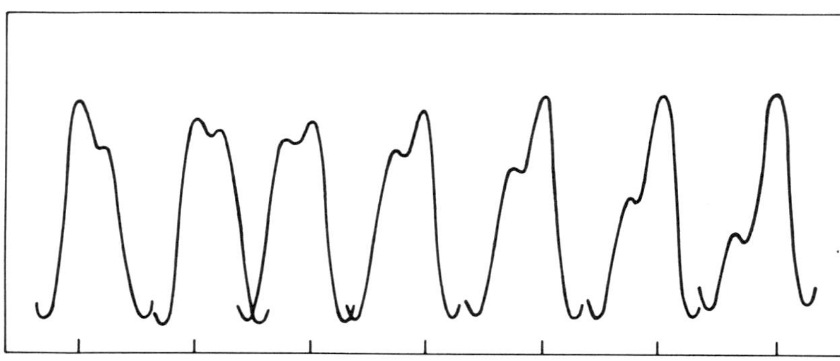

Figure 10.3 Measured surface profiles at successive positions (left to right) along a wave flume (Chandler and Sorensen, 1972).

coefficient. The reflected waves propagate back to the wave generator and rereflect back toward the structure. Thus, depending on the length of the wave flume, there is only a short time period after the generator is started before the waves attacking the structure are different from those being generated, owing to the superposition of the rereflected waves.

In earlier days the rereflection problem was dealt with by constructing long wave flumes and generating waves in bursts so that the generator could be stopped for a period of time between bursts to allow reflected waves to dissipate. This was unsatisfactory for several reasons. Testing took an excessive amount of time, the wave group phenomena caused lower waves at the beginning and end of each burst, and generator blade inertia meant that the initial and final waves in the burst often had longer wave periods than the desired waves. An alternate way to deal with this rereflection problem is to use a wide wave basin that is subdivided by vertical walls to form a narrow test channel with wave-dissipating beaches on either side (Fig. 10.4). The reflected waves propagate back into the wider basin and most of the energy from the rereflected waves is dissipated at the beach sections.

Recently, wave generators have been designed that detect the reflected waves and adjust the blade motion to cancel out these reflected waves (Bullock and Murton, 1989; Salter, 1981). The reflected waves are detected by a thin wire wave gage on the blade face or by sensing the load in the arm that drives the generator blade. From this, the blade motion is continually adjusted to generate waves that are the same amplitude and 180° out of phase with the reflected waves to cancel them out.

Another reflection problem that is occasionally observed in a wave flume occurs when the wave period being generated equals the fundamental or a harmonic resonant period of the flume. Lengthwise oscillations of long waves in a flume can be troublesome as they take a long time to dissipate owing to their high reflectivity. Lateral oscillations across the flume width are also sometimes seen as the waves propagate along the flume. The wave crest oscillates at the resonant period being alternately higher on one flume wall and then the other. This problem is best resolved by testing with wave periods that are different from the lateral resonant periods of the flume.

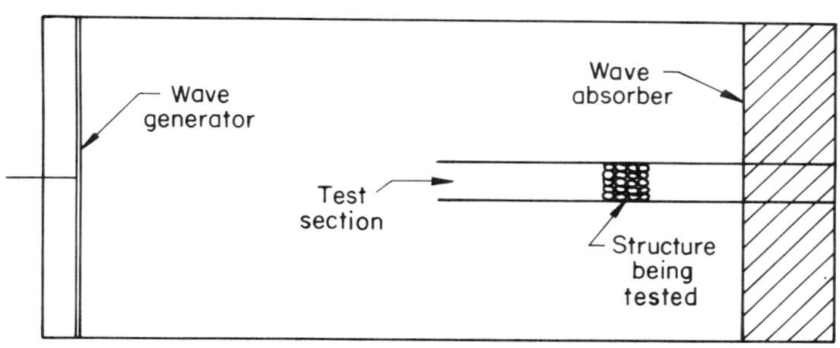

Figure 10.4 Subdivided flume to minimize wave rereflection problems.

10.2.2 Irregular Waves

Our discussion of the generation of irregular waves is broken into three segments: mechanical generation of two-dimensional waves, wave generation by wind, and the generation of directional wave spectra. Most laboratory investigations employing irregular waves are carried out in two-dimensional flumes where the waves are generated mechanically. We describe the generic type of generator used to generate irregular waves and the three basic procedures for producing the signal required to drive the generator. Over the years laboratory waves have also been directly generated by wind, but scale problems are encountered. The procedures used and the nature of these scale problems are discussed. Recently, very sophisticated generators capable of generating directional wave spectra in basins have been developed. These are discussed briefly.

Mechanical Wave Generation. Figure 10.5 is a schematic depiction of the generic type of device commonly used to generate irregular waves. Often these generators have a dry well behind the blade. This reduces the power requirements of the generator and eliminates the need for a wave absorber to dampen water motion behind the blade. To reduce leakage, pressure seals are placed in the gaps between the blade and flume walls and bottom.

An appropriate electrical input signal must be prepared (see discussion below) and sent to the generator to drive the piston/blade by a hydraulic, pneumatic, or mechanical device. The servo senses the piston motion and sends a proportional voltage feedback to the signal control. The feedback and input signals are continuously compared to adjust the piston motion to the desired form.

This type of wave generator can be used to generate monochromatic waves by simply imputing a sinusoidal signal. A pair of sinusoidal signals can be input to eliminate harmonic free waves as discussed above, or nonsinusoidal input signals can be used to generate any desired wave form such as cnoidal

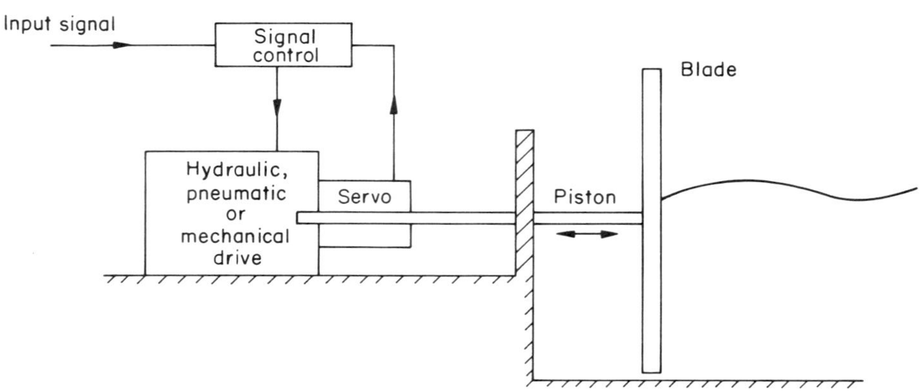

Figure 10.5 Schematic of irregular wave generator.

or solitary waves. Input of appropriate complex signals, in turn, will generate the desired irregular waves.

The input signal for irregular wave generation is commonly produced in one of three ways:

1. By the superposition of a large but finite number of sine waves of different amplitudes and periods with random phasing.
2. By the filtering of white noise to form the desired irregular wave spectrum.
3. By creating a signal that replicates a measured or artificially constructed surface elevation time history.

The purpose of the first two procedures is just to generate a desired wave spectrum, either the spectrum derived from a measured wave record or a theoretical spectrum model (e.g., JONSWAP). The individual wave phasing is not specified nor are the wave grouping characteristics (although certain levels of wave grouping may be caused by the shape of the selected spectrum). A particular wave train cannot be reproduced on demand, as can be done by the third procedure. The third procedure is most desirable, for example, when a particular sequence of extreme or grouped waves is desired to test a moored or fixed structure or the stability of rubble mound structures under attack by select groups of high waves. It also allows us to use exactly the same wave sequence in successive tests.

Procedure 1: Two steps are required to produce a wave generator input signal by the superposition of sine waves. First, the appropriate sine components must be determined and added; then a transfer function is required to convert the summed sine components to the desired blade motion. Several approaches have been used to determine the sine components. Borgman (1969) used equal amplitude sine components where the frequency range represented by each component was inversely proportional to the spectral amplitude for that frequency range. Consequently, there were many components with frequencies around the spectral peak and the components away from the spectral peak represented a much wider frequency range. This will lead to a poorer representation of the spectral shape at the high and low frequency ends. Borgman used a total of 200 sine components.

Goda (1970) employed sine components whose amplitudes were not constant but were proportional to the area of the spectrum segment represented by the frequency band covered by the selected frequency. The selected frequencies are taken at random, but with the spacing between frequencies increasing as the frequency increases. After the amplitude and frequency of the individual components are selected, the phasing of each component is randomly selected from a uniform distribution of phases. Seelig (1980), in a study of the wave reflection and transmission characteristics of breakwaters exposed to irregular waves, used Goda's approach with 50 sine components. His resulting wave spectra were rough, so more than 50 sine components are typically required to produce a smooth spectrum with sufficient accuracy in matching the desired spectrum. Carvahlo (1989) recommends that between 75 and 150 components be used.

The frequency and phase of each since component translate directly to the wave generator input signal. The amplitude of each component must be transformed to give the appropriate blade stroke amplitude for the desired component wave amplitude by using a transfer function such as given by Eq. 10.6 or 10.7. Note that the transfer function is frequency dependent so that the shorter frequencies in the spectrum require less blade stroke than do the longer frequencies to generate the same amplitude component. The addition of the sine components is typically done digitally by computer so the signal must be run through a digital-to-analog converter to develop the desired analog input signal for the wave generator. The spectrum of waves then has to be generated in the flume, measured, and analyzed to compare the resulting spectrum to the desired spectrum. Commonly, adjustments have to be made to achieve the desired result.

Procedure 2: The input signal to an irregular wave generator may be formed by filtering the signal from an analog or digital white noise generator. The white noise has the frequency range for the desired spectrum. A set of adjustable filters allow the white noise signal to be modulated to form a signal having the desired spectral shape. An electronic filter can be used for analog white noise (Carvahlo, 1989) and a finite Fourier transform algorithm can be used for the digitally produced white noise (Hudspeth et al., 1985). The appropriate transfer function then converts the signal to yield the necessary blade stroke for each wave. If the white noise is generated by a digital computer, a digital-to-analog converter is also needed to produce the final input signal for the wave generator.

Procedure 3: The input signal for the generation of a specified wave train may be derived by the Fourier decomposition of the wave record into its individual sine components (see Lundgren and Sand, 1978 and Carvahlo, 1989). This yields the amplitude, frequency, and phase of each sine component which can (as with Procedure 1) be transformed by Eqs. 10.6 or 10.7 to the desired blade motion input signal.

Gravesen et al. (1974) applied a procedure for generating specified wave trains that is based on a calculation of the resulting water particle motion in the waves. The horizontal particle velocity can be related to the time-dependent water surface elevation by employing the linear wave theory. This time-dependent particle velocity is then integrated to yield the desired time-dependent blade motion. A piston-type wave generator was used so this approach was most appropriate for shallow water irregular waves.

It should be remembered that the desired spectrum may be generated, but that this is done at the generator. As the waves propagate down the flume, wave dispersion causes the lower-frequency waves to move ahead and reach the test area first. With time, the total desired spectrum of waves will appear at the test area. This is so, with the caveat that the spectrum may be modified by Benjamin–Feir type transfers of energy to side-band frequencies. Funke (1974) shows a plot of a JONSWAP spectra measured at 5-m intervals along a wave flume. As the waves propagated down the flume, the spectra flattened and the peak frequency shifted slightly to successively lower values.

Wave Generation By Wind. Several decades ago, before the development of mechanical irregular wave generators, studies of two-dimensional irregular wave runup and attack on coastal structures were made in flumes where the waves were generated by wind. These flumes had a lid over their entire length to contain the air, an exhaust fan at the downwind end, and an air intake to smoothly guide the incoming air to the upwind water surface. Studies of this type continued into the 1960s (see Prins, 1960 and Plate and Nath, 1968). With the development of effective mechanical irregular wave generators, laboratory studies employing waves generated solely by wind are no longer commonly done.

To scale the wind wave generation process correctly requires that (see Eq. 6.45)

$$\left(\frac{gF}{U^2}\right)_{\mathrm{m}} = \left(\frac{gF}{U^2}\right)_{\mathrm{p}} \tag{10.8}$$

This assumes that the waves are fetch limited. For even the longest wave flumes (having lengths on the order of 100 m or more), the required wind speed for similitude is so small that waves of incredibly small amplitude and period are generated. These waves are of little use for engineering studies of wave runup or structure stability. Also, serious surface tension scale effects are likely to occur. There are other physical problems that develop, such as the growth of boundary layers along the flume walls and ceiling that are unrealistic and the development of a significant pressure gradient along the flume that distorts the wave generation process (Harris, 1976).

In order to have sufficient wave amplitudes and periods for the study of engineering effects of wind waves, the flumes were run with exaggerated wind speeds. This produced unrealistically small values of the parameter gF/U^2 for even the longest flumes. One consequence of this procedure was that unnaturally narrow wave spectra were generated (D'Angremond and Van Oorschot, 1969). However, some worthwhile studies of air–sea interaction and of the initiation and growth mechanism involved in the wind wave generation process have been carried out. Worthwhile studies have also been conducted on the wind setup of water and of oil floating on water.

With the development of mechanical irregular wave generators that can generate the desired spectrum size and shape as well as the desired wave record, there is little need for the use of wind-generated waves in laboratory studies of coastal and ocean engineering problems. Wind has been effectively used, however, by blowing the wind over irregular waves that have been generated by a mechanical wave generator (D'Angremond and Van Oorschot, 1969). The effect of the wind is to more realistically steepen the fronts of the individual waves, as happens in the field during a storm.

Generation of Directional Spectra. The procedures discussed above can be used to generate straight long-crested monochromatic waves or one-dimensional wave spectra in wide basins. But many three-dimensional wave studies require that directional wave spectra be used. Examples include the stability of offshore oil drilling structures (particularly when freak waves are of concern), the investigation of realistic nearshore refraction–diffraction problems, and studies of wave action inside coastal harbors (see Elgar et al., 1992; Goda, 1985; Salter, 1981; and Vincent and Briggs, 1989). As a consequence, during the last two decades a number of the major laboratories have constructed basins with wave generators capable of generating directional wave spectra.

Consider the wave generator shown schematically in plan view in Figure 10.6. It consists of a row of individually actuated wave generators. If all of the generators are run with the same blade stroke amplitude, frequency, and phasing, a straight long-crested monochromatic wave having its crest axis parallel to the row of generators would be produced. However, if there is a progressive phase shift ϵ of each of the blades as shown in Figure 10.6, a straight long-crested monochromatic wave will be generated with its axis at an angle α with the axis of the generators. It can be shown that this angle is given by (see Salter, 1981)

$$\sin \alpha = \frac{\epsilon L}{2\pi W} \tag{10.9}$$

where W is the blade width, L is the generated wave length, and the phase shift is in radians. The amplitude and frequency for each generator blade is established in the normal way. For basin experiments with monochromatic waves, where it is desired to frequently change the incident wave angle, a ''snake''-type generator as depicted in Figure 10.6 is more efficient than a generator with a single long blade that has to be rotated each time the incident wave angle has to be changed.

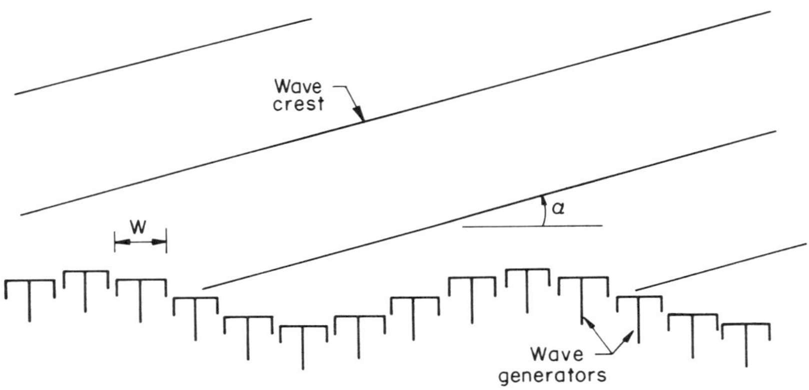

Figure 10.6 Multicomponent wave generator (after Salter, 1981).

The generator depicted in Figure 10.6 can be used to generate directional wave spectra by adding sine components (as done above for one-dimensional spectra generation). The directions of the individual components are also varied to develop the desired directional spreading distribution for the spectrum. It becomes apparent that the computational task required to produce the individual input signals for all of the generators is immense. The development of directional spectra wave generators has essentially followed the development of adequate computers to run the generators.

A directional spectra wave generator at the University of Edinburgh has 80 individual wave generators (Salter, 1981). It can employ up to 75 component waves to develop the desired spectrum. A spectral wave generator at the U. S. Army Waterways Experiment Station in Vicksburg, Mississippi has 61 blades, each 0.46 m wide, but the blades are driven by 60 piston arms that are connected to the attachment points between the blades (Vincent and Biggs, 1989). Thus there is a phase shift between each piston arm, but no gaps between individual blades since the blades can rotate as well as translate. This provides for smoother wave generation.

10.2.3 Generation of Long Waves

Monochromatic Waves. In some instances, such as for the investigation of tsunami wave effects, it is necessary to generate shallow water waves of true form. Solitary and cnoidal form waves can best be generated by a vertical piston blade generator. Solitary waves are generated by moving the blade forward a certain distance and stopping it. When Goring and Raichlen (1980) used a linear forward blade displacement, the generated solitary wave was trailed by a group of lower-amplitude, slower-moving frequency dispersive oscillatory waves. Similarly, when the blade was moved periodically with a frequency that would produce cnoidal waves, a nonpermanent wave form that did not fit the cnoidal theory surface profile was generated.

Goring and Raichlen were able to generate essentially "true" solitary and cnoidal waves of permanent form by matching the blade velocity with the calculated time-dependent depth-averaged water particle velocity in the solitary or cnoidal wave that was desired. That is,

$$\frac{ds}{dt} = U(s, t)$$

where s is the horizontal displacement of the blade from its mean position and $U(s, t)$ is the position and time-dependent depth-integrated average particle velocity in the wave. It was important to determine the value of U at the instantaneous position s of the blade in order to have a true match of blade and particle velocity.

Group-Induced Waves. Figure 10.7 shows a typical set of grouped irregular waves and the associated mean water level. Owing to radiation stress, the mean

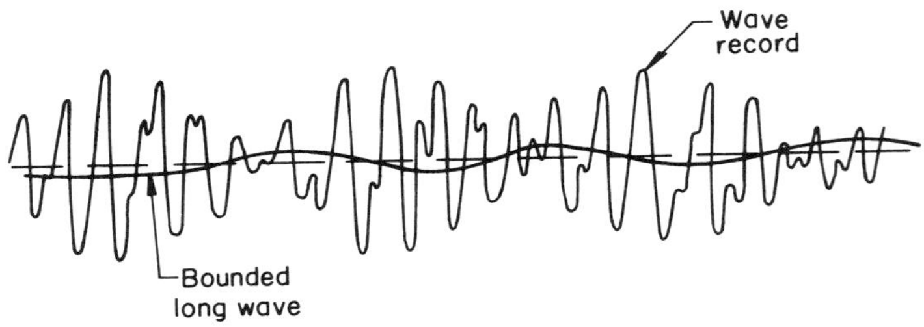

Figure 10.7 Grouped irregular wave and bounded long wave.

water level will set down in the vicinity of higher waves and set up in the vicinity of relatively lower waves. This produces a bounded long wave that propagates with the wave group and has the same period as the group. In irregular wave studies of harbor resonance, mooring loads for floating vessels, and wave runup on beaches, it is important that these group-induced long waves be correctly reproduced.

The bounded long waves behave as normal waves in that there is a forward water motion under the set up (crest) and a reverse motion under the set down (trough). When one of the procedures discussed in the previous section is used to generate irregular waves, these forward and reversing water motions tied to the bounded long waves are not reproduced at the generator blade. The bounded waves develop as the wave groups propagate down the flume, but additional free long waves are generated since the required boundary conditions are not fully satisfied at the generator blade. The undesired free long waves propagate up and down the flume until they are dissipated and, consequently, may cause undesirable effects. They also increase in amplitude as the water depth decreases.

These undesired free long waves can be eliminated, as discussed before, by adding to the generator input signal a long wave component that is equal in amplitude and opposite in phase to the undesired component. For a more detailed discussion of this phenomenon and techniques for its elimination see Ottesen Hansen et al. (1980), Sand (1982), Barthel et al. (1983), and Mansard and Barthel (1984).

10.3 WAVE ABSORBERS

In wave flumes, experiments on basic wave characteristics or on structures that do not completely intercept the incident wave require that an efficient wave absorber be installed at the end of the flume. In wave basin experiments, wave absorbers may also be required along the sides of the basins. The effectiveness of any wave absorber can easily be tested by measuring the reflection (see Section 2.5.3) of a range of incident wave height/period combination monochromatic waves for the expected range of water depths.

The ideal wave absorber would be a rough porous flat slope (e.g., a flat-sloped beach made of pea gravel). The flat slope induces repeated wave breaking and the porosity allows the remaining runup to seep into the beach rather than run down the slope to generate a reflected wave. Svendsen and Jonsson (1976) suggest that wave reflection will be negligible if the value mL/d (where m is the beach slope) is of the order of unity. This could require a rather long wave absorber. Svendsen (1985) notes that the most efficient absorber would keep the value mL/d small at all points along the slope. As the wave propagates up the slope, the slope depth decreases, but the wave length also decreases by the square root of the depth. This leads to the most efficient slope being a parabola of the form $d = ax^2$ where x is measured seaward from the still water line. With a parabolic slope of length equal to $10d$–$15d$ and the effective value of mL/d around unity, Svendsen (1985) indicates that the wave reflection coefficient will be below 0.05.

Of course, long flat slopes take up much space and are not easy to move should they need to be relocated. So shorter wave absorbers with steeper slopes are commonly used. The slopes are covered with wire mesh, plastic-coated hair mats, or other materials that will significantly dissipate wave energy as the wave moves up the slope.

Jamieson and Mansard (1987) report on a wave absorber developed for the Canadian National Research Council Hydraulics Laboratory. They wanted an efficient wave absorber that took up less space and was effective for a wide range of water depths. The result was a series of perforated vertical metal sheets placed so the face of each sheet is parallel to the incident wave crest. The sheets progressively decrease in porosity toward the rear of the absorber. The change in porosity is used because sheets with high porosity are most effective for high steepness waves, and sheets with lower porosity are more effective for lower steepness waves. As the waves propagate through the absorber energy dissipation results in decreased wave heights (i.e., wave steepness), so progressively lower porosity sheets optimize energy dissipation. If lower porosity sheets were placed in the front of the absorber, there would be excessive reflection of the higher waves. Wave energy reflected from the lower porosity interior sheets must pass through more sheets to exit the front of the absorber and is thus effectively dissipated. There is a progressive decrease in the spacing of the sheets toward the rear of the absorber because the lower wave heights cause shorter horizontal displacements of water particles in the waves. Wave energy that reaches the rear wall of the flume will reflect so optimum location of the absorber would place the lowest porosity sheets at the nodal position ($L/4$) of the partially reflected wave.

Another effective wave absorber that would take up less space would involve blades that articulate to absorb the incoming wave (like the technique discussed above for eliminating reflected waves from wave generators). The blade would have a mechanism for detecting the incoming wave surface elevation versus time and cause the blade to move so that it just follows the water particle motion in the wave (see Milgram, 1970). The blade is then essentially invisible to the waves and they would pass through the blade if there was water on the other side. If the other side is dry, the incident wave energy is absorbed by the motion of the blade.

REFERENCES

Barthel, V., Mansard, E. P. D., Sand, S. E., and Vis, F. C. (1983), "Group Bounded Long Waves in Physical Models," *Ocean Eng.*, **10**, 261–294.

Bascom, W. (1980), *Waves and Beaches*, Anchor/Doubleday, Garden City, NY.

Benjamin, T. B. and Feir, J. E. (1967), "The Disintegration of Wave Trains in Deep Water, Part 1, Theory," *J. Fluid Mech.*, **27**, 417–430.

Borgman, L. E. (1969), "Ocean Wave Simulation for Engineering Design," *J. Waterw. Harbors Div., Am. Soc. Civ. Eng.*, November, 557–583.

Briggs, M. J., Borgman, L. E., and Outlaw, D. G. (1987), "Generation and Analysis of Directional Spectral Waves in a Laboratory Basin," *Proceedings, Offshore Technology Conference*, Houston, pp. 495–502.

Buhr Hansen, J. and Svendsen, I. A. (1974), "Laboratory Generation of Waves of Constant Form," *Proceedings 14th International Conference on Coastal Engineering*, American Society of Civil Engineers, Copenhagen, pp. 321–339.

Bullock, G. N. and Murton, G. J. (1989), "Performance of a Wedge-Type Absorbing Wave Maker," *J. Waterw. Port Coastal and Ocean Eng. Div., Am. Soc. Civ. Eng.*, January, 1–17.

Carvahlo, M. M. (1989), "Sea Wave Simulation," *Recent Advances in Hydraulic Physical Modelling*, R. Martins, Ed., Kluwer Academic Publishers, Dordrecht, pp. 447–502.

Chandler, P. L. and Sorensen, R. M. (1972), "Transformation of Waves Passing a Submerged Bar," *Proceedings, 13th International Conference on Coastal Engineering*, American Society of Civil Engineers, Vancouver, pp. 385–404.

D'Angremond, K. and Van Oorschot, J. H. (1969), "Generation of Irregular Waves on Model Scales," *Proceedings, Research on Wave Action Symposium*, Vol. 1, Paper 2, Delft Hydraulics Laboratory, Delft.

Dai, Y. B. and Kamel, A. M. (1969), "Scale Effects Tests for Rubble Mound Breakwaters; Hydraulic Model Investigation," Report H-69-2, U. S. Army Engineer Waterways Experiment Station, Vicksburg, MS.

Dalrymple, R. A., Ed., (1985), "Introduction to Physical Modelling in Coastal Engineering," *Physical Modelling in Coastal Engineering*, A. A. Balkema Publisher, Rotterdam, pp. 3–9.

Dean, R. G. and Dalrymple, R. A. (1984), *Water Wave Mechanics for Engineers and Scientists*, Prentice-Hall, Englewood Cliffs, NJ.

Elgar, S., Guza, R. T., Freilich, M. H., and Briggs, M. J. (1992), "Laboratory Simulations of Directionally Spread Shoaling Waves," *J. Waterw. Port Coastal Ocean Eng. Div., Am. Soc. Civ. Eng.*, January/February, 87–103.

Flick, R. E. and Guza, R. T. (1980), "Paddle Generated Waves in Laboratory Channels," *J. Waterw. Port Coastal Ocean Eng. Div., Am. Soc. Civ. Eng.*, February, 79–97.

Funke, E. R. (1974), "Random Wave Signal Generation by Minicomputer," *Pro-*

ceedings, 14th International Conference on Coastal Engineering, American Society of Civil Engineers, Copenhagen, pp. 352–371.

Galvin, C. J. (1964), "Wave Height Prediction for Wave Generators in Shallow Water," Technical Memorandum 4, U. S. Army Coastal Engineering Research Center, Washington, DC.

Gilbert, G., Thompson, D. M., and Brewer, A. J. (1971), "Design Curves for Regular and Random Wave Generators," *J. Hydraul. Res.*, **9,** 163–196.

Goda, Y. (1970), "Numerical Experiments on Wave Statistics With Spectral Simulation," Report 9-3, Port and Harbor Research Institute, pp. 7–57.

Goda, Y. (1985), *Random Seas and Design of Maritime Structures*, University of Tokyo Press, Tokyo.

Goring, D. and Raichlen, F. (1980), "The Generation of Long Waves in the Laboratory," *Proceedings, 17th International Conference on Coastal Engineering*, American Society of Civil Engineers, Sydney, pp. 763–783.

Gravesen, H., Frederiiksen, E., and Kirkegaard, J. (1974), "Model Tests with Directly Reproduced Nature Wave Trains," *Proceedings, 14th International Conference on Coastal Engineering*, American Society of Civil Engineers, Copenhagen, pp. 372–391.

Harris, D. L. (1976), "Wind-Generated Waves for Laboratory Studies," Technical Paper 76-12, U. S. Army Coastal Engineering Research Center, Ft. Belvoir, VA.

Hudspeth, R. T., Nath, J. H., and Sollitt, C. K. (1985), "Digital to Analog Wavemaker Simulations," *Physical Modelling in Coastal Engineering*, R. A. Dalrymple, Ed., A. A. Balkema, Publisher, Rotterdam, pp. 81–103.

Hudson, R. Y., Herrmann, F. A., Sager, R. A., Whalin, R. W., Keulegan, G. H., Chatham, C. E. and Hales, L. Z. (1979), "Coastal Hydraulic Models," Special Report 5, U. S. Army Coastal Engineering Research Center, Ft. Belvoir, VA.

Hulsbergen, C. H. (1974), "Origin, Effect and Suppression of Secondary Waves," *Proceedings, 14th International Conference on Coastal Engineering*, American Society of Civil Engineers, Copenhagen, pp. 392–411.

Jamieson, W. W. and Mansard, E. P. D. (1987), "An Efficient Upright Wave Absorber," *Proceedings, Coastal Hydrodynamics Conference*, American Society of Civil Engineers, Newark, DE, pp. 124–139.

Kamphius, J. W. (1985), "On Understanding Scale Effect in Coastal Mobile Bed Models," *Physical Modelling in Coastal Engineering*, R. A. Dalrymple, Ed., A. A. Balkema Publisher, Rotterdam, pp. 141–162.

Kamphius, J. W. and Readshaw, J. S. (1978), "A Model Study of Alongshore Sediment Transport Rate," *Proceedings, 16th International Conference on Coastal Engineering*, American Society of Civil Engineers, Hamburg, pp. 1656–1674.

Kreibel, D. L., Dally, W. R., and Dean, R. G. (1986), "An Undistorted Froude Model for Surf Zone Sediment Transport," *Proceedings 20th International Conference on Coastal Engineering*, American Society of Civil Engineers, Taipei, pp. 1296–1310.

Lundgren, H. and Sand, S. E. (1978), "Natural Wave Trains: Description and Reproduction," *Proceedings, 16th International Conference on Coastal Engineering*, American Society of Civil Engineers, Hamburg, pp. 312–319.

Mansard, E. P. D. and Barthel, V. (1984), "Shoaling Properties of Bounded Long Waves," *Proceedings, 19th International Conference on Coastal Engineering*, American Society of Civil Engineers, Houston, pp. 798–814.

Milgram, J. H. (1970), "Active Water-Wave Absorbers," *J. Fluid Mech.*, **43,** 845–859.

Noda, E. K. (1972), "Equilibrium Beach Profile Scale-Model Relationship," *J. Waterw. Harbors Div., Am. Soc. Civ. Eng.*, November, 511–528.

Ottesen Hansen, N.-E., Sand, S. E., Lundgren, H., Sorensen, T., and Gravesen, H. (1980), "Correct Reproduction of Group-Induced Long Waves," *Proceedings, 17th International Conference on Coastal Engineering*, American Society of Civil Engineers, Sydney, pp. 784–800.

Plate, E. J. and Nath, J. H. (1968), "Modeling of Structures Subjected to Wind Generated Waves," *Proceedings, 11th International Conference on Coastal Engineering*, American Society of Civil Engineers, London, pp. 745–760.

Prins, J. E. (1960), "Model Investigation of Wind-Wave Forces," *Proceedings, 7th Conference on Coastal Engineering*, Council on Wave Research, The Hague, pp. 766–777.

Quinn, M.-L. (1977), "The History of the Beach Erosion Board, U. S. Army, Corps of Engineers, 1930–1963," Miscellaneous Report 77-9, U. S. Army Coastal Engineering Research Center, Ft. Belvoir, VA.

Salter, S. H. (1981), "Absorbing Wave Makers and Wide Tanks," *Proceedings, Directional Wave Spectra Applications Conference*, American Society of Civil Engineers, Berkeley, pp. 185–202.

Sand, S. E. (1982), "Long Wave Problems in Laboratory Models," *J. Waterw. Port Coastal Ocean Eng. Div., Am. Soc. Civ. Eng.*, November, 492–503.

Seelig, W. N. (1980), "Two-Dimensional Tests of Wave Transmission and Reflection Characteristics of Laboratory Breakwaters," Technical Report 80-1, U. S. Army Coastal Engineering Research Center, Ft. Belvoir, VA.

Snyder, C. M., Wiegel, R. L., and Bermel, K. J. (1958), "Laboratory Facilities for Studying Water Gravity Wave Phenomena," *Proceedings, 6th Conference on Coastal Engineering*, Council on Wave Research, Engineering Foundation, University of California, Berkeley, pp. 231–251.

Svendsen, I. A. (1985), "Physical Modelling of Water Waves," *Physical Modelling in Coastal Engineering*, R. A. Dalrymple, Ed., A. A. Balkema Publisher, Rotterdam, pp. 13–48.

Svendsen, I. A. and Jonsson, I. G. (1976), *Hydrodynamics of Coastal Regions*, Technical University of Denmark, Lyngby, Denmark.

Ursell, F., Dean, R. G., and Yu, Y. S. (1960), "Forced Small Amplitude Water Waves: A Comparison of Theory and Experiment," *J. Fluid Mech.*, **7,** 33–52.

Vennard, J. K. and Street, R. L. (1982), *Elementary Fluid Mechanics*, 6th ed., Wiley, New York.

Vincent, C. L. and Briggs, M. J. (1989), "Refraction–Diffraction of Irregular Waves over a Mound," *J. Waterw. Port Coastal Ocean Eng. Div., Am. Soc. Civ. Eng.*, March, 269–284.

INDEX